TS 253 .P75 1994

NEW ENGLAND INSTITUTE
OF TECHNOLOGY
LEARNING RESOURCES CENTER

Progressive Dies

Progressive Dies

Principles
and Practices
of Design
and Construction

Second Edition

Society of Manufacturing Engineers
in cooperation with the Forming
Technologies Association of SME
One SME Drive, P.O. Box 930
Dearborn, Michigan 48121

Based on
Progressive Dies: Design and Manufacture
By Daniel B. Dallas

Copyright © 1994 by Society of Manufacturing Engineers

Second Edition
98765432

All rights reserved, including those of translation. This book, or parts thereof, may not be reproduced in any form or by any means, including photocopying, recording, or microfilming, or by any information storage and retrieval system, without permission in writing of the copyright owners.

No liability is assumed by the publisher with respect to the use of information contained herein. While every precaution has been taken in the preparation of this book, the publisher assumes no responsibility for errors or omissions. Publication of any data in this book does not constitute a recommendation or endorsement of any patent, proprietary right, or product that may be involved.

Library of Congress Catalog Card Number: 94-065575
International Standard Book Number: 0-87263-448-5

Additional copies may be obtained by contacting:

Society of Manufacturing Engineers
Customer Service
One SME Drive
Dearborn, Michigan 48121
1-800-733-4763

SME staff who participated in producing this book:

Donald A. Peterson, Senior Editor
Dorothy M. Wylo, Production Secretary
Frances M. Kania, Production Secretary
Rosemary K. Csizmadia, Operations Administrator
Robert A. Ankrapp, FTA/SME Association Manager
Judy D. Munro, Manager, Graphic Services
Cover design by Barbara L. Sierp

Printed in the United States of America

Table of Contents

Preface vii

Introduction 1

Chapter 1 - Punches and Dies 3

Chapter 2 - Stock Guides, Strippers, and Pilots 31

Chapter 3 - Cam Stages 55

Chapter 4 - Press Selection 65

Chapter 5 - Grinding Operations 87

Chapter 6 - Blank Development 119

Chapter 7 - Basic Types 141

Chapter 8 - Strip and Stamping Design 163

Chapter 9 - Conventional Progressive Dies 183

Chapter 10 - Progressive Transfer Dies 199

Chapter 11 - Carbide Progressive Dies 215

Chapter 12 - Materials Selection 225

Chapter 13 - Die Engineering — Planning and Design 255

Chapter 14 - Design Practice 277

Chapter 15 - EDM and Progressive Dies 307

Chapter 16 - Progressive Die Mathematics 327

Chapter 17 - Progressive and Transfer Die Lubrication 353

Chapter 18 - Electronic Sensors and Die Protection 381

Chapter 19 - Quick Die Change 417

Glossary 443

Bibliography 459

Index 461

PREFACE

Against a backdrop of steadily increasing advances in manufacturing technologies, certain fundamental processes continue to prevail as the foundation on which industry depends to make things efficiently. Progressive die stamping is such a process. But though the process itself is mature and almost universally adopted, advances in the way progressive dies are designed and built have outpaced the texts to get the word out about them.

The intent of this book is to fill that void — to make known the technological advances in progressive die work in a way that will aid the designer and diemaker in developing dies that meet the quality, agility, and cost-efficiency demands of twenty-first-century global markets. New technologies are presented within the context of classic diemaking fundamentals: advances in sensor technology that prolong die life and produce higher precision parts; durable new materials that extend the length of part runs; and the latest in electrical discharge machining technology that speeds die cutting while ensuring a much greater level of accuracy. These and other developments in the diemaking business have had a profound impact on both the designer and the diemaker and their ability to respond to increasing customer demands.

This book serves as a guide for the contemporary diemaker: throughout, emphasis is placed on those practices, techniques, and materials that produce the greatest economies in design and the highest degree of productivity in construction. Compiled and written by experts in the field, the book reflects the collective body of knowledge of practitioners who routinely push the state of the art with each diemaking challenge they encounter.

Acknowledgments

With much gratitude we at SME commend the following team of die design and diemaking specialists who generously contributed their intellectual energies to the creation of this work.

Special acknowledgment to Howard Bender for his perseverance in shepherding the book through the several authors and reviewers: to James Albrecht, E. Lee Bainter; Vernon Eads; James Finnerty; Jeffrey Fredline; Harry Hastilow; Joseph Ivaska, Jr.; Robert Johnson; Paul Miner; David Smith; William Turner III; and Robert Wilson for taking responsibility for writing, reviewing, and ensuring the technical accuracy and currency of the book.

Donald A. Peterson
Editor

|||| Introduction ||||

A progressive die performs a series of fundamental sheet-metal operations at two or more stations during each press stroke in order to develop a workpiece as the strip stock moves through the die. Each working station performs one or more distinct die operations, but the strip must move from the first station, or stage, through each succeeding station to produce a complete part. One or more idle stations may be incorporated in the die to locate the strip, facilitate interstation strip travel, provide maximum-size die sections, or simplify their construction.

The linear travel of the strip stock at each press stroke is called the "progression," "advance," or "pitch" and is equal to the interstation distance.

The unwanted parts of the strip are cut out as it advances through the die, and one or more ribbons or tabs are left connected to each partially completed part to carry it through the stations of the die. The operations performed in a progressive die could be done in conventional, single dies as separate operations but would require individual feeding and positioning. In a progressive die, the part remains connected to the stock strip which is fed through the die with automatic feeds and positioned by pilots with speed and accuracy.

When parts are made from individual blanks moved from die to die by mechanical fingers in a single press the dies are known as transfer dies.

When total production requirements for a given stamping are high, a progressive die should be considered. The savings in total handling costs by progressive fabrication compared with a series of single operations may be great enough to justify the cost of the progressive die.

The present application of computer-aided die design, together with the general use of wire-burn EDM for making die sections, has greatly simplified the design and construction of progressive dies.

The quality of stampings made on progressive dies is often higher than that produced on individual dies. There is less chance for off-gage conditions due to part locating problems, and the human factor has less influence on part quality. Often the savings in labor costs, together with the more consistent quality of progressive die stampings, have been the deciding factors in justifying the greater material cost of coil stock over producing stampings from offal in recovery dies.

The fabrication of parts with a progressive die should be considered when:

- Stock material is not so thin that it cannot be piloted;
- Coil handling equipment, stock straighteners, and feeders are available or can be justified;
- The overall size of the die, as determined by part size, number of stations, and strip length is not too large for available presses;
- The total press tonnage capacity required is available;
- The press is level and in good condition — problems with worn gearing, loose gibbing, and worn bearings can result in alignment problems that can damage precision tooling;
- Quality and part consistency requirements are high;
- Quick die change and flexible manufacturing requirements exist.

1

Punches and Dies

Introduction

The words "punch" and "die" have accrued a number of meanings in the industry that have caused no small amount of confusion. The term "dies," for example, has been used to refer to the entire press tool. It also has been used to refer to the lower half of the press tool or the unit clamped to the bolster plate of the press.

The word "punch" has come to refer variously to the entire upper half of the press tool and, more specifically, to any so-called male section that punches or forms metal within a female section, the female being called the die section. Although this array of definitions stands up quite well in most die work, it falters when specifics are described. For example, in a compound die, the punch section is mounted in the die and the die section is mounted in the punch.

In the interest of clarity, throughout this book the punch is a male member, a die section a female member, and the press tool itself will be referred to as a progressive die. A progressive die is a press tool which performs a series of sequential operations on a piece of metal. Each part of the die performing an operation is referred to as a stage or as a station, terms used interchangeably.

Part Size

One of the fundamental rules of diemaking practice is *the punch determines the hole size and the die determines the blank size.* This is because a part tapers according to the clearance, as shown in Figure 1-1. The clearance is the amount of space between the punch and the die on one side and is dependent on the type of material being perforated, as well as on punch-to-hole ratio.

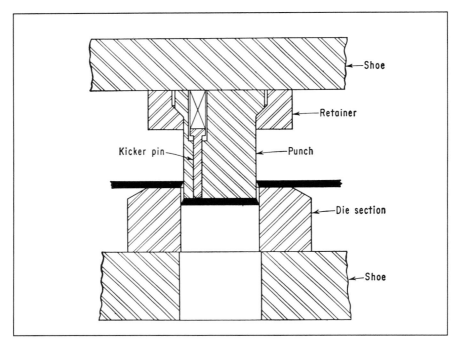

Figure 1-1. *Angular-edge part breakage in a blanking operation. Off-center ejector pin breaks vacuum between punch and stamping.*

Vacuum

Slugs of all sizes have a tendency to stick to their punches, creating a hurdle for all die designers to overcome. The most universally accepted way to "shed" the slug is with an ejector, or "kicker," pin, shown in Figure 1-1. Ejector pins work very well, are simple and inexpensive, and are difficult to improve upon.

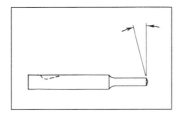

Figure 1-2. *Pointed piercing punch designed to keep slugs in lower die. Angle is dependent on material hardness and thickness.*

Punches that are too small to contain an ejector pin are sometimes pointed as shown in Figure 1-2. This design is effective in keeping the slug from returning with the punch, but sharpening time is greatly increased. However, it is recommended that specification of this type of punch be left to the discretion of the diemaker. If, in tryout, slugs begin to pull, grinding an angle on punches will likely rectify the problem. The grinding of this angle is simplified by use of a device familiarly called a vertical "whirligig." The whirligig is set at one degree to the back rail, and the punch is fed into the side of the wheel.

Tapered Buttons

Some firms specify that die buttons be ground with a half-degree taper, as shown in Figure 1-3. This forces the slug into an ever-decreasing diameter which holds it from returning on the punch. Although effective, such a design should only be incorporated as a last resort. It is expensive, the buttons must be internally ground, and punch breakage is high.

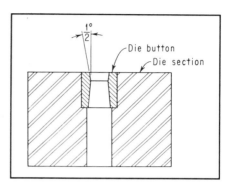

Figure 1-3. *Tapered die button designed to retain slugs. Taper relief in most cases reduces punch breakage as it controls the slug.*

Breaking Slug Columns

Small slugs often present a problem in the die itself because of their inherent tendency to stick together. These columns of slugs should be broken up. Although rare and quite costly, one of the best ways to do this is by angling the escape hole through the backing plate, as shown in Figure 1-4. Slugs emerging from the die set should be kept out of the holes in the bolster plate by means of a sheet-metal guard. Clearance can be milled or shaped into the risers and the guard secured by means of flat-headed screws.

Shear

A punch with shear operates more smoothly than a flat punch and requires less tonnage. It also minimizes problems of slugs sticking to the punches. Because in most progressive dies the punch removes metal that is not wanted, shear is put on the punch — important to bear in mind because a punch with shear distorts the part it cuts out of the strip.

If a progressive blank die is being built, the situation is reversed because the finished panel is the part being cut out. In such a case the shear must be put in the die, generally equal to metal thickness as shown in Figure 1-5.

Progressive Dies

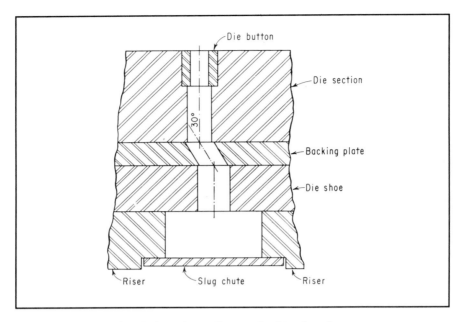

Figure 1-4. *Angular hole in backing plate is effective in breaking slug columns.*

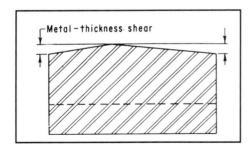

Figure 1-5. *Sectional view of a blanking punch showing shear.*

Push-off Pins

Push-off, or ejector, pins have been mentioned with respect to slug removal from a punch. But yet another important use of these pins is shown in Figure 1-6. Illustrated is a typical cutoff punch for a two-per-stroke progressive blank die. The outer die section (detail 1) is angled at about 30 degrees for blank removal. The push-off pin ensures that the part will start down the slide.

Push-off pins are occasionally used to complete a "stripping" operation, as illustrated in Figure 1-7. On a panel such as the one shown, the designer must ensure that every portion of the stock can be stripped. The stripper, or knock-off, pin on this die takes care of the entire strip except for the final form stage.

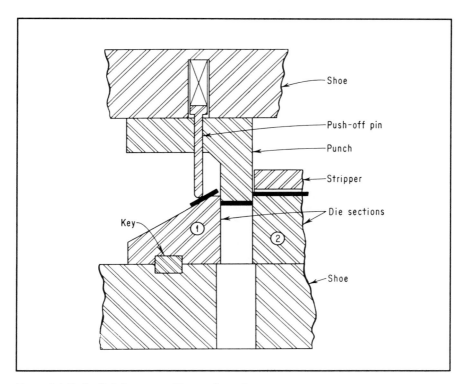

Figure 1-6. *Push-off pin incorporated for part disposal.*

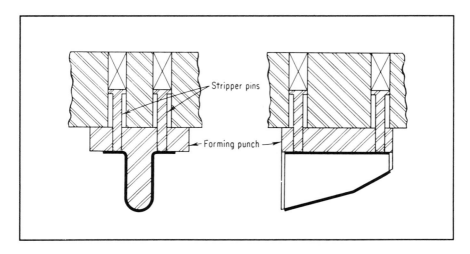

Figure 1-7. *Push-off pins used for stripping. When a forming punch covers the entire panel, the stripper cannot free the part from the punch.*

Die Life

The term "die life" refers to a dimension, specifically, the length of the land in a cutting edge. This dimension is governed by the number of stampings expected from the die. Progressive dies are usually built to produce millions of parts, and it is reasonable to expect that cutting sections will have to be replaced when they have been ground and shimmed over a period of time. However, the sections should be designed for maximum possible use and, generally speaking, each cutting edge should have at least a 0.125-inch (3.175-mm) land ending at a two-degree draft.

In truth, it is not important whether the angle is one, two, or three degrees because any open angle will permit the slugs to fall through. However, the designer should specify a certain angle so the diemaker will not be tempted to undercut the cutting edge, as illustrated in detail 2 of Figure 1-8. Undercutting the edge is much easier for the diemaker than grinding an angle, but it steals an eighth of an inch (3.18 mm) or so from the die life of the section. The designer specifies a two-inch (50.8-mm) angle merely to make sure he gets a draft and not an undercut. If the die is built for a comparatively short run, however, the cutting sections should be designed with an undercut instead of a draft as an economy measure.

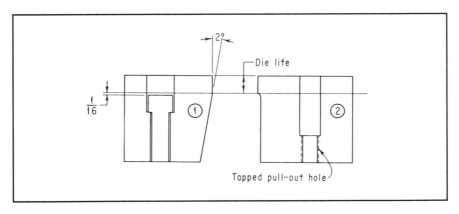

Figure 1-8. *Two methods of indicating die life. Design at left is preferable in long-run progressive dies.*

Counterbored Holes

By no means the rarest but certainly one of the saddest sights in a die shop is a cutting section that is completely finished without provision for die life above the screw heads. Diemakers should always remember that holes for socket-head screws must be counterbored to a depth equal to the head thickness, plus die life, plus 0.062 inch (1.57 mm), as shown in Figure 1-8.

Pull-out Holes

Many progressive die sections are pocketed in the die shoe. This is to hold location of the details without the use of dowels and to back up the cutting steels. If, for instance, the cutting steels shown in Figure 1-8 were mounted on the surface of the die shoe, the slugs piling up in the die would start to spread open the sections in spite of the heaviest dowels possible. Once this opening action starts, it is impossible to stop because each successive slug is slightly larger than the previous one. Therefore, the sections must be pocketed to hold location.

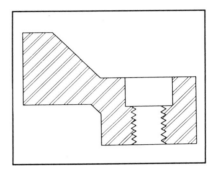

Figure 1-9. *Tapped counterbored hole. Thread dimension is one size larger than screw that fits counterbored hole.*

Once the sections are pocketed, provision must be made for their removal when desired. Most companies specify that the designer incorporate one or two tapped holes of a diameter equal to the screws holding the section down whose purpose is to remove the die sections. In operation, it is often necessary to remove a section from a die without removing the die from the press. In these cases tapped pull-out holes are mandatory.

Many smaller sections and inserts are so crowded that tapped holes are not feasible. In these cases tapping the counterbored screw holes is recommended, as shown in Figure 1-9.

Pocket Depth

Prior to shipping an assembled progressive die, the finished height of the sections should be stamped — in decimals — on the bottom sides. This practice makes it easier for the die repairman to sharpen the section and shim it back to the correct height. This indication of size will not be made unless the designer so specifies.

Punch Heels

All notching punches should be backed up by a heel block (Figure 1-10) before contacting the strip. Even large notching punches, with more powerful footing and stability, must be heeled if for no other reason than good mechanical design. An example appears in Figure 1-11. The machine-steel block holding the heel should be keyed and doweled in place.

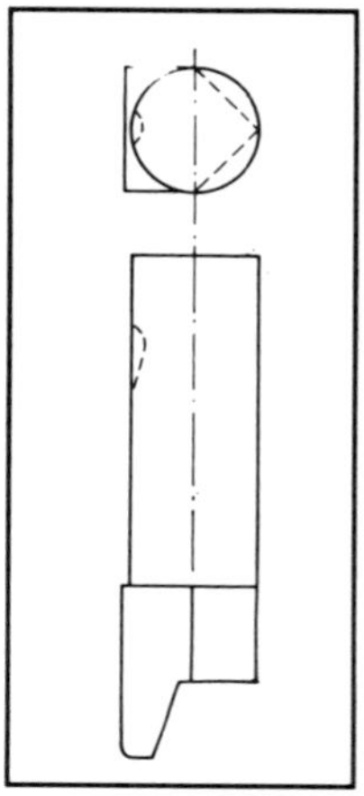

Drawing Dies

Drawing is a process of changing a flat, precut metal blank into a hollow vessel without excessive wrinkling, thinning, or fracturing. The various forms produced may be cylindrical or box-shaped with straight or tapered sides or a combination of straight, tapered, or curved sides. The size of the parts may vary from 0.250 inch (6.35 mm) in diameter or smaller, to

Figure 1-10. *Heeled notching punch.*

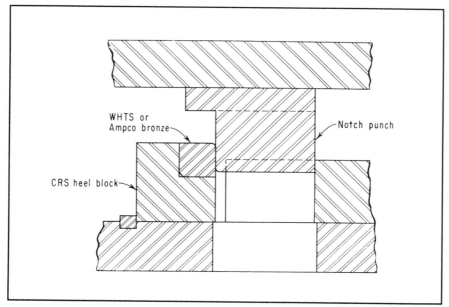

Figure 1-11. *Another method of heeling a punch. Punches that do not cut around their entire periphery must be heeled.*

aircraft or automotive parts large enough to require the use of mechanical handling equipment.

Metal Flow

When a metal blank is drawn into a die, a change in its shape is brought about by forcing the metal to flow on a plane parallel to the die face, with the result that its thickness and surface area remain about the same as the blank. Figure 1-12 shows in schematic form the flow of metal in circular shells. The units within one pair of radial boundaries have been numbered and each unit moved progressively toward the center in three steps. If the shell were drawn in this manner, and a certain unit area examined after each depth shown, it would show (1) a size change only as the metal moves over the die radius and (2) a shape change only as the metal moves over the die radius. Note that no change takes place in area 1, and the maximum change takes place in area 5.

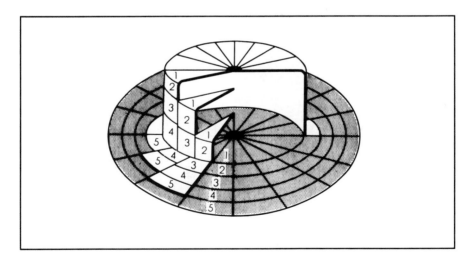

Figure 1-12. *A step-by-step flow of metal.*

The drawing of a rectangular shell involves varying degrees of flow severity. Some parts of the shell may require severe cold working and others, simple bending. In contrast to circular shells, in which pressure is uniform on all diameters, some areas of rectangular and irregular shells may require more pressure than others. True drawing occurs at the corners only; at the sides and ends metal movement is more closely allied to bending. The stresses at the corner of the shell are compressive on the metal moving toward the die radius and tensile on the metal that has already moved over the radius. The metal between the corners is in tension only on both the side wall and flange areas.

The variation in flow in different parts of the rectangular shell divides the blank into two areas. The corners are the drawing area, which includes all the metal in the corners of the blank necessary to make a full corner on the drawn shell. The sides and ends are the forming area, which includes all the metal necessary to make the sides and ends full depth. To illustrate the flow of metal in a rectangular draw, the developed blank in Figure 1-13B has been divided into unit areas by two different methods. In Figure 1-13A the corners of the shell drawn from the blank in view B are shown. The upper view is the corner area which has been marked with squares, and the lower view is the corner area marked with radial lines and concentric circles. The

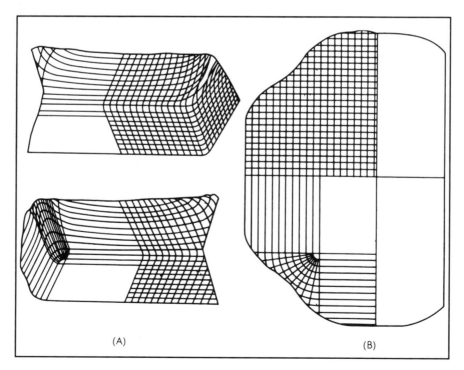

Figure 1-13. *Metal flow in rectangular draws: (A) blank marked before drawing; (B) corner areas after drawing.*

severe flow in the corner areas is clearly shown in the lower view by the radial lines of the blank being moved parallel and close together and the lines of the concentric circles becoming farther apart the nearer they are to the center of the corner of the blank. The relative parallel lines of the sides and ends show that little or no flow occurred in these areas. The upward bending of these lines indicates the flow from the corner area to the sides and ends to equalize the height where these areas on the blank were blended to eliminate sharp corners.

In the compound blank and draw die of Figure 1-14, a round blank is cut out of the strip by the action of detail 7 entering detail 6. Once this disk of metal has been cut from the strip, it is held firmly against the draw ring (or binder, as some prefer to call it) and the drawing action begins. The metal actually flows between detail 7 and detail 3 into the shape of the form punch. The metal is not pulled over the form punch like a sheet of rubber being

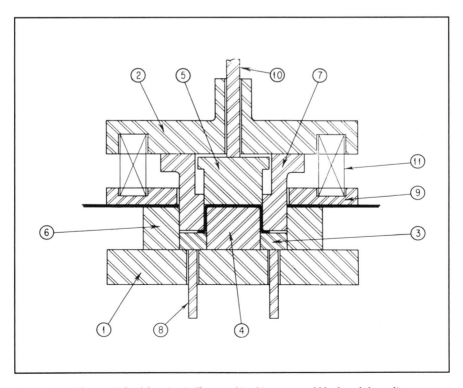

Figure 1-14. *The principle of drawing is illustrated in this compound blank-and-draw die.*

pulled over a cylinder, but rather it flows from a flat form into a cupped shape as if it were molten and being cast.

To pick out the most important detail in this die would be as difficult as selecting the most important link in a chain, but it is the draw ring that actually controls the plastic flow of metal. The binding action of the draw ring is determined by the air cushion and can be regulated for optimum results. If the pressure is too high, the metal cannot flow, and so it will fracture. If the air pressure is too low, the draw ring cannot hold the metal enough to start the plastic flow, and wrinkles and folds will appear.

Draw Stages

Drawing action in a progressive die requires the same components as those in a single-operation die: a form punch (male) and a form die section (female) as well as a binder or draw ring.

If the part is formed down into the die, the binder must be built into the stripper. Although this works well on most shallow draws, in deeper draws the draw ring binds the metal by spring or nitrogen cylinder pressure, which is uneven. Thus, at first contact, there is little or no pressure, depending, of course, on the initial compression of the springs. This means wrinkles will start immediately. As the die closes, however, the compressing springs rapidly build up an enormous pressure, so that the metal which originally started to wrinkle from lack of binding pressure now starts to fracture because of an excess. If, as stated, the draw is shallow, the stripper can be mounted with initial spring or gas pressure and a reasonably even amount of pressure maintained throughout the draw.

If the part does wrinkle, greater pressure is required. This can be provided by the addition of more gas or spring units as close as possible to the draw stage.

Draw Beads

One of the best devices for forestalling wrinkle formation is the draw bead illustrated in Figure 1-15. The bead is designed to control the rate of metal slippage and, although effective, more metal stretch and less plastic flow is found than in a true draw operation. The bead can be made tight enough to lock the metal, allowing no flow at all, and subsequent grinding

Punches and Dies

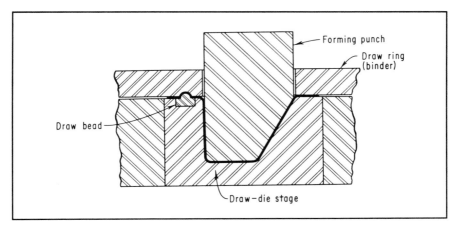

Figure 1-15. *Inserted draw bead. This design decreases the rate of metal slippage during plastic flow.*

operations performed on a trial-and-error basis will determine optimum size and shape. This, of course, is experimental, and the designer who uses this device must specify maximum contact.

Difficulties in Drawing Stages

A draw operation is far more sensitive than many diemakers realize until they become involved with one.

A case in point is illustrated in Figure 1-16. It was intended that this part be drawn in the first draw stage of a progressive die, but because of the large diameter of the blank and the small diameter of the cup, the metal simply would not flow into the shape. There was too much resistance to plastic flow in the design. At a later stage of the die, however, a hole of 1.75-inch (44.5-mm) diameter was pierced in the cup. So, as an experiment, several blanks were turned on a lathe and a 1.5-inch (38.1-mm) hole was drilled in the center of

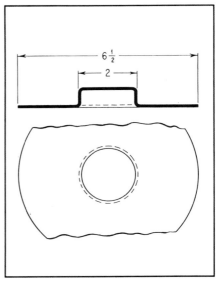

Figure 1-16. *A shallow but exceedingly difficult draw operation. The difficulty arises from too large a flange for the relatively small cup.*

15

each. These blanks were placed in the die and hit, the experiment resulting in several perfect draws.

The subsequent success of the operation was due to a law of plastic flow, i.e., plastic flow in steel is much more easily accomplished away from a center than toward a center. By piercing a preliminary hole in a panel, the diemaker allowed the metal to move away from the center. The axiom is clear: If a hole-piercing operation is to be performed on a drawn panel, a smaller hole placed in a preliminary stage will facilitate the draw operation.

Draw Radii

An important consideration in sheet-metal draw dies is the size of the radii R and R_1 detailed in Figure 1-17. For the best results, R_1 should be equal to four times metal thickness.

$$R_1 = 4 \times \text{metal thickness}$$

If possible, R should be equal to or greater than R_1.

$$R \gtreqless R_1$$

If R_1 is less than four times metal thickness, it is more difficult to draw the metal without fracturing it. The more R_1 is reduced, the more difficult the problem becomes until, in theory, one arrives at a square edge over which any draw action becomes impossible.

If, on the other hand, R_1 is increased above four times metal thickness, too much metal is free of the restriction of the draw ring and a wrinkling action will begin. Hence, $R_1 = 4 \times$ metal thickness represents a delicate balance between two impossible conditions.

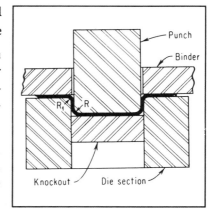

Figure 1-17. *The radii illustrated are of paramount importance in a drawing operation.*

R_1 should be as near to a perfect radius as is possible and on round sections it should be ground with an ID grinder. R_1 should also be a complete radius, for if a sharp edge is left, as in Figure 1-18, it will hinder plastic flow.

Punches and Dies

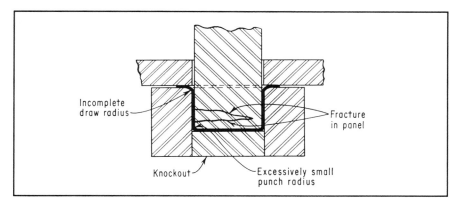

Figure 1-18. *Fracture resulting from an incomplete draw radius and a far-too-small punch radius.*

Finally, whether this radius is machine-ground, as is possible in a circular contour, or hand-ground on an irregular contour, its final polish must be put on in the direction of the metal flow. This seemingly small step often spells the difference between the success and failure of a draw section.

As stated, R should at least equal R_1, for if it is less, the wall will thin down adjacent to R (Figure 1-18). If R should be reduced to zero radius, a fracture would result, as a draw die would then tend to act like a blanking die.

As far as increasing the size is concerned, there is no mechanical limitation on R.

Of course the optimum sizes of R and R_1 may not coincide with the desired sizes, although many part designers could make occasional concessions at these two points. If this is impossible, the die designer should plan on a restrike stage to shorten these radii to the desired size. Often, however, subsequent operations in the die can be utilized to shorten the radii, and the designer should investigate this possibility before designing a separate restrike stage.

Wrinkles

Much has been said about wrinkle formation, and one immutable law of mechanics can be codified: Wrinkles in a steel stamping cannot be entirely removed from that stamping by any subsequent restrike operation.

Even though the part may go through one or more restrike stages, diemakers should remember that when wrinkles form in a panel, they are there to stay. Many progressive dies have been permanently damaged by bringing the ram down a little bit more in an effort to coin out wrinkles.

The implication is obvious. Isolate the cause of the wrinkles and cure it or report its incurability to those responsible.

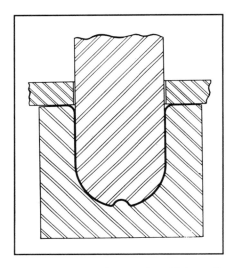

Figure 1-19. *A costly way of machining a die section for the panel illustrated.*

Metal Follows a Punch

Many designers and diemakers unknowingly drive up the cost of a draw stage by matching the die section exactly with the outside of the shell, as shown in Figure 1-19. This expense can easily be avoided by bearing in mind that the metal must by the laws of mechanics follow the contour of the punch. The punch and die section shown in Figure 1-20 has a male characteristic incorporated into its design. Female characteristics of the part end at the radii so clearance only is machined. The die section is further simplified by the use of an insert, which, at the end of the draw, forces metal up into the punch. Although the die may be designed as shown in Figure 1-19, the simpler and more cost-effective solution is that shown in Figure 1-20.

The fundamental rule here is that in all draw operations the metal follows the punch.

Ensuring Part Integrity

Almost any draw stage can be made to work one way or another. As an example, the panel shown in Figure 1-16 was successfully drawn without a hole. Powdered graphite was used as a lubricant and the press was inched over, giving the metal a chance to slip. As a result, a part was made which could have been "sold" to an affable inspector. However, the strip was not representative of the conditions under which the die was to operate, i.e., ordinary lubricant at 60 strokes per minute. The honest diemaker saws the drawn cup in two and examines the wall to see how much it has thinned down. Although it may not have fractured, it may be stretched to the point of weakness. If this is the case, the diemaker should improve the draw stage if it is within his authority to do so. If he cannot improve the panel, it is possible that he is fighting a bad design; and that being the case, he has isolated the trouble and relieved himself of the responsibility.

Blank Stages

The strip of the blank stage of a progressive die normally follows the pattern shown in Figure 1-21. The dimensions at A and B are significant in that if they are too large, they cause resistance to the drawing action and fractures occur. The smaller A and B are, the easier the metal can flow into the draw; and although these dimensions will vary because of metal thickness and tensile strength, the designer should strive for the smallest possible size. Reduction of these dimensions aids the draw action, but the resulting strip is very weak, particularly in those dies with more than four or five stages. The designer should consider supporting the strip as far as possible by the inclusion of bar-type lifters especially suited to fragile strips.

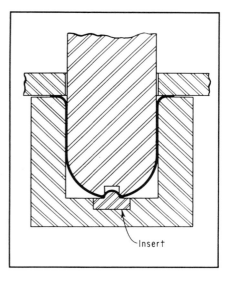

Figure 1-20. *The drawn panel in Figure 1-19 can be made more cost-effectively by omitting the female configurations. Design is further simplified by an insert in the die cavity.*

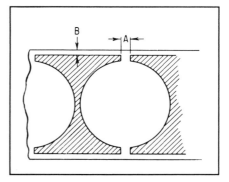

Figure 1-21. *Hourglass blank design typical of a progressive draw die. Dimensions A and B are critical because they can restrict drawing action.*

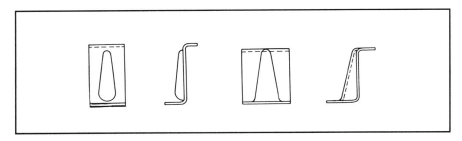

Figure 1-22. *"Teardrops" — typical embosses for strengthening formed panels. Teardrop at left stiffens one surface only. Teardrop at right stiffens entire panel.*

Embossing Stages

An embossing operation is a form of metal stretching. A part may be embossed for any number of reasons, though the majority of embosses are used as metal stiffeners, as shown in Figure 1-22 and Figure 1-23. In these examples, the emboss gives the stamping greater strength and resistance to deformation, but will draw some metal from the sides of a panel. Therefore, where possible, it should be placed in the early stages of the die or in some stage that precedes the final determination of the part's outline.

Inserts are often useful in embossing stages in that they effect economies in machining time. Their use also makes it much easier to grind the die section.

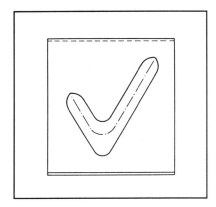

Figure 1-23. *V-type emboss — an ideal design for stiffening large flat areas.*

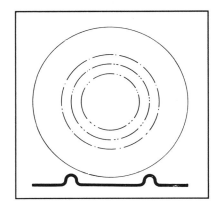

Figure 1-24. *Circular emboss frequently used in progressive die designs.*

One of the most frequent and difficult embossing operations is illustrated in Figure 1-24. If the die layout permits, this emboss is ideally performed in the two operations detailed in Figure 1-25. Although such an emboss can be accomplished in one stage, much more press tonnage is required than in two stages. It is also much more difficult to hold the height of the emboss to a three-place decimal, as is often required. However, should such an emboss be required in one stage, the design of Figure 1-25 should be followed, with the punch made 0.015 inch (0.38 mm) higher than the mean dimension. As in the case of draw cavities, clearance only is machined in the die insert. No attempt should be made to put a female configuration in this section. In tryout, the diemaker can reduce the excess height of the insert to establish a close tolerance dimension.

Designers and diemakers should be cautioned not to forget to provide emboss clearance in the stripper in subsequent stages. The word "emboss" should be a visual signal meaning "clearance in the stripper will be

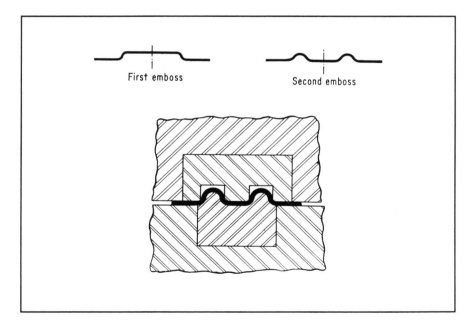

Figure 1-25. *Two-stage method of stamping a circular emboss (top). Creation of the emboss in one stage (bottom) increases tonnage requirements.*

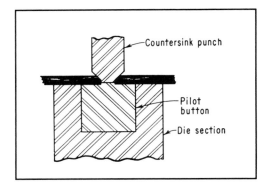

Figure 1-26. *Design for a countersinking stage.*

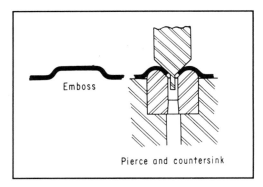

Figure 1-27. *Two-stage countersinking operation.*

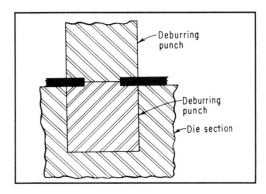

Figure 1-28. *Deburring can be accomplished in progressive dies by this design.*

required." Countless dies have been pulled out of tryout presses to complete this job — a simple enough operation but one often forgotten.

If the emboss is down instead of up, the omission of clearance is much more serious, because hardened die sections must be cleared.

Countersinking and Deburring Stages

Countersinking and deburring are still regarded by some in the trade essentially as secondary operations. However, many progressive die designs now provide stages to countersink and deburr holes. A typical countersink stage as appears in Figure 1-26 is a commercial punch ground to the requisite size and length.

The diemaker should not attempt to grind this detail until the rest of the die is finished and the exact shut height of the die is established in decimals. The reasoning is obvious, for if the countersink is too short, it will not function, and if it is too long, it will be broken.

The button in the die section is an insert with a small pilot which contacts the countersink when the die is closed. This is essentially a swaging operation, and the hole must be supported if its diameter is to be maintained.

Another form of countersinking is shown in Figure 1-27. This countersink is for small-gage stock and is accomplished first by embossing and then by piercing and countersinking in one operation. Although this two-stage operation works, it has a number of drawbacks, first of which is that the pierce portion of the countersink must be developed. Second, the piercing operation is done without any die support, as the piercing countersink will depress the metal only a short distance before it pierces through.

Deburring is accomplished in a similar manner, as shown in Figure 1-28. Here the burrs are swaged back into the part and a clean hole results. As in countersinking, the deburring punch should not be ground until the die is otherwise complete.

Shearing a Punch to Die Size

When a punch with a difficult contour is made, an operation known as shearing in the punch is useful in simplifying an otherwise difficult operation. Shearing, no matter how carefully done, will dull the edges of the die section. If these sections have already been brought down to height, they will have to be ground again and shimmed back up to die level. Thus, the new die leaves the shop with shims in it, a practice not acceptable in first-class die shops.

Second, after the punch has been roughed out, it should be screwed and doweled to the shoe. The diemaker should dowel the punch in place with dowels one size smaller than specified. Some diemakers will attempt to shear in a punch without doweling it, and invariably will end up in trouble. When the punch is finally sheared in, the diemaker simply opens the dowel holes to the correct size. After heat-treatment, the punch can be redoweled in location.

Third, after the impression of the die has been made on the rough punch, the punch is milled or shaped as close as possible to the line and then doweled in place. This may take four or five more trips to the mill or shaper

Progressive Dies

at this point to force the punch home. Actually, the diemaker cannot take off very much metal by a shearing operation, but he or she can get an excellent impression of the die for about 0.063 inch (1.59 mm). This is the metal sheared; the remainder is machined.

Flange Lines

An important point often overlooked, not only in progressive die work, but in other types of die work as well, is that a flanging section should always have a contour that follows the part bend line.

Figure 1-29 shows a panel and both the correct and incorrect designs for the flanging sections. In progressive work the contour of the panel is generally made in a forming stage, after which the panel advances through an idle stage to the flanging stage. In single-operation die work, the panel

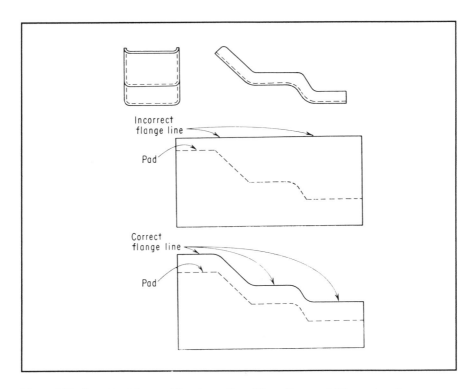

Figure 1-29. *Correct and incorrect flanging sections. Flange lines must follow pad contours.*

is formed on a pad and against air pressure. When the pad is forced to bottom, the flanges are wiped.

If a flanging steel is made as shown in the incorrect view, the lower part of the flange is wiped to vertical before the remainder begins to turn. This, of course, means that the partially formed flange will have a 90-degree twist in it. Although this condition can be tolerated in panels with comparatively short flanges, the metal in those panels having longer flanges will tear at the first radius.

The metal, having torn at this point, will then tend to overlap itself as the inclined portion of the flange begins to wipe. Because flange outlines generally must be developed, the blank development phase cannot proceed until this condition is corrected.

The "incorrect" flanging section of Figure 1-29 presents an impossible condition. The "correct" flanging section, as shown, can also be very difficult, depending on the sharpness of the angles. The difficulty arises at the radius where metal is forced toward a common center as this portion of the flange is wiped. In cases where a comparatively long flange is wiped, the metal will buckle and actually fold over at this point. The product designer wise in the ways of diemaking will generally allow a shorter flange at these points, knowing that flanging at an inside radius is actually a draw operation that must be performed without a binder and, as such, has definite limitations.

To help the operation, the diemaker must experiment with the flanging contour until he finds its optimum shape (Figure 1-30). A higher contour at the points where metal tends to gather, i.e., the radii, has to be included. This higher contour causes the flange to start wiping at the critical points first, thus holding back the excess metal moving toward the common center. Again, finding the correct contour is a matter of experimentation, for if the cor-

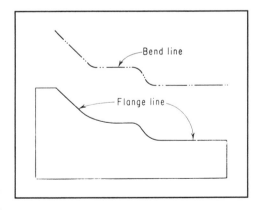

Figure 1-30. *Internal radii are trouble spots when flanging. Raising these portions of the flange line, as shown above, assists this operation by forming the flange at the radii slightly ahead of the straight part.*

rected contour is too high, the problem is not solved; it is merely shifted to other points in the flange.

Restrike Stages

Many mechanical functions can be performed in a progressive die, e.g., blanking, piercing, notching, lancing, drawing, forming, coining, swaging, burnishing, etc., but none is as misunderstood as restriking.

A restrike stage in a progressive die or in a separate restrike die in a series of line dies has only one purpose: to rehit a finished panel, setting its various members to exact part-print size and location. As an example, a drawn panel may have radii that must be shortened to part-print size. This is often necessary when the radii, as specified by the part print, are not large enough to allow a successful draw. A restrike stage is then necessary to shorten one radius or more.

Such an operation as this is no problem, and the restrike stage can be built to predetermined dimensions. Similar examples are embosses or extrusions which must be defined by a sharp line. As a general rule, such embosses or extrusions are made with a small radius to permit metal to flow (Figure 1-31). A restrike stage then gives definition to the emboss. The restrike stage is comparatively simple and can be made from predetermined dimensions.

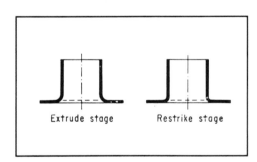

Figure 1-31. *Extrusions are best made in two stages. Radii are shortened by restriking.*

When a restrike stage is used to correct a springback or overbend condition, it cannot be dimensioned until a panel has gone through the final form stage.

For illustration, consider a piece of sheet metal (as in Figure 1-32) with a slight belly that needs to be flattened. It would be possible to place the part on a perfect surface plate and then place an anvil on top of it, but regardless of how long the anvil rests on it, the part would immediately return to its convex shape.

Similar and equally futile are attempts to flatten an elastic material by squeezing it in a vise. When the vise is opened, the irregularity returns. The combination of two flat surfaces and enormous pressure must always fail as a flattening device. It follows, then, that a restrike stage designed and built to part-print dimensions cannot correct overbend and springback. Similarly, wrinkles cannot be removed by subsequent restrike stages.

Conversely, however, a convex piece of metal can be straightened by supporting it on two dowel pins and exerting pressure against the belly (Figure 1-33). By pushing the convexity to a point somewhere past the elastic limit of the metal, the part can be flattened.

It is an immutable law of mechanics (Hooke's law) that to change the shape of metal by pressure, it must be moved past its limit of elasticity whether we are trying to flatten a convex surface or put curvature in a flat piece. The limit of elasticity is a variable which, of course, is dependent upon the analysis of the material being stamped.

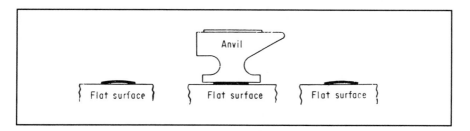

Figure 32. *Why restrike stages built to part-print sizes fail. Subjected to extreme pressure, the curved piece of sheet metal will always return to its natural curvature because of its elasticity.*

Figure 1-33. *Principle of restrike stages. The weight of a finger pressing this part against two pins will correct the curvature.*

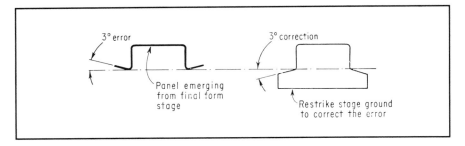

Figure 1-34. *Design of restrike stage for correcting three-degree springback.*

The tabs on the panel of Figure 1-34 must be formed at 90 degrees ±30 minutes for spot welding, so the form stage will be ground to a perfect square. The restrike stage should be roughed out with plenty of grind stock remaining at the working surface. Its other dimensions should be ground, allowing the section to be placed in the otherwise finished die. The part is then advanced through the forming stage and examined for size. If it consistently shows a three-degree springback, and the springback constant for this particular metal is 3 degrees in 90 degrees, the diemaker needs only grind the restrike die section and its mating punch to 93 degrees and put them in the die. The sizes will fall well within part-print tolerance. The rule is simple: Determine the amount of overbend or springback in the final form stage. Make a corresponding correction in the restrike stage.

Above all, it is important to shun the philosophy that the restrike stage is built to part-print dimensions and then subjected to enormous pressure by leaving its punch a few thousandths (mm) long.

Lance Stages

The operation known as lancing is found frequently in all types of sheet-metal die work. Its most common use is in the formation of short tabs wherein one punch is used to both cut and form the metal. Such a punch is shown in Figure 1-35, and though it is not difficult to design or build, certain rules must be observed to create a successful design.

First, the lance must be ground at an angle from the point where the cutting action begins. If this portion of the punch is ground square, the tab will be badly stretched and deformed.

Second, the radius R must be calculated and designed into the die to prevent scoring of the tab by the sharp edge. This means that this portion of the lance should be treated for what it is — a forming punch.

The action of this punch in its descent is a truly unique operation. First it cuts like a blanking punch, then bends the area it has cut for a short distance, and finally cuts again and forms simultaneously. Again, this is not a difficult operation, but it is critical that the radius and the angle shown be incorporated into the punch.

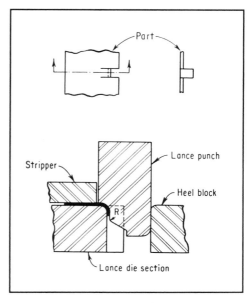

Figure 1-35. *Design of a lance-form stage for tab formation.*

A note to the diemaker: It is not unusual for the designer to omit showing the radius or even the angle in this punch design. Often, such an omission is made in the interest of cutting costs by cutting design time. Whatever the case, the diemaker should make the correction.

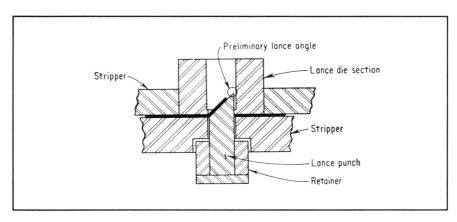

Figure 1-36. *Lance-station design for forming a tab upward. This is one of the more difficult lance operations.*

Lancing Upward

Another basic lance operation is cross-sectioned in Figure 1-36. Because of the formation of the flange there is no question that the lance punch will be ground on an angle. If, however, this angle is continued to the upper edge of the lance, a feather edge will result. To eliminate this, a short 10- to 15-degree angle is ground at the upper portion which makes the initial breakthrough.

A flange lanced and formed in this manner will not be spanked flat, and this is often one of the weaknesses of the operation. It is possible, however, to put an insert in the lance die section for the flange to form spank against, or a restrike stage can be utilized later in the sequence.

The restrike stage generally is the best solution, since there is often a certain amount of distortion of the flange caused by the stripping action. Hence, spanking the flange against an insert makes it perfect only at the moment of its formation.

It will be noted that two strippers are specified in this design. The lower stripper is generally a small pad used only to strip the stock off the lance. The upper stripper serves to strip the flange out of the lance die section.

To facilitate the latter stripping action, the die section is relieved to a very short die life. Although this smaller die life shortens the interval between die replacements, it minimizes part distortion by giving the flange less area in which to "hang up."

||||2||||

Stock Guides, Strippers, and Pilots

Stripper Wear

Of the many components of a progressive die, none shows the evolution of this type of press tool as much as the stripper. In the less advanced stage of its development, the stripper was regarded solely as a means of stripping the stock from the punches and pilots. Though this remains its primary purpose, it is now also used either to flatten the part or to give it a desired contour. It may act either as a retainer or as a guide for small-diameter pilots and punches. In some dies, particularly large automotive progressive dies, the stripper may actually be a retainer for large die sections; in this type of die it almost loses its identity as a stripper and actually becomes a third die shoe.

When the stripper is used to flatten or spank the strip, it should be hardened or it should contain a hardened insert, as shown in Figure 2-1. If the stock is 30 gage or less, a hardened stripper is unnecessary, but if heavy-gage stock is being stamped, the strip will make an impression in the stripper before the die leaves the tryout press. Water-hardening steel is generally selected for hardened strippers and it is heat-treated to Rockwell C 45-50. This degree of hardness is sufficient to keep the stripper from wearing without causing crystallization or fracture when double-headers are hit. Final determination of the hardness as well as the type of steel used is made by the size and shape of the part itself, and it is a decision which should be left to the designer.

While it is generally accepted practice to increase the stripper hardness in direct proportion to the thickness of metal being stamped, there are too many variables to make this a hard-and-fast rule. For example, if

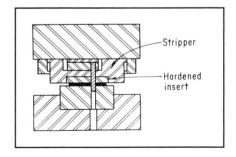

Figure 2-1. *Sectional view of hardened insert in stripper.*

31

a comparatively short-run die is being designed, wear and abrasion are not important considerations.

Some designers, knowing that the stripper is the most complicated part of the die, leave it soft enough to be machined after hardening. Being machinable, it can also be welded without fracture; in this way, all toolroom errors, whether positive or negative, can be corrected. There is no unanimity on the subject, and where one large automotive concern specifies Rockwell C 60-62 on strippers, many others leave a similar stripper soft.

Stripper Support

Because almost all strippers are now used to spank the work, it is necessary that the stripper bottom against the die shoe, retainers, and flanges. Too often designs call for the stripper to bottom against the shoe only, leaving clearance under the retainers and flanges, as illustrated in Figure 2-2. The resulting lack of support eventually causes a bow in the center of the stripper and, obviously, a stripper that is not flat cannot spank the strip. Figure 2-3 shows the correct way for a stripper to bottom, and if the designer has failed to show it this way, the diemaker should make the necessary correction.

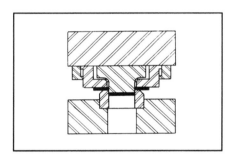

Figure 2-2. *Unsupported stripper. If stripper bottoms against die shoe, it should also bottom on flanges and retainers.*

Many progressive dies designed with retainers to hold small punches and pilots are not manufactured to exact size, and may have as much as 0.015 inch (0.38 mm) variation in thickness. One of the first tasks of a diemaker, when starting a progressive die, should be to grind all retainers to a uniform thickness. When the smallest one cleans up, all retainers are uniform in size. This size is then marked on the print in decimals and is considered the depth of the finished pocket. The pocket is later machined to this depth plus whatever grind stock the diemaker considers necessary, usually 0.025 inch to 0.050 inch (0.64 mm to 1.27 mm). After heat-treatment, the back of the stripper is ground until the pocket depth is equal to, or 0.000 inch to 0.001 inch (0.00 mm to 0.025 mm) greater than, the retainer dimension.

Stock Guides, Strippers, and Pilots

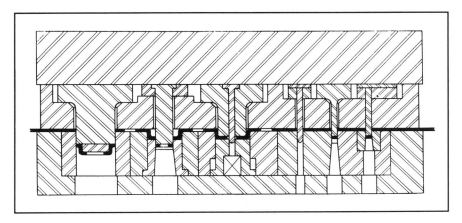

Figure 2-3. *Sectional view of properly supported stripper in a five-stage progressive draw die.*

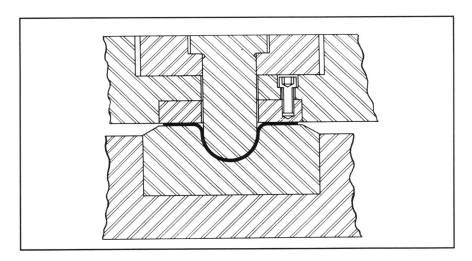

Figure 2-4. *Hardened insert used as a binder (draw ring) in a progressive draw die. Insert should be several thousandths above surface of stripper plate.*

Shoulders on all punch sections should be assigned a decimal dimension when the job is started and their pockets machined accordingly. In this way the stripper receives maximum support.

It is of particular importance that the stripper be kept flat when progressive dies that have drawing stages are built. In such dies a hardened insert, which acts as a binder or draw ring, is placed in the stripper to hold the metal while it is being drawn, as shown in Figure 2-4. This insert generally is

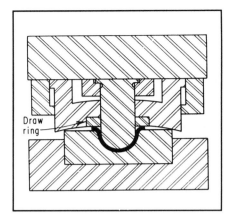

Figure 2-5. *A source of wrinkle formation in progressive draw dies. Because the unsupported stripper has developed a concavity, the draw ring no longer binds the metal.*

ground from 0.001 inch to 0.002 inch (0.025 mm to 0.05 mm) higher than the stripper to ensure that it will seat properly on the stock. If the insert does not sit down tightly against the metal — for instance, if it should be below the level of the stripper — wrinkles will form as the stock is drawn into the form. If the stripper is not fully supported by the retainers and shoulders on the punch sections, a concavity will develop in its surface and the binding action of the draw ring will be lost, as illustrated in Figure 2-5.

Stripper Balance

The importance of proper seating of a binder insert in the stripper has been pointed out. Also important is the necessity of a balanced stripper. A balanced stripper is one which cannot tilt when stock is first introduced to the die. Balanced strippers are of particular importance in progressive draw dies, if draw stages are to perform without causing wrinkles.

Frustration arises for many diemakers when, although their dies may be functionally correct, wrinkling action at the draw stage continues. The wrinkling is caused by an unbalanced stripper. The sequence of events begins as the stock is introduced to the die and a wrinkled panel is drawn because of the angle of the unbalanced stripper (Figure 2-6). The binder, of course, cannot function at this angle — hence the wrinkle formation. As the strip carrying the wrinkled panel moves through the die, the wrinkles prevent the stripper from seating, so that still more wrinkled panels result. At this point the cause of the trouble lies not in the die, but in the strip.

The only way out of this situation is to start over again with a fresh piece of stock. Two small pieces of stock should be placed at the last station as a means of holding the stripper in a parallel plane. Assuming that the die is functionally correct, an unwrinkled panel can be drawn. When this panel has reached the last station, the strip itself serves as an effective balancing device and good panels can be produced. The best solution, however, is to

Stock Guides, Strippers, and Pilots

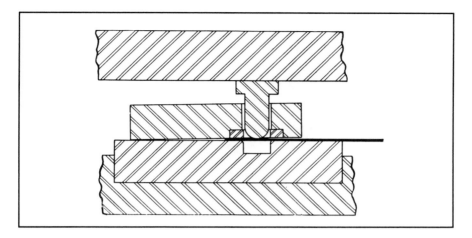

Figure 2-6. *An unbalanced stripper is another source of wrinkle formation. In this example the binder cannot act upon the stock.*

screw two or more pieces of support stock at appropriate locations in the lower die.

Stock Guide Design

In many progressive dies the strip is raised as it advances between stations, being lowered into place by the action of the stripper on the stock guides. Actually, it is necessary to raise the stock only that amount necessary for proper clearance. As the stock is introduced to the die, two or four round stock guides which also act as lifters are generally used. These stock guides (Figure 2-7) are simple enough but many diemakers have trouble when making them by failing to observe the mathematics involved. There is a definite relationship between *A*, *B*, and *C*. In most progressive die designs, the designer does not bother to dimension anything at this point except *D*, which is the height the stock must be raised above die level to successfully clear all die sections. *C* is generally accepted as being 1.5 times stock thickness, although twice stock thickness is permissible on heavy-gage stock.

$$C = 1.5 \times \text{stock thickness}$$

The dimension *B*, which may be scaled, is generally a nominal size, usually 0.5 inch, 0.63 inch, or 0.75 inch (12.7 mm, 16.0 mm, or 19.1 mm). However, it must be assigned a decimal equivalent which is marked on the

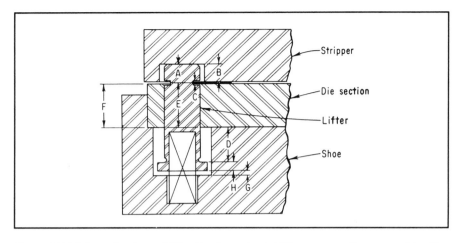

Figure 2-7. *Mathematics of a stock lifter. Poorly constructed lifters are an endless source of trouble.*
A = B + (stock thickness + 2)
B = arbitrary dimension
C = stock thickness x 1.5
D = travel
E = die section + (stock thickness + 2)
F = die section
G = estimated die life x 1.5
H = arbitrary

print by the diemaker. To maintain interchangeability of the lifters, their respective pockets in the stripper are machined to this dimension plus grind. When the lifter-stock guide is machined, dimension A should be equal to B plus one-half of metal thickness.

$$A = B + \frac{\text{metal thickness}}{2}$$

The nominal size F of the die section must also be marked on the die drawing in decimals. Total length of the lifter can then be calculated, and this dimension should also be marked on the print.

Length (above shoulder) = A + D + E

Length (above shoulder) = B + die section + stock thickness + travel

When building dies which stamp heavy metal, the diemaker often is able to use scale sizes, but on thin stock he or she can get into trouble unless these calculations are made. If A, for example, is too long with respect to B, the stripper will force notch C below die level and nip the stock, as shown in

Figure 2-8. If B is too long with respect to A, the work of forcing the stock guide down against spring pressure will be performed by action of the stripper on the stock itself (Figure 2-9). As a result, thin stock will buckle and tear out of the lifter.

The "stock-guide lifters" shown in Figure 2-7 are practical only when there is no break or interruption in the edge of the stock. Obviously, they can be used in any progressive die before the stock has been notched and in cases where the stock edge remains unbroken and, as in the progressive draw die illustrated in Figure 2-10, they can be used throughout the die.

When stock width is known, dimension D can be calculated immediately, but D' must be developed if all lifters are to be interchangeable. (In well designed progressive dies all lifters must be interchangeable.) If, however, interchangeability is not necessary, the location can be approximated and the lifters made with

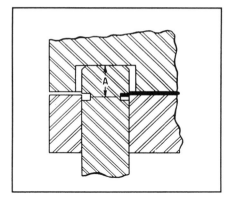

Figure 2-8. *Dimension A is too long, and the lifter nips the stock.*

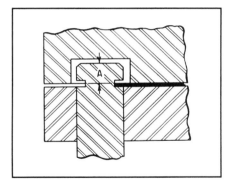

Figure 2-9. *Dimension A is too short. The work of depressing the lifter is performed by the stock instead of the stripper.*

predetermined dimensions. If such a lifter is on the tight side (and the astute diemaker will see that it is), the necessary correction can be made in tryout. If it is on the loose side, it can still be used, provided it is not also being used as a stock guide. If it is being used as a stock guide, it should not be made until the strip has been run through the die and the diemaker has accurately measured the amount the stock has narrowed down. There are several ways these lifters can be designed; the most satisfactory way by far is shown in Figure 2-11a.

Many designers persist in using the lifter shown in Figure 2-11c. Such a lifter will work successfully only when held in perfect alignment with the

Progressive Dies

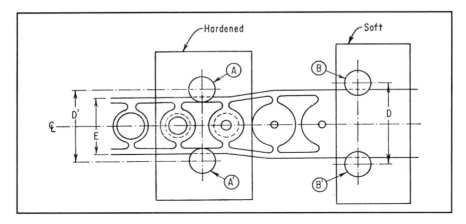

Figure 2-10. *If interchangeability is required of lifters, a problem is posed as D' must be developed. One good solution is to "cold develop" D' in the hardened section; in tryout, E will be established. Lifters A and A' can be finished (from their dimensions D can be calculated and their holes bored in the soft section) and then B and B' can be made identical to A and A'.*

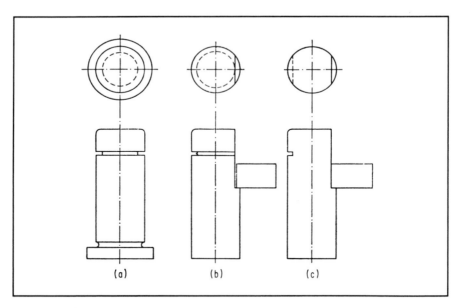

Figure 2-11. *Three types of stock lifters: (a) excellent; (b) good; (c) poor.*

strip. If it is allowed to rotate the slightest amount, it will bind the strip. However, if it is held in perfect alignment by the clamp, it will be so tight that it will tend to stick when in the down position, as shown in Figure 2-12.

Another drawback to the lifters shown in Figures 2-11b and c is that additional parts must be made in the form of small clamps. Clearance for these clamps must then be milled in the stripper. The lifter shown in Figure 2-11a requires no extra parts or machining, and it can rotate without any ill effect.

The slot shown in these lifters is easily made with a parting tool and, if made in a good lathe, the diemaker need only calculate the depth required, add to this one-half the grind stock, and plunge the cut. If carefully done, the slot will be within a few thousandths of correct size and this is closer tolerance than most stock is sheared to.

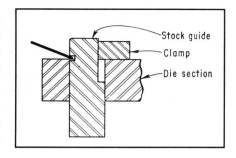

Figure 2-12. *This is the drawback to the lifter shown in Figure 2-11c. The stripper has left the bottom die but the lifter is still down. Either the clamp is too tight or some foreign matter has worked in between the clamp and stock guide. But if the clamp is loosened too much, the stock guide can rotate and bind the stock as it is being fed into the die.*

Bar Lifters

In cases where the edge of the stock is notched in the first or second stage (and more often than not this happens), an entirely different type of lifter must be used. In some cases, a series of pins, as shown in Figure 2-13, is satisfactory, provided there are no interfering cuts or notches and stock

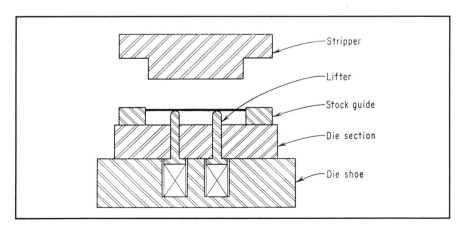

Figure 2-13. *An inexpensive lifter suitable for low-production progressive dies.*

Progressive Dies

alignment is not a requisite. This lifter will mark the part if a heavy spring is required and it is seldom used, although it is practical and inexpensive under certain conditions.

A better lifter is shown in Figure 2-14. In this example a solid bar of machine steel is used to raise the stock. It is ground to fit between the die sections, and holes for liner bushings are jig-bored. These liners are for pilot entry, thus ensuring perfect alignment of the strip before the die closes.

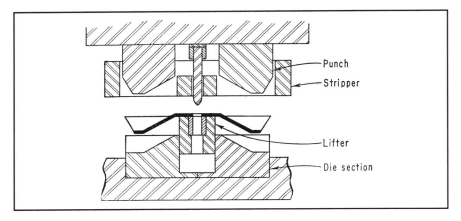

Figure 2-14. *Sectional view of a bar-type lifter moving between die sections.*

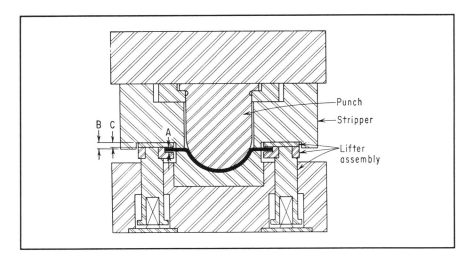

Figure 2-15. *Bar-type lifters. This design is especially good when the strip is relatively fragile.*

Still another example of a bar lifter is illustrated in Figure 2-15. This lifter is easily machined but it requires clearance in the stripper as well as in the die sections. Extra clearance can be cut in the die sections under the lifters. A should be considered 1.5 x stock thickness; B, the depth of clearance in the stripper, is then marked on the print as equal to C + 0.25 stock thickness. C is an arbitrarily established dimension.

Stock Lifters

To assist stock guides in raising the stock, many designers utilize the bar lifter shown in Figure 2-16. This lifter has no stock-guiding function whatsoever, its purpose being to raise that portion of the stock outside the die. Thus, the only lifting function of the posts or internal bar lifters is to raise the strip itself.

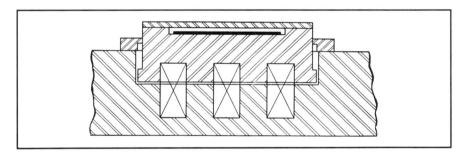

Figure 2-16. *This lifter is excellent for raising the stock from die level to feed level.*

Because this type of lifter floats freely in a pocket in the shoe, no attempt should be made to utilize it as a guiding device. It should have an opening width of three or four times stock thickness and an opening length of stock width plus 0.25 inch (6.4 mm).

Lifters of this type present the same timing problem as the bar- and post-type lifters already discussed. Formulas already discussed are, of course, applicable in this instance.

Stripper Keepers

Two methods of holding a stripper in place are illustrated in Figure 2-17 and Figure 2-18, the design of Figure 2-17 being the most popular. In die designs where strippers should be guided as well as supported by keeper blocks, hardened tool-steel inserts should be used as a means of preventing wear. The keeper blocks should also be made of hardened tool steel or carburized and hardened machine steel. If the function of the keeper is support only, inserts and hardened keepers are unnecessary.

The round counterbored retainer shown in Figure 2-18 is an effective stripper-holding device in crowded dies which do not have sufficient room

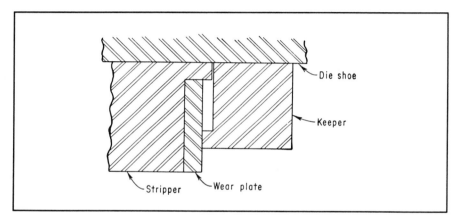

Figure 2-17. *Conventional stripper keeper. Hardened insert is required when the keeper guides the stripper.*

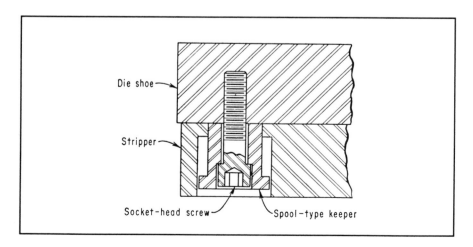

Figure 2-18. *Spool-type stripper retainer. This design is useful when die space is too limited for conventional keepers.*

for keeper blocks. Screws no smaller than 5/8-11 should be used to prevent spring pressure from stripping the threads.

An advantage of this type of retainer is that the tapped holes can be used to depress the stripper for purposes of punch sharpening. By removing the retainer and substituting a heavy washer in its place, the diemaker can draw the stripper down past the level of the punches. It is then possible to place the entire unit on a large surface grinder and sharpen the punches without dismantling the die.

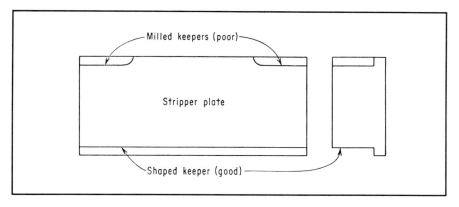

Figure 2-19. *Retaining lips should be cut through the entire length of the stripper.*

When designing the stripper, the designer should specify a keeper which runs the entire length of the plate (Figure 2-19). Many designers will show a milled pocket as a means of support. This milling operation, which must be repeated at least four times, is expensive. The better design calls for a shaper cut all the way through.

Stripping Pressure

In designing progressive dies with long cutting perimeters, it is necessary to calculate the die stripping pressure in order to determine conventional spring or nitrogen cylinder requirements. To determine stripping pressure in pounds, it is necessary to multiply the length of the cutting perimeter by the thickness of the metal in decimal inches by 2,500 pounds (1125 kg).

Thus, the formula reads

$$P = L \times T \times 2{,}500$$

where P = stripping pressure
  L = total length of all cutting perimeters, inches
  T = thickness of stock, decimals

Having determined the stripping pressure, the designer need only divide this quantity by the pressure rating of the springs.

$$N = \frac{P}{R}$$

where N = number of springs (or nitrogen cylinders) required
  P = stripping pressure
  R = pressure rating of individual springs or cylinders

Spring/cylinder layout is an extremely important aspect of die design. Designers should always try to place them as close as possible to the punches which must be stripped. In this way maximum mechanical advantage is secured.

If it is not inconsistent with company standards, spring pockets for stripper springs should be drilled completely through the die shoe. This is done in many plants for two reasons. First, it enables the die to contain longer springs, which in many cases are necessary to obtain a long travel dimension or a more even stripper pressure at draw stages. Second, it facilitates the loading of springs in an assembled die.

The latter reason is especially important in larger dies where stripper-spring pressures are calculated in tons. To draw a stripper plate down against such spring pressure requires great strength and ingenuity. If, however, spring holes go through the shoe, the springs can be inserted in the die after stripper assembly.

Some companies which do not approve of through-holes for stripper springs use threaded plugs to convert through-holes into pockets (Figure 2-20). In this design, as in the through-hole design, the diemaker can insert the springs after die assembly. Designers should be aware, however, that when the spring is at its fully expanded length, it must be far enough below the die-shoe surface to permit the diemaker to start the plug in its thread. In this respect, the threaded plug design does not allow the travel permitted by a through-hole, but it does facilitate die assembly.

Punch Contours

In some progressive dies, particularly those in which an intricately shaped punch is used to cut thin stock, a problem arises in holding the contour of the punch in the stripper. If it is a comparatively small stripper and if it remains soft, the problem is minimized, but if it is hardened into the higher Rockwell numbers, the heat-treatment involved usually distorts the punch openings of the hole. Figure 2-21 shows such a punch and the solution to this problem of heat-treatment distortion. The diemaker should remember that the maximum clearance permitted around the punch is one-half stock thickness. This condition, more often sought after than attained, is easily established by cutting out a large clearance hole, as shown by the broken line in the figure, and then milling a pocket for the inserts. Holes are then tapped in the pocket and after heat treatment, the stripper is located on the die and its keepers doweled in position. At this point inserts can be machined to the punch contour and the tapped holes transferred back to the inserts. After the inserts have been

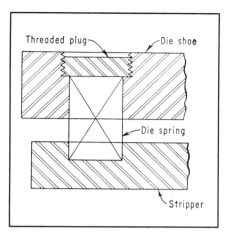

Figure 2-20. *Threaded plug design for spring pockets.*

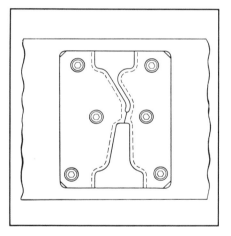

Figure 2-21. *Difficult contours in hardened stripper are "held" by means of inserts.*

hardened, they can be screwed in position and ground at assembly to the stripper. Oil-hardening gage stock is ideal for inserts of this type.

Stripper and Pilot

One basic problem sometimes ignored in the design of progressive dies is the work done by the stripper in conjunction with the pilots. A pilot should enter its full diameter into the stock before the stripper makes contact with the strip. This is particularly important in progressive dies where the stock is not raised by stays at die level. If the pilot does not make its full entry before stripper contact, the stock is held securely by the stripper and the pilots pull and distort the hole. As a result, piloting is only approximate and pilot breakage becomes excessive. If the designer has used a spring-loaded pilot (Figure 2-22), true gaging becomes impossible.

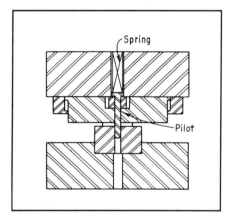

Figure 2-22. *Spring-loaded pilot. In the event of a mis-hit or broken punch, this pilot retreats against spring pressure into the upper shoe.*

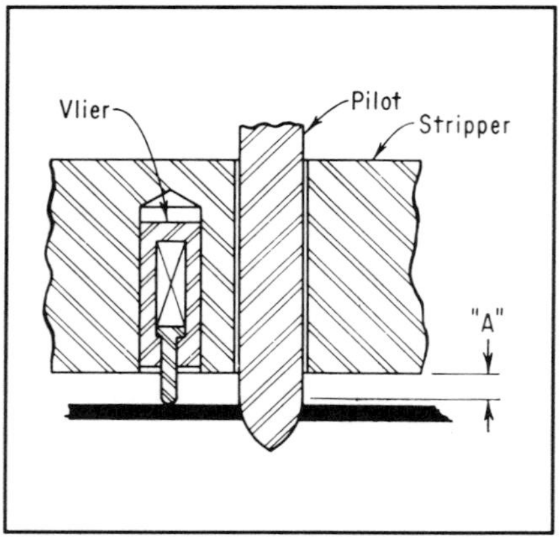

Figure 2-23. *Method of stripping a pilot. Dimension A equals 1.5 x stock thickness.*

Opposing this design necessity is that if pilots are long enough to make complete entry before stripper contact, the stock cannot be completely stripped and will tend to rise on the pilots. Some designers are now incorporating small spring plungers to complete the stripping operation. As shown in Figure 2-23, the stripper removes the stock from everything but the tip of the pilot, from which the stock is removed by the spring plungers.

The hole should be tapped as close as practical to the pilot clearance hole to obtain the maximum possible thrust from the plunger.

Spring-loaded Pilots

Figure 2-22 shows the design of a spring-loaded pilot. Its one great advantage over stationary pilots is that if, for one reason or another, the stock has advanced a distance greater or less than the progression and the hole is off location so far that the pilot cannot possibly register it, the pilot will retreat against spring pressure into the upper shoe. It is nearly impossible to break this type of pilot but it does have its disadvantages. First of all, it will not gage the strip with the accuracy of a solid pilot. The very fact that it is spring-loaded makes it necessary for the pilot hole to be close to the final position before pilot entry. If the die is being fed by an automatic hitch feed, this condition is realized and pilots of this type are ideal. Here, they serve as excellent safeguards against pilot breakage due to the breaking of a punch which pierces the pilot hole. But if the die is being fed manually and the stock has a tendency to stick in the guides, this pilot will spend as much time in the upper shoe as it does in the stock, to the detriment of accurate gaging.

Second, in the event this pilot does break, it is much more difficult to replace than a commercial pilot held securely in a retainer. A third objection to this type of pilot is that it is more expensive than the commercial pilot blanks and retainers on the market. Finally, this pilot can be used only in a die with a heavy upper shoe, as a great deal of room is necessary for it to make a full retreat from stock level. However, when designing and building heavy-duty progressive dies to be run automatically, this pilot should be considered.

Microswitches

Spring-loaded pilots are often used to actuate microswitches which cause the press to stop in the event of a mis-hit. The action here is almost instantaneous. As soon as the pilot is raised the slightest amount, the microswitch shuts off the electric power, causing the brake to stop the motion of the ram. This microswitch setup is a must on many dies. Progressive dies with shave stations and progressive lamination dies are both examples of dies which can be seriously damaged by a mis-hit. The addition of this design to any progressive die is a wise precaution. It does, of course, require a press which can be stopped instantly.

Immovable Pilots

Another way to minimize pilot breakage due to errors in gaging or a broken punch is shown in Figure 2-24. In this design, pilot holes in the die sections are ground or lapped to the pilot diameter plus double stock thickness.

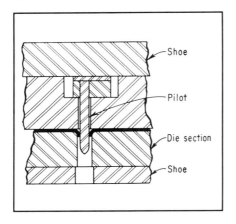

Figure 2-24. *In this example, hole diameter equals punch diameter plus double metal thickness.*

In the event the stock is off location as the dies close, the pilot pierces the strip flanging the metal over the edge of the pilot hole. This design is regarded by many as the optimum in pilot design.

In most automotive progressive dies the clearance hole in the stripper is simply a drilled hole laid out with a scale or a height gage if the stock is thin. The diameter of the hole should not exceed the pilot diameter plus metal thickness. However, on small high-precision dies operating at high speeds pilots must be guided and supported by the stripper itself. This can be accomplished as shown in Figure 2-25.

In this design the pilots are guided and supported by liner bushings pressed into the stripper. The stripper itself must then be guided, for if there is the slightest lateral movement the pilots will snap. The necessary alignment is secured by the use of four guideposts moving in liner or shoulder bushings pressed into the stripper. As these dies often operate at enormous speeds, it is the duty of both designer and diemaker to ensure proper lubrication. Once the pins and bushings become galled, the accuracy of the die is permanently impaired.

Another good safety measure is to use shoulder bushings instead of liner bushings for guidepost movement. Liners have been known to get pushed out for one reason or another and when this happens the die is usually completely ruined.

Stock Guides, Strippers, and Pilots

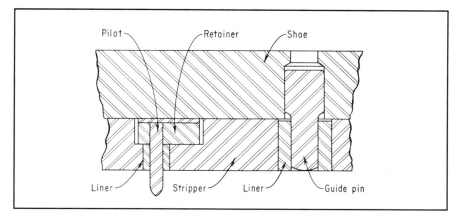

Figure 2-25. *Stripper design for small high-precision dies. The pilots and punches are stabilized by the stripper, which in turn is stabilized by four guide pins.*

In the past, piloting was kept at a minimum and in most cases still is, but today, with the more intricate pieces that are being made, piloting has become a problem of major importance. To pilot a strip through a die, a designer must have one of the following three conditions: a round hole in the part, an opening in or between the parts, or scrap that is removed somewhere past the third stage. In the first two conditions diemakers need only pilot in holes or openings; in the last case, only piercing one or more holes in the scrap and piloting them in is needed.

If there is a large hole in the part, as in Figure 2-26, a smaller hole of nominal size can be pierced in the first stage and used for piloting in subsequent stages. At a later stage, the desired hole is pierced in the strip.

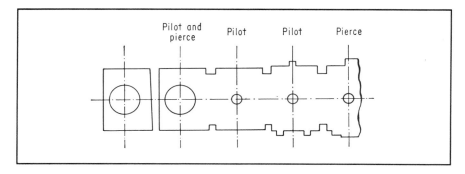

Figure 2-26. *Piloting in the scrap. This is easily accomplished when large holes are blanked in terminal stations.*

49

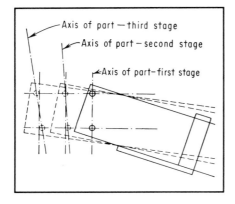 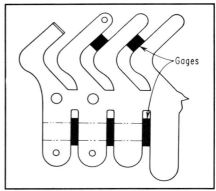

Figure 2-27. *Rotation of a part in a strip caused by a faulty stock guide.*

Figure 2-28. *Gages used for part location.*

The diemaker constructing a manually fed progressive die must take one important precaution that is often overlooked. The outer stock guide of such a die must run absolutely parallel to the line of pilots. As shown in Figure 2-27, the amount that the outside stock guide is out of parallel to the line of pilots tends to multiply with each successive progression. If the error is 1 degree at the second stage, it will be 2 degrees at the third. This means that in an eight-stage die, the part will have rotated nearly 7 degrees on its axis and will, of course, be far outside of tolerance. This out-of-parallel condition also causes an enormous lateral thrust on the pilots. Pilots shown in Figure 2-22 will retreat to the upper shoe, and those shown in Figure 2-24 will soon break. The diemaker should mount the stock guide as close as possible to the correct position, omitting dowels. In tryout he can make any final corrections necessary, pull the lower half of the die out, and dowel it.

Gages

On some dies conventional piloting is impossible or impractical and other solutions to problems of strip guidance must be found. This is particularly true on strips having long extended "leaves," as shown in Figure 2-28. On these dies, small gages mounted on the die sections will locate the strip as accurately as any form of piloting.

There are holes in this part which are pierced after a forming operation has taken place — which brings up another consideration. All these hole locations could have been developed and placed in the preliminary stages of the die. Subsequently, they could have been used for piloting. However, it is extremely poor practice to pilot in a hole whose location has shifted because of forming operations.

Positive Strippers

Positive strippers can often be used on progressive die designs in which stock stays at die level. They are comparatively simple to build, are 100 percent effective, and eliminate the need for springs and spring pockets. Whenever a designer can employ this type in lieu of a spring-actuated stripper, it is recommended to do so. Such instances are rare because nine out of ten progressive dies include forming stages, and when steel is formed progressively, its feed level is higher than the die level—in which case spring-actuated strippers must be used.

Lamination dies are examples of dies in which feed level and die level are identical, but ones in which positive strippers cannot be used. The reasoning here is that a spring-actuated stripper is necessary to guide the punches. In any die in which punches require support — and this includes the great majority of the carbides — spring-actuated strippers are mandatory.

An example of a conventional positive stripper is seen in Figure 2-29. The precision member of this assembly is the back rail, which must be hardened and ground. Water-hardening tool steel is generally used for this component, although carburized and hardened machine steel is acceptable. Because the function of this component is to maintain stock alignment while holding dimensions from the die center line, carburized and hardened cold-rolled steel should never be used. Many designers prefer cold-rolled steel for this member because it is finished to nominal size and no machining is required. However, cold-rolled steel is an unstable material which should not be used for precision details. The stripper plate itself should be constructed of cold-rolled steel, as should the other members of the assembly.

Ample clearance should always be provided in this assembly. Dimension T should be at least twice stock thickness to provide clearance for warped or bent stock. Dimension W should be at least 0.25 inch (6.4 mm) wider than stock width. Here again, the new diemaker gets into trouble. Having been trained to work in three- and four-place decimals, he or she often tries to reduce W to a "nice sliding fit" on the stock. This would be all right if stock were always sheared to exact size — which it is not. To try to

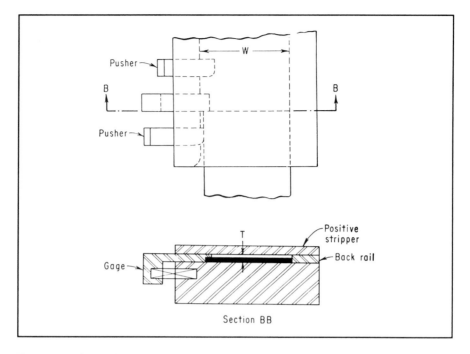

Figure 2-29. *Sectional view of a positive stripper, backrail, and pushers.*

fit W to stock width plus 0.010 inch (0.25 mm) is an invitation to trouble and it accomplishes nothing. The moral here is simple — "do not be afraid to open it up."

In operation, stock is held securely against the back rail by means of two "pushers" which are actuated by springs. It will be noted that the pushers are well chamfered to permit easy stock entry, as are the back and front rails. Thirty degrees is often specified as an entry angle. Pushers can be made of almost any material as long as their tips are hardened. Cold-rolled steel or machine steel, cyanided on the end, works very well.

Section BB of Figure 2-30 is a view through the gage which is used to locate the stock for the initial hit. When a spring-actuated stripper is used, the initial hit is gaged visually.

Visual gages are not feasible when positive strippers are used, although some designers specify a drilled peephole in the plate.

The gage shown is quite effective and sometimes a number of them must be used in a single die. For example, if the strip is gaged on an end gage and there are eight stations in the die, seven such gages are necessary to guide the strip through the length of the die. After the seventh hit, the end gage picks up the strip and performs the gaging function.

In designing this gage, the designer should specify a 0.015-inch (0.38-mm) pull-back if the die utilizes pilots.

3

Cam Stages

Design

Cam-action stages are used infrequently in modern progressive die work but on occasion they are useful for certain forming, piercing, and stripping operations. Important in their use is that, where possible, they have positive returns. In single-operation work, spring-returned cams are acceptable since the panel generally cannot be removed from the die in the event of spring failure. Progressive dies, however, normally operate at high speeds and are fed automatically. If spring failure leaves a cam in its forward position, the result may be a ruined die and quite possibly a damaged press.

Figure 3-1 is an example of a good design for piercing using a positive return cam. The cam itself can be made of water-hardening tool steel, as are inserts 2 and 3 in the figure. The use of inserts such as these effects an economy in machining time and allows for ground working surfaces within

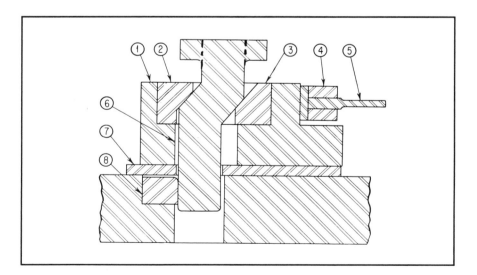

Figure 3-1. *Positive-return cam. Use of inserts simplifies building and timing of cam.*

the cam itself. To make such cams a solid piece results in a tool of inferior quality that generally costs more.

The optimum driving angle for cams is generally accepted as 40 degrees from vertical, although it can vary, depending on the travel required. The closer the driving angle is to vertical, the greater its mechanical advantage. The driver, detail 6, should be made of 0-6 material and constructed with a mild steel base for economy and low cost of replacement if broken.

If the cam is to pierce or form heavy material, the driver may soon fail if it is unsupported. To prevent such failure, designers incorporate an insert (number 8 in Figure 3-1) to share the strain which otherwise is placed on the driver itself. The designer's decision to use this added support should be carefully weighed, inasmuch as it is expensive and entirely unnecessary on dies which stamp lighter gage material. Use of standard, purchased carbon steel inserted keepers adds to economy of construction.

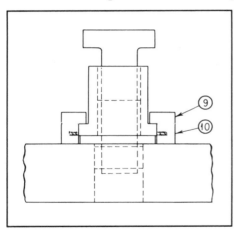

Figure 3-2. *Simplified design for cam guide rails.*

In Figure 3-2, the two details (10) on each side of the cam govern its position, so they must be set close enough to allow no sideplay. Therefore, the dowel holes in these details should be lapped to a gage fit. If they are any tighter than a gage fit, the dowels, when driven in, will cause a swelling, bind the cam, and cause cam "freeze-up." If alignment requirements dictate a tight dowel fit in the keepers, the diemaker should drive the dowels into the keeper and grind its sides, thus eliminating the swelling. The pins are then driven out and the keepers set and doweled in position.

Cam Forming Die Sections

Cam-action stages are sometimes useful for completing a forming operation in a progressive die. Figure 3-3 shows the first, second, and third forming stages of such a strip as well as a cross section of the third forming station in the die. Obviously, this part could be formed over a stationary die section in the third forming stage. However, the strip could not be raised from die level, as the "hook" would make contact and be deformed by the die section. To solve this problem, the forming section is made movable, receding enough to clear the hook as the stock rises to feed level.

In such designs, proper timing of the cam is critical. The cam should reach its forward position just before the stripper brings the strip down from feed level. This is important because the cam must start its return as soon as the strip leaves die level. Important as well is that the driver be well established in its period of "dwell" before the flanging action begins. This ensures that the cam is well backed up against the thrust caused by this operation. In addition, it is good practice to provide for positive return of the cam slide.

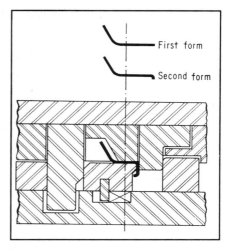

Figure 3-3. *Complex cam forming stage. Timing in such designs is critical.*

All in all, before beginning the design of the driver, the designer must consider:
- Cam movement required,
- Length of flanging section,
- Period of dwell,
- Stripper thickness,
- Stripper travel.

This cam-section design is not used frequently and should be considered only when other options have been discounted. If the strip can possibly be kept at die level, stationary sections can be used for this operation as long as clearance for the hook is machined in the subsequent stations.

Consistent with this objective, it is important to remember that die designs should always be as simple as possible. Many highly sophisticated mechanisms and motions are in place that lend themselves well to die design, but they are costly, and a primary concern of designers and diemakers should be cost-effectiveness. With a finite amount of money budgeted for tooling, a master mechanic or chief engineer will not be entranced with the mechanical superfluities of designers.

Cam Forming Punches

Cams are occasionally used to carry forming punches in both single operation and progressive work. The forming punch can be made an integral part of the cam, but experience teaches that a separate punch designed as an insert in the cam is preferable. One reason for designing the punch as an insert is that it can be made of a higher quality steel. But this comes with a higher price tag as well.

If a cammed punch is to be subjected to excessive wear and abrasion, as are the inserts shown in Figure 3-4, a high-carbon, high-chrome steel should be used. The cam carrying the punch should be made of water-hardening tool steel or carburized and hardened machine steel.

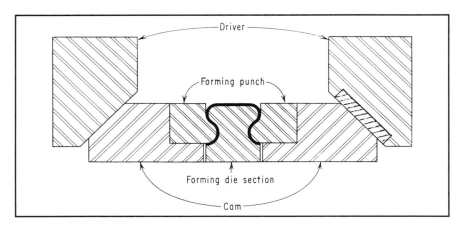

Figure 3-4. *Cam forming stage. Inserts provide easy correction of errors in part formation.*

Dwell Period

Cams carrying piercing punches may or may not have a dwell period, but those carrying forming and restriking punches should not. If such cams and drivers dwell, there is no quick way of increasing or decreasing the forming pressure. In the case where dwell in a forming punch is absolutely necessary, an insert, either in the cam or in the driver, should be incorporated that can be shimmed out to increase the forming pressure. If the punch insert is not doweled vertically in the cam, it can be shimmed forward when necessary. From this, another useful rule emerges: *Always* use a heel on the cam when *forming*.

Cammed Curling Stages

The designer or diemaker who has had experience with single-operation curl dies will have no difficulty in understanding Figure 3-5. Those new to the profession might feel instinctively that it will not work.

The first form is a conventional "form at 45-degree stage" with a small amount of the curl formed into the ends of the flange.

The second stage forms the flanges to the desired 90 degrees, after which a restrike stage may be designed, if desired.

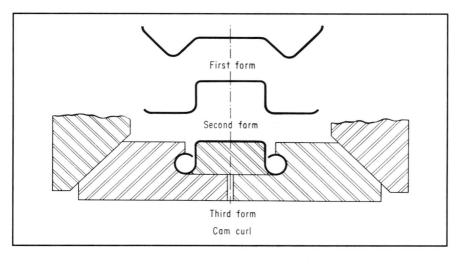

Figure 3-5. *Cam curl formation — the third forming stage of a progressive die.*

The final form stage is the cam curl, in which the metal follows the form contour in the advancing cams.

When curls of relatively small diameter are required — or when forming thick metal — it is wise to build the cams and the form section first. By mounting them in a separate die set, the diemaker can "prove" the operation. If it will not function, it is obviously impractical as a progressive operation and a single-operation die with mandrels must be built.

Note that part of the curl is formed in the first forming stage, the purpose of which is to act as a lead in the cam operation. The designer should strive to obtain the maximum amount of curl possible in preliminary form stages — 90 degrees if possible. The more precurl, the more certainty of a successful cam operation. Another necessary precaution is proper application of polish in the direction of metal flow. The smoother the curling surface, the more certain the diemaker is of success with this portion of the die.

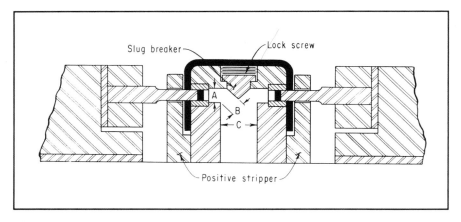

Figure 3-6. *Positive stripper for cam-pierce operation.*

Strippers for Cam Piercing

An inexpensive way of stripping a cam-pierce punch is with a positive stripper, as illustrated in Figure 3-6. In this example the stripper is simply a piece of machine steel screwed to the side of the die block. It is machined to allow passage of the advancing strip, and a punch clearance hole is drilled after its location has been determined at assembly. Although this type of stripper is preferable from a cost standpoint, it has limited applications in progressive dies. It is primarily for use on heavy material, and when piercing holes opposite each other, a burr forms which will cause holdup of feed and possible mis-hit. Also, if the surface to be pierced has a number of projections such as tabs, embosses, etc., or if its plane is not parallel to the die center line, this type of stripper cannot be used.

Spring Strippers

If there is sufficient room in the die, a spring-actuated stripper can be used to strip a cam-pierce operation (Figure 3-7). A shoulder screw can be used, to both guide and support the stripper. Generally, four springs are used, properly placed for balance. Designers should calculate stripping pressure requirements, then double them as a safety precaution. They should also take care to ensure that screw heads do not extend beyond the surface of the stripper when the cam is in its forward position.

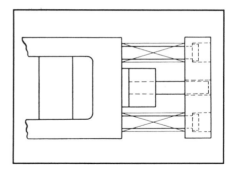

Figure 3-7. *Spring-actuated stripper for cam piercing.*

Rubber Strippers

In operations where comparatively light-gage material is being cam-pierced, it is feasible to use a piece of rubber as a stripper. With this method, a hole slightly smaller than the punch is drilled through the block of rubber

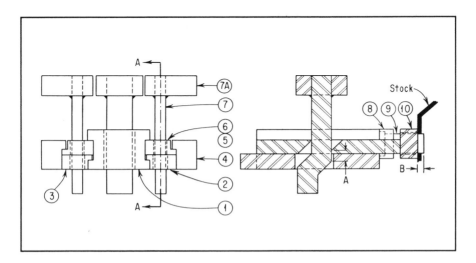

Figure 3-8. *Cam-actuated stripper for cam piercing. This design is unique since it provides positive stripping action from a stripper that moves.*

61

which is then forced over the punch. Now commercially available, these devices effectively shed the stock. Today they are in fairly widespread use on less expensive dies.

Cam-actuated Cam Strippers

Cam-actuated cam strippers are unique in that they are the only examples of positive strippers that move, providing positive stripping action although they advance and recede. Such strippers, illustrated in Figure 3-8, are foolproof and highly efficient. They effectively strip long-perimeter punches out of heavy-gage metal with minimal potential for mechanical failure.

A major drawback of this device, however, is the large size of its complete assembly which limits its use to larger progressive dies. If used in smaller progressives, it may require two or more additional idle stations.

The major cam (detail 1 of Figure 3-8) and the two minor cams (details 5 and 6) have reached their maximum forward position and the die is closed. The punch (detail 9) has advanced through the stripper (detail 10) and has pierced the stock. It is now ready to withdraw and begins the stripping action. As the die opens (refer to Figure 3-1), the piercing cam (detail 1) returns by positive action but the two stripping cams maintain their position. They cannot begin to recede until the drivers have raised distance A, but by this time the piercing cam has receded a distance greater than B, thereby stripping the punch.

If the driving angle of the piercing cam is 45 degrees, it will return at exactly the same rate of speed that the die is opening. And if A is equal to B, the stripping cam begins its return at precisely the moment the punch is flush with the stripper. Good practice, however, is to leave at least 0.25 inch (6.35 mm) more so that A equals B plus 0.25 inch.

If the driving angle deviates from 45 degrees, the problem of determining A is slightly altered.

The distance a positive-return cam advances or recedes is a function of the tangent of its driving angle. This means that the distance the die must open to produce a movement of A is equal to B divided by the tangent of the driving angle.

$$A = \frac{B}{\tan \text{driving angle}} + 0.25 \text{ inch (6.35 mm)}$$

The safety factor of 0.25 inch (6.35 mm) is, of course, subject to change by the designer, varying according to size and strength of the cam assembly.

It is well to point out that this assembly, like the cam-action die section of Figure 3-3, is still another example of a relatively complicated mechanism performing a simple function. It should be used with discretion and only when other stripping devices will be ineffective.

Slug Disposal

In cam-piercing stages in progressive dies, as in single-operation cam-piercing dies, there exists the problem of slug disposal. Slugs have to turn a corner to drop out through the die shoe, and the diemaker as well as the designer has a definite responsibility to ensure that the die is foolproof against slugging up.

A rule that simplifies this disposal problem is that the slug, when changing direction, must always enter a larger area (Figure 3-6). Thus, dimension B at the corner is larger than dimension A but less than half the size of C.

Slugs have a tendency to stick together in columns, and these columns, whether vertical or horizontal, must be broken up. For example, if slug columns were not broken up in the die section illustrated in Figure 3-6, they would eventually produce disastrous consequences. The simplest possible insurance against such an eventuality is the conical slug breaker (separator) shown. This breaker, which is held securely in place with a setscrew, can be turned out of cold-rolled steel and cyanided for maximum effectiveness.

Press Selection

Proper selection of a press is essential for successful and economical operation. The purchase of a press represents a substantial capital investment, and return on investment depends upon how well the press performs the job required. No general-purpose press exists that can provide maximum productivity and economy for all applications. Compromises usually have to be made to permit a press to be employed for more than one job. Careful consideration should be given to both present and future production requirements.

Important factors influencing the selection of a press include size, force, energy, and speed requirements. The press must be capable of exerting force in the amount, location, and direction, as well as for the length of time needed to perform the specified operation(s). Other considerations must include the size and geometry of the workpieces, operation(s) to be performed, number of workpieces to be produced, production rate needed, accuracy and finish requirements, equipment costs, and other factors.

Size Requirements

Bed and slide areas of the press must be large enough to accommodate the dies to be used and to provide space for die changing and maintenance. Space is required around the dies for accessories such as keepers, pads, cam return springs, and gages; it is also needed for attaching the dies to the press. Shutheight of the press, with adjustment, must also be suitable for the dies.

Presses with as short a stroke as possible should be selected because they permit higher speed operation, thus increasing productivity. Stroke requirements, however, depend upon the height of the parts to be produced. Progressive blanking can be done with short strokes, but some forming and drawing operations require long strokes, especially for ejection of the parts.

Size and type of press to be selected also depends upon the method and direction of feeding; the size of sheet, coil stock, blank, or workpiece to be formed; the type of operation; and the material being formed and its strength. Material or workpiece handling and die accessibility generally determine whether the press should be of gap-frame or straightside construction (discussed later in this chapter) and whether it should be inclined or inclinable.

Physical size of a press can be misleading with respect to its capacity. Presses having the same force rating can vary considerably in size depending upon differences in length of stroke, pressing speed, and number of strokes per minute.

Force required to perform blanking and forming operations determines the press capacity, expressed in tons or kilonewtons (kN). The position on the stroke at which the force is required and the length of stroke must be considered.

Mechanical presses are generally rated near the bottom of the stroke. It is customary to provide the torque necessary to exert the rated press force at some given point above the bottom of the stroke and capacity decreases above this point. Operations requiring force application higher in the stroke should be performed on presses with greater torque in their drives and more flywheel energy.

Energy or work (force times distance), expressed in inch-tons or joules (J), varies with the operation. Blanking and punching require the force to be exerted over only a short distance; drawing, forming, and other operations, over a longer distance. The major source of energy in mechanical presses is the flywheel, the energy varying with the size and speed of the flywheel. Energy available increases with the square of the flywheel speed.

Possible problems are minimized by selecting a press having the proper frame capacity, drive-motor rating, flywheel energy, and clutch torque capacity.

Types of Presses

Presses are classified by one or a combination of characteristics which include the source of power and the number of slides. Other classification methods are the types of frames and construction, types of drive, and intended applications.

Source of Power

Power for progressive die press operations is generally mechanical or hydraulic.

Mechanical presses use flywheel energy which is transferred to the workpiece by gears, cranks, eccentrics, or levers. Mechanical presses can be nongeared or geared, with single or multiple-reduction gear drives, depending upon the press size and force requirements.

Hydraulic presses provide working force through application of fluid pressure on a piston by means of pumps, valves, intensifiers, and accumulators. While mechanical presses are still the predominant type in use, hydraulic presses are becoming increasingly popular because of their improved performance and reliability.

Number of Slides

In terms of function, presses may be classified by the number of slides incorporated and are referred to as single, double, and triple-action presses.

Single-action presses are most commonly used with progressive dies. They have one reciprocating slide (tool carrier) acting against a fixed bed. Presses of this type, which are the most widely used, can be employed for many different metal stamping operations, including blanking, embossing, coining, and drawing. Depending upon the depth of draw, single-action presses often require the use of a die cushion for blankholding. In such applications, a blankholder ring is depressed by the slide (through pins) against the die cushion, usually mounted in the bed of the press.

Press Speeds

Press speed is a relative term that varies with the point of reference. High speeds are generally desirable, but they are limited by the operations performed, the distances above stroke bottoms where the forces must be applied, and the stroke lengths. High speed, however, is not necessarily the most efficient or productive. Size and configuration of the workpiece, the material from which it is made, die life, maintenance costs, and other factors must be considered to determine the highest production rate at the lowest cost per workpiece. A lower speed may be more economical because of possible longer production runs with less downtime.

Speed Ranges. Simple blanking and shallow forming operations can be performed at high speeds. Mechanical presses have been built that operate to 2000 strokes per minute (spm) with a 1-inch (25.4-mm) stroke, but applications at this maximum speed are rare. Speeds of 600 spm to 1400 spm are more common for blanking operations, and thick materials are often blanked at much slower speeds. For drawing operations, contact velocities are critical with respect to the workpiece material and presses are generally operated at slide speeds from 10 spm to 300 spm, with the slower speeds for longer stroke drawing operations.

Requirements for High-Speed Operation. High-production presses that operate to about 200 spm generally do not require special dynamic balancing. Those operating from 200 spm to 600 spm, with force ratings of 25 tons to 400 tons (222 kN to 3558 kN), usually have crankshaft counterweighting to minimize vibration. Presses that operate at speeds over 600 spm, with force ratings generally less than 200 tons (1779 kN), are available with adjustable or fixed stroke lengths. Adjustable types have a maximum speed for each stroke length; fixed-stroke presses are dynamically balanced for the maximum speed.

All high-speed presses require rigid frames and beds to minimize deflection and increase shock-absorbing capabilities. Alignment of the slide to the bed of the press is critical and requires minimum-clearance gibbing or antifriction bearings. The presses are usually furnished with automatic recirculation systems to provide lubricant to all wear surfaces, with the systems interlocked with the press drives to stop the presses if the oil pressure falls below a safe limit.

Automatic, high-speed, accurate means for feeding and unloading the presses and fast, reliable safety systems are essential. Most high-speed presses are equipped with variable-speed motor drives. It is generally recommended that high-speed presses be mounted on inertia blocks or isolation mounts to isolate them from the plant foundation.

Limitations of High-Speed Operation. Press speeds above about 700 spm increase the amount of noise generated and may require the use of sound enclosures. High-speed operation also increases the amount of heat generated and decreases the shutheight because of stretching of the drive connections. This can generally be controlled, however, by maintaining the recirculation lube at a constant temperature and carefully monitoring shut height and adjust, if needed. Other possible limitations include decreased accuracy, repeatability, and die life.

Mechanical versus Hydraulic Presses

Mechanical presses are the most predominant type used for blanking, forming, and drawing of sheet metal, but the use of hydraulic presses is increasing. There are applications for which hydraulic presses offer certain advantages, and in some cases, are the only machines that can be employed. For example, very high force requirements can only be met with hydraulic presses. A comparison of characteristics and preferred uses for both mechanical and hydraulic presses is presented in Table 4-1.

Press Systems

Presses integrated with material-handling equipment, feeding and unloading devices, and other manufacturing equipment for automated, synchronized operation are being used more extensively. The reasons include improved quality, increased productivity, lower costs, and reduced inventories. Press systems are not limited to high-production applications; the development of means for making quick die changes has increased their flexibility and made short runs economical.

Mechanical Presses

Mechanical presses have a slide or slides actuated by mechanical means. They are sometimes referred to as power presses to differentiate them from hydraulic presses. All mechanical presses employ flywheel energy which is transferred to the workpiece by gears, cranks, eccentrics, or levers. They are available in many different types and sizes and with various drives.

Types of Frames and Construction. Basic functions of a press frame are to contain the loads imposed with a minimum of deflection, which requires ample rigidity. The two major types of press configurations are gap-frame and straightside. Straightside presses are sometimes constructed with column-type frames. Important criteria for selecting the type of frame to be used include accessibility and operating characteristics, convenience of feeding and unloading, stiffness, and profile.

Table 4-1
Comparative Characteristics of Mechanical and Hydraulic Presses

Characteristic	Mechanical Presses	Hydraulic Presses
Force	Variable (depends upon slide position).	Relatively constant (does not depend upon slide position).
Stroke length	Limited.	Capable of long strokes (100 inches [2540 mm] or more).
Slide speed	Higher speed capability. Highest at mid-stroke. Can be variable.	Slower pressing speeds with rapid advance and retraction. Variable speeds uniform throughout stroke.
Capacity	Under 1500 tons (13.4 MN) for progressive dies.	For progressive dies, generally below 500-600 tons (4.5-5.4 MN).
Control	Full stroke generally required before reversal.	Adjustable, can reverse slide at any position.
Preferred uses	Operations requiring maximum pressure near bottom of stroke. Cutting operations (blanking, shearing, piercing). Forming and drawing to depths of about 4 inches (100 mm). High-production applications. Progressive and transfer die operations.	Operations requiring steady pressure throughout stroke. Deep drawing. Die tryout. Flexible die forming. Drawing irregular shaped parts. Straightening. Hubbing of mold and die cavities. Operations requiring high and variable forces. Operations requiring variable or partial strokes.

Press frames are made of cast iron, cast steel, or welded or bolted-steel construction. Some frames are made from machined posts or pillars. Many press builders are now using computers for frame design to optimize material utilization and ensure maximum stiffness, strength, and performance. The five most common types of construction are:

- One-piece frame of cast iron, cast steel, or steel weldments.
- Four-piece, steel, tie-rod frame. This construction consists of the bed, two uprights, and the crown held together by steel tie rods (usually four) which are preshrunk in excess of the rated force.
- Bolted frame. This construction consists of the bed, two side members, and the crown keyed and held together by bolts. This type of construction is used frequently for single- and double-crank, gap-frame presses.
- Modified tie-rod frame. This construction combines the tie-rod and bolted types of frames.
- Solid frame with tie rods. With this type of construction, steel tie rods are shrunk into the solid frame.

Gap-frame Presses. The housings of a gap-frame (also called C-frame) press are cut back below the gibs to form the shape of a letter C. Presses of this type are the most versatile and common in use and are lower in cost than straightside presses. They provide unobstructed access to the dies from three sides, and their backs are usually open for the ejection of stampings and/or scrap. Press feeding can be done conveniently from the side or front (on open-back presses). Gap-frame presses generally have a lower overall height than straightside presses of the same capacity, which is important when overhead clearance is limited.

Gap-frame presses are available in several different designs, some of which are shown in Figure 4-1. The types include permanently upright presses, such as the adjustable-bed stationary (ABS) and open-back stationary (OBS) presses illustrated; permanently inclined presses; and open-back inclinable (OBI) presses, which are the most common. The inclined presses often facilitate feeding and permit finished stampings to fall out by gravity or to be blown out by air at the rear of the press.

Gap-frame presses are made with either one or two points of suspension. Large gap-frame presses are generally equipped for mounting cushions for workpiece liftout or for shallow-draw operations. Powered slide-adjustment systems are available for faster die changing.

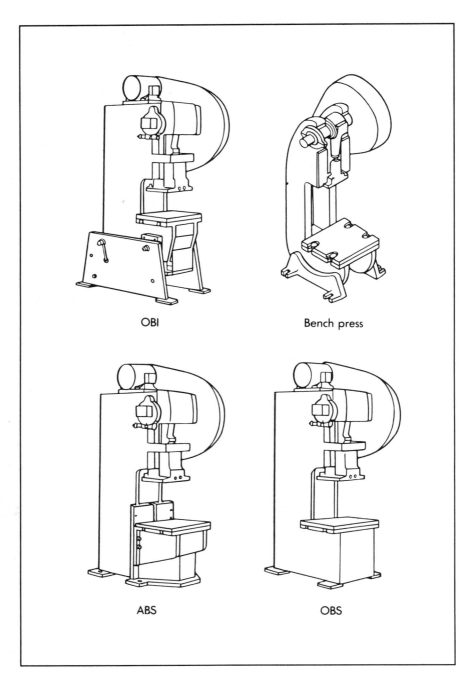

Figure 4-1. *Several types of gap-frame presses: OBI — open-back inclinable, ABS — adjustable-bed stationary, OBS — open-back stationary; and bench press.*

These presses are commonly arranged with their crankshafts extending from right to left of the die space. They are also available with the crankshaft extending from front to back, with the flywheel or gear at the rear of the press.

Gap-frame presses are available from small bench types of 1-ton (8.9-kN) force capacity to OBIs of 300-ton (2669-kN) capacity. Such presses with two-point suspension are made with capacities to 300 tons (2669 kN) or more.

While gap-frame presses have many desirable features, they have the disadvantage of more deflection under load than straightside presses of the same capacity. Also, due to the geometry of the gap construction, the deflection results in an out-of-parallel condition between the top surface of the press bolster and the bottom surface of the slide. Proper location of the dies, especially progressive dies, is critical because single-point, gap-frame presses are not generally designed for off-center loading unless plunger guides are used. Excessive deflection can damage the dies or press and can result in the production of unacceptable parts. The degree of misalignment is proportional to the force required. If punch and die alignment are critical, either from the standpoint of workpiece accuracy or excessive die wear, gap-frame presses should not be used.

Straightside Presses. Presses with straightside frames consist of a crown, two uprights, a bed which supports the bolster, and a slide which reciprocates between the two straightsides or housings. The crown and bed are connected with the uprights by tie rods or by bolting and keying together, or all members can be cast or welded into one piece. Continuous, welded, box-type construction is commonly used to minimize twisting, especially when the press is to be subjected to off-center loading. Each construction method has certain advantages and limitations. Solid-frame, straightside presses are generally less expensive than tie-rod presses, but their size is limited because they must be transported from builder to user in one piece.

Tie-rod construction provides several advantages. If the press slide becomes stuck at the bottom of its stroke, it is possible to remove the stress applied to the rods by heating them, thereby enabling the cause of the problem to be corrected. To facilitate this procedure, hydraulically operated tie-rod nuts are available, generally as original equipment.

Another advantage of tie-rod construction is that tie rods provide some overload protection. Figure 4-2 shows an example of deflection curves for solid frames and for tie-rod frames with cast iron and steel uprights. With

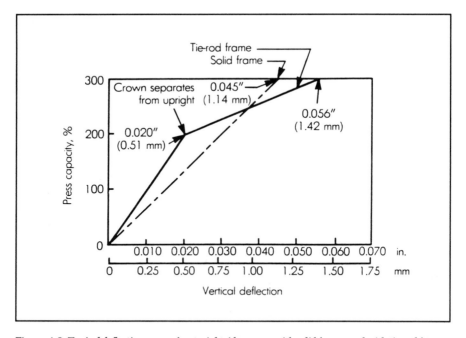

Figure 4-2. *Typical deflection curves for straightside presses with solid frames and with tie-rod frames.*

a tie-rod frame, the crown separates from the uprights at about 200 percent of press capacity and the stiffness of the frame under additional load is determined by the rods alone. Rate of deflection changes from about 0.010 inch (0.25 mm) to 0.036 inch (0.91 mm) per 100 percent of press capacity, and the frame is only 28 percent as stiff after crown separation. As a result, load buildup because of accidental interference increases at a slower rate, thus decreasing the possibility of a catastrophic accident. Use of a hydraulic overload device, however, provides more positive protection for short displacements.

Straightside presses are available for single or multiple-action operation and in a number of sizes from small-capacity, special-purpose machines to those having capacities of 2000 tons (17 792 kN) or more.

Dies mounted in straightside presses are accessible primarily from the front and rear. Openings are generally provided in the uprights, giving some limited access to the right and left ends of the die space, and are used for automatic feeding and unloading. The wide range of bed sizes available with these presses allows selecting the narrowest front-to-back dimension commensurate with the dies to be used. Moving bolsters are often used on large, straightside presses to facilitate die changing.

Straightside presses are also stiffer than gap-frame presses. They have no angular deflection, and vertical deflection of their uprights under load is practically symmetrical when they are symmetrically loaded. As a result, usually no problems are created with respect to punch and die alignment.

Vertical stiffness of a press frame depends upon the cross-sectional area of the uprights on solid or bolted-and-keyed frames. With tie-rod frames, stiffness depends upon the cross-sectional areas of the uprights and tie rods and the proportion of one to the other. Adequate stiffness is necessary for the die to function properly and to prevent excessive snap-through during blanking or severe punching operations. Too much stiffness, however, can cause a proportionally greater load on all press members, including the drive, in cases of excessive stock thickness or misfeeds.

Round-column Presses. Column-type presses are similar to straightside presses, but they have round columns, pillars, or posts instead of the side uprights. Most column-type presses have four columns, but some have two or three columns. Bushings of the bronze-sleeve or ball-type surround the columns for good guidance. Column presses are available for horizontal operation, permitting gravity ejection of stampings and/or scrap.

Press Slide Connections. Mechanical straightside presses are generally arranged so that the rotary motion of their drives is transmitted to their reciprocating slides through one, two, or four connections (pitmans). Depending on the number of connections, the presses are referred to as being of single-point, two-point, or four-point suspension.

Presses with small, nearly square slide-face areas, usually 42 inches^2 (27 097 mm^2) or less, are generally of *single-point* suspension. The parts of the one connection are sized to carry the rated press capacity.

When a single-point slide is loaded at any position other than directly under the connection, the slide tilts. The amount of tilt depends on the clearance between the slide guides and the gibs and on the amount of load. Single-point presses are not recommended, however, when significant off-center loading is to be applied.

Presses with single-point suspension are also not generally recommended for precision, long-run operations because die wear can be excessive. When progressive dies are used, the loading at the various stations should be balanced.

Presses with *two-point* suspension generally have rectangular slides with a front-to-back dimension not exceeding about 54 inches (1372 mm). Each connection is designed for a maximum of 50 percent to 60 percent of the rated press load, depending on the press manufacturer. The maximum load should ideally be placed at the midpoint between the connections to avoid overloading. Also, it should be remembered that the slide tilts in the front-to-back direction if the resultant load on the slide is not in the same plane as the connections. Workpiece quality and die life are both improved if the dies are set in line with the connections.

Presses with *four-point* suspension generally provide the highest accuracy and longest die life. The rating for each connection is usually either one-fourth or one-third of the press capacity, depending on the press builder.

The four pitmans on these presses are better able to reduce tilting in either direction when off-center loading occurs. Press builders, however, recommend that the loads be centered as much as possible or that, if off-center loading is necessary, the loads be distributed as evenly as possible, making sure that the load on any connection is within its capacity.

Force, Energy, and Torque Considerations. The force rating of a mechanical press, expressed in tons or kilonewtons, is often the major consideration in the selection and application of the press. Torque and energy (work) capacities, however, are also critical, especially for deep-drawing operations, and should be given careful attention.

The *force capacity* of a mechanical press is the maximum force that should be exerted by the slide or slide-mounted dies against a workpiece at a specified distance above bottom of stroke. It is a relative measure of the torque capacity of the press drive and an actual measure of the structural capacity (size and physical strength) of the press components to withstand the applied load and resist deflection within the specified tolerances.

Distance Above Bottom of Stroke. Proper selection of a press for a specific application depends upon where in the stroke the maximum force is to be applied. Only forces less than the capacity of the press should be applied further up on the stroke than the distance specified in the press rating.

The reduced force capacities of straightside presses (rated at full capacity 0.5 inch [12.7 mm] above bottom of stroke) for force applications at various distances above bottom of stroke are shown in Figure 4-3. For example, if a stroke of 6 inches (152 mm) is required, with maximum force applied 2 inches (50.8 mm) above bottom of stroke, only 60 percent of the rated force capacity of the press should be applied. Since many presses are rated at different distances from the bottom of the stroke, a chart such as the one shown should be obtained from the builder for each press.

Torque Capacity. The drives of mechanical presses provide a constant torque, but owing to the mechanical advantage of the linkage, the forces transmitted through their clutches to rotating members and reciprocating slides vary from a minimum at midstroke to infinity at the bottom of the stroke. The torque value equals the force times the perpendicular distance from the force to the axis about which the force is applied.

Torque limitations of a press are determined by the size of the drive components (shaft, clutch, gears) that transmit flywheel energy in the form of torque to the slide, by the stroke length, and by the distance above bottom where the force is applied. Presses rated higher in the stroke (above bottom dead center) require greater torque capacity in the drive components and more flywheel energy. Exerting too high a torque causes press members to fail.

Energy Requirements. In addition to knowing that a press and its drive can provide adequate force at a certain distance above stroke bottom, it is equally important to determine whether the press has sufficient work (energy) capacity. This is the ability to deliver enough force through the distance required to make a particular part. The product of force times distance (working stroke) is the energy load on the press, usually expressed in inch-tons or joules (J) — 1 inch-ton equals 226 J.

As energy is expended in forming a part in the die, the flywheel slows down. If the amount of energy used is within the design limitations of the press drive, the motor returns the energy used to the flywheel, bringing it back to its starting speed before the press ram reaches the top of its stroke, and starts the next stroke. For intermittent press operation, 20 percent is considered the maximum the flywheel may be slowed in removing energy from it. The motor must be of sufficient size to restore the energy in time for the next stroke.If not enough energy is available to form a specific part on the press, slowdown of the flywheel during the working portion of the press stroke becomes excessive. The results could be loss of press speed on each successive stroke, belt slippage and wear, and/or overloading of the main

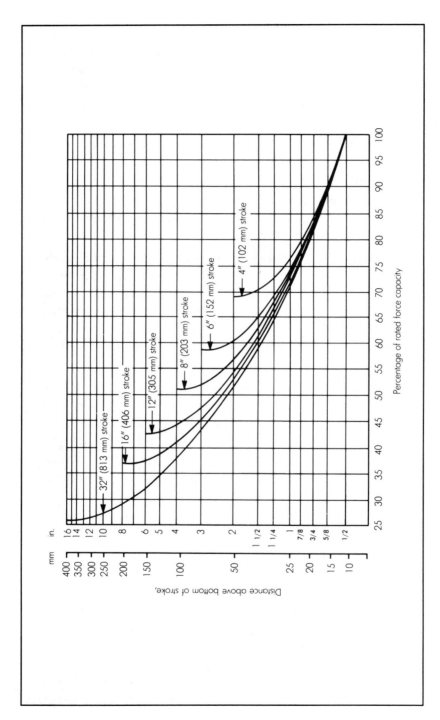

Figure 4-3. Percentage of rated force capacity for straightside presses at various distances above bottom of stroke.

drive motor. If flywheel slowdown is severe enough due to the use of most or all the energy available during a stroke, motor overheating can cause thermal overloads in the motor starter sufficient to break the circuit and stop the press. Even more critical is the potential for the press to stick at the bottom of the stroke. If this happens, considerable work may be required to get the press unstuck.

A general rule of thumb is that a standard press should have work capacity at least equal to its force rating times the distance from the bottom of stroke at which it is rated. For example, a 100-ton (890-kN) press rated at 14 inches (6.3 mm) from stroke bottom would usually be capable of providing only 25 inch-tons (5650 J) of energy to do work on a continuous basis. For presses using multistation dies, the energy requirements at all stations must be added.

Motor Selection. Determining the proper motor power for a specific application requires careful consideration of energy requirements, stroking mode and rate, and press inertia. Empirical methods based on data that has given satisfactory results for similar applications are sometimes used to select motor power, flywheel energy, and other design criteria. Analytical methods require solving exponential equations involving many independent variables. More recently, the use of computer programs for determining duty cycles has facilitated the process, with less chance for errors.

Controls. The subject of press controls deals with the electrical, electromechanical, pneumatic, hydraulic, electronic, and associated equipment used to control power presses. These controls range from a simple, single-motor starter and disconnect switch to a sophisticated multimotor, multifunction control incorporating a variety of sensors, control systems, and devices. But regardless of the type, the resulting operation must yield consistent, predictable high production rates without jeopardizing the safety of either the operator, the press, or the dies. Proper design application and installation are critical factors, whereas easy accessibility, efficient operation, minimum downtime, and limited maintenance are desirable features for press controls. Self-checking and diagnostic capabilities are becoming increasingly important.

Systems used for the control of mechanical power presses are predominantly either electromechanical or solid state or a combination of the two. More recently, programmable controllers (PCs) and computer numerical control (CNC) systems have been popular, particularly in conjunction with more complex automated systems.

Hydraulic Presses

Hydraulic presses use one or more cylinders and pressurized fluid to provide the motion and force to form or blank workpieces. While mechanical presses are still the predominant type for progressive die work, hydraulic presses are sometimes used.

An increase in use of hydraulic presses in recent years is attributed to their improved performance and reliability. Improved hydraulic circuits and new valves with higher flow capacities and faster response times are major factors.

Advantages of Hydraulic Presses. The greatest advantage of a hydraulic press is its adjustability, which increases the versatility of the machine. With a nonfixed cycle and full force availability at any point in the stroke, a hydraulic press is compatible with various dies and operations.

Variable Force. The force exerted by a hydraulic press is infinitely adjustable from about 20 percent of its maximum rated capacity by simply varying the pressure relief valve setting. This is important in protecting dies designed for limited capacities and for various operations on different workpiece materials. The preset force is relatively constant throughout the stroke. It is virtually impossible to overload a hydraulic press because the press only operates to the preset force, regardless of variations in stock thickness, inaccurate dies, doubleheaders, or other factors.

Full force capacities of hydraulic presses can be applied at any point in their strokes, regardless of stroke lengths, thus permitting their use for both short- and long-stroke applications. This capability allows the use of a hydraulic press with a lower force rating than that needed for a mechanical press on which the operation requires the application of force high above bottom of stroke.

Variable Stroke and Speed. Compared to the fixed stroke of a mechanical press, the stroke of a hydraulic press can be easily adjusted to stop and reverse the slide at any position in the stroke. The speed of hydraulic presses is also variable; slide speed can be slowed from rapid advance to pressing speed just prior to contacting the workpiece. This can lengthen die life by reducing shock loads. Variable speed also permits selecting the optimum speed for each operation and workpiece material, thus ensuring high-quality parts and reducing setup time.

Other Advantages. Variable capacity is another feature of hydraulic presses. Bed size, stroke length, press speed, and force capacity are not necessarily interdependent. Hydraulic presses are available with large beds and low force ratings, small beds and high force ratings, and a wide variety of stroke lengths. While standard-size hydraulic presses are available, many are purchased especially designed to suit user requirements. Hydraulic presses are more compact than mechanical presses of comparable capacity.

Since hydraulic presses have fewer moving parts than mechanical presses, they generally have less downtime and require less maintenance. Hydraulic presses are essentially self-lubricating; the only additional lubrication required is that needed for the slide gibbing or column bearings. With properly designed and mounted hydraulic systems, the presses provide quiet operation.

Limitations. Hydraulic presses still have slower speeds than mechanical presses. The speed, however, depends on stroke length, and many applications exist in which hydraulic presses can outperform mechanical presses. There are small, automatic-cycling hydraulic presses operating at 900 spm for short-stroke applications. Pressing speeds range to about 600 ipm (15 240 mm/min), with approach and return speeds as high as 2000 ipm (50 800 mm/min).

Even with slower speeds, hydraulic presses are ideally suited for many applications, particularly hand-fed operations and small lot sizes. Automatic feeds require the use of an external or auxiliary power unit, integrated with the press control system.

Frame Construction. Hydraulic presses are available in a number of frame configurations, including straightside, column-type, C-frame, open-back stationary (OBS), and special-purpose designs. They are made in horizontal and vertical-acting models, with the vertical type being predominant.

Straightside Presses. Straightside hydraulic presses are available with single or two-point (connection) suspensions having one or two cylinders and may be single-, double-, or triple-acting. Force capacities generally range from under 50 tons to 500 or more tons for progressive die operations.

Straightside hydraulic presses are used extensively in applications for which good slide and bed alignment are essential for accuracy. Their rigid construction permits some off-center loading.

Operations in which these presses are used encompass the full range of metalforming and blanking operations. Presses with large bolster and slide areas permit producing large parts which may require blanking, punching, forming, and often, deep drawing of the same part. Controlled and adjustable speed stroke length and force facilitate the use of progressive dies for producing complex parts.

C-Frame Presses. Hydraulic presses with C- or gap-frames (see Figure 4-4) are a more recent development. The two solid steel housings on each press are cut back below the gibs to form the shape of a letter C similar to gap-frame mechanical presses. The ample throat and open sides of these presses provide easy access to the die area from three sides and permit loading/unloading wide or irregular-shaped workpieces. The force ratings of C-frame hydraulic presses generally range from 1 ton to 1000 tons (8.9 kN to 8896 kN).

Controls. Properly designed and applied controls are essential for the maximum protection of personnel, presses, and dies and to ensure optimum production rates with a minimum of downtime. Reliability of the

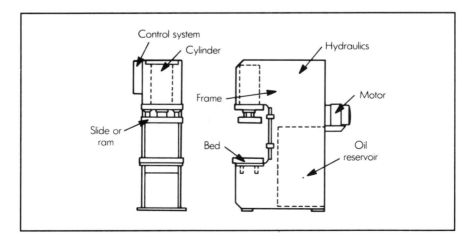

Figure 4-4. *Hydraulic press with C- (gap-) frame.*

controls is critical, and self-checking and diagnostic capabilities are important advantages.

The controls used for hydraulic presses are similar to those employed for mechanical presses with the addition of pressure gages and controls for varying the force applied, stroke length, and ram speed.

Types of controls employed include electromechanical systems, programmable controllers (PCs), solid-state controls, and microprocessors. Some electronic control systems provide for digital setting of all pressures and movements. The PCs permit fast setting of forming or blanking pressure, stroke length, dwell time, pull-out pressure, and other functions, as well as the sequence of functions for different applications.

Some microprocessor controls provide a data entry keyboard for simplified programming of slide stroke positions, forces applied, and auxiliary functions. Data can be stored in the control memory for future use on the same job. The controls are available with a visual display of data for setup, operation, and production counts; fault indications; and diagnostic information.

The maximum anticipated working pressures in any hydraulic system must not exceed the safe working pressure rating of the lowest rated component in that system. Mechanisms operated by hydraulic cylinders should have provisions for shutting off and bleeding pressure from all components that might cause unexpected operation or motion.

Dieing Machines

Dieing machines, sometimes called die presses, are used for high-speed precision stamping operations on small workpieces, often with progressive dies and coil feeds. These machines are actually inverted mechanical presses with their drive mechanisms located under the beds. Advantages include accurate die alignment and a low center of gravity. No floor pits are required, and the machines can be installed in rooms with low ceilings. The machines are available with capacity ratings ranging from 25 tons to 200 tons (222 kN to 1779 kN), speeds of 100 spm to 600 spm, and stroke lengths of 1 inch to 2 inches (25 mm to 51 mm). A high-volume application for progressive die applications is high-speed perforating work.

Lamination Presses

So-called "lamination presses" are high-speed, automatic, mechanical presses frequently used for stamping laminations, but also employed for shallow-drawing operations, progressive die operations, and other operations. The short-stroke presses are generally of straightside, tie-rod construction. Variable speed drives are often standard. Coil stock reels, straighteners, and feeders are offered as optional equipment.

The presses are available with rated capacities of 30 tons to 300 tons (267 kN to 2669 kN), strokes of 1 inch to 3 inches (25.4 mm to 76.2 mm), and speeds of 133 spm to 1500 spm.

Four-slide Machines

Four-slide or multi-slide machines are versatile presses equipped with progressive dies for automatically producing small, intricately shaped parts to close tolerances. Originally developed to form wire, these machines are now also used to progressively form sheet-metal parts. The machines straighten the metal as it is taken from a coil, feed it in exact lengths, form and cut off the parts, and form them further if required, thus minimizing or eliminating the need for secondary operations. Substantial material savings are generally obtained compared to other pressworking methods.

Parts formed on four-slide machines are comparable to those formed in progressive dies on presses, but more difficult operations can be accomplished without the expense of providing slides in the dies. Slides that can be timed independently are a standard part of four-slide machines, and they can be tooled inexpensively. Forms can be completed within the dies or over a block at the forming end of the machine. The form tools can dwell at closure while other operations are performed by auxiliary slides. If the part cannot be completed in one cycle, it can be transferred further out on the mandrel and completed in succeeding cycles (called multistage forming). Four-slide presses are generally eccentric operated, but toggle presses are available for heavy coining operations.

Four-slide machines are available in a range of sizes and for operation in the horizontal or vertical plane. All working components are mounted on the machine table. Camshafts, driven by a variable-speed motor through a flywheel and gears, actuate the press section and slides in synchronized movements.

Stock entering the press section goes through a progressive sequence in the die(s). Parts are then cut from the coil, and if additional forming is required, the parts are held for forming at the kingpost. Forming action at the kingpost is accomplished with two, three, four, or more of the machine slides.

||| 5 ||||

Grinding Operations

In medium and large die shops, there are usually two key persons involved in the building of a progressive die. First, of course, is the die leader, who bears the responsibility for the success or failure of the die. The grinder is the second. A good grinding specialist can be the pivotal person in the competitive equation.

Because a top-flight grinder can completely grind a progressive die in about one-third the time required by a diemaker, his or her value is easily understood. Moreover, the specialist generally knows more about progressive dies than a diemaker knows about grinding and can often apply that knowledge to making certain die sections which may prove difficult to grind.

Smaller job shops, with their smaller scope of operations, generally do not have a grinding specialist on the payroll because there is not enough work to keep one permanently employed. These shops often have one or more "all-around" machinists who grind with competence and can fill in on a drill press, milling machine, shaper, or lathe. Such persons often need help in grinding the more complicated sections.

Jig Grinding

Very few shops have a jig grinder, a machine essential to the building of extremely close-tolerance progressive dies. However, in most industrial areas, there are usually a number of grinding job shops which depend for their existence upon the grinding overflow from tool and die shops. Many die shops, even the larger ones, find it more economical to let grinding shops handle this work than to invest in a full-time jig borer.

However, there are many jig-grinding jobs that can be handled in the average die shop by use of an ID grinder if tolerances no finer than ±0.0005 inch (0.013 mm) are required. Pilot holes and button-die locations are generally held no closer than this, meaning that the great majority of ground holes can be ID-ground by the method that follows.

Progressive Dies

In the die section shown in Figure 5-1, five holes are specified to be jig-ground to a tolerance of ±0.001 inch (0.03 mm). The pilot hole P is used as a construction hole and the section is dimensioned from its center lines. To jig grind this section (or any other section) on an ID grinder, a hardened solid square is necessary. This square must be made in the shop because it is not commercially available. Machine steel, carburized and hardened, oil-hardening, and air-hardening steels are all satisfactory for this tool.

The square must be ground as closely as possible to 90 degrees, inasmuch as any error in its squareness will, of course, be reflected in the work performed.

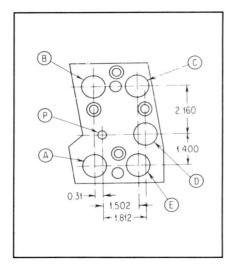

Figure 5-1. *Typical detail of a progressive die piercing stage.*

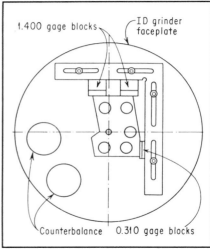

Figure 5-2. *Setup for jig grinding die button holes on an ID grinder. Setups of this type are more economical than jig grinding.*

The three preliminary steps in making the jig-grinding setup are:

1. Set the work so that the construction hole is running on dead center,
2. Study the moves which will have to be made, then chart them,
3. Set the square to the work.

The section is screwed to the faceplate of the ID grinder and the pilot (construction) hole P is indicated. It is always wise to ream this hole 0.002 inch to 0.003 inch (0.05 mm to 0.08 mm) undersize and then true it up on the ID grinder. With this done, a perfectly round construction hole running exactly on dead center is produced. The secret to jig grinding this section is to make five exact moves; and this is done by following the three steps mentioned, as shown in Figure 5-2.

In calculating chart dimensions (Figure 5-3), there are few rules to memorize. Common sense dictates that the die section be moved so that the center line of the desired hole falls in the center line of the grinder. Study the section to determine the most favorable two sides to place the square. Find the maximum hole spacings from the construction hole on the sides opposite the legs of the square. Place this dimension in gage blocks between each arm of the square and the section.

The square is positioned so that the distance from the construction hole to A represents the greatest distance on the opposite side of the construction hole. Therefore, the square is set with 0.310 inch and 1.400 inches (7.87 mm and 35.56 mm) in gage blocks between its arms and the die section. With this setup, when hole A is ground, the die section will be resting against both arms of the square.

Hole	Arm 1	Arm 2
A	0.000	0.000
B	3.560	0.000
C	3.560	1.812
D	1.400	2.122
E	0.000	1.812

Figure 5-3. *Chart of movements for ID grinding the section illustrated in Fig. 5-1.*

Many grinding specialists make all setups with small pieces of paper between the gage blocks and the arms of the square. In this way when hole A is positioned and all gage blocks are removed, the paper alone remains to indicate whether the die section is properly positioned.

Another important precaution — in fact an essential one — is to thoroughly demagnetize the die section and make sure it is clean and free of burrs. Any small obstruction or foreign matter will cause inaccuracies in the work.

The Chart

With the construction hole trued up and the position of the square established at 0.310 inch and 1.400 inches (7.87 mm and 35.56 mm), the square can be set. This is a two-person operation requiring great care, in that the square must be set exactly in location with the gage blocks tight, without disturbing the position of the workpiece itself. When the square is properly set, the faceplate should be able to rotate without any of the gage blocks flying out. At this time the construction hole should again be indicated to ensure that the workpiece has not been moved during the setup operation.

The square, having been set, must not be moved or disturbed until the jig-grinding operation has been completed. It is now possible to make a chart of the required moves by referring to Figure 5-1.

Loosening the workpiece and moving it on the faceplate according to the chart dimensions will bring each hole to the axis of the grinder where it can be ID-ground.

The only limitations on accuracy in such a series of moves are the accuracy of the square and the level of the grinder's craftsmanship.

Hole Sizes

When a hardened die section such as that shown in Figure 5-1 is ground to receive a hardened die button, the hole size should be ground to between 0.0003 inch and 0.0005 inch (0.008 mm and 0.013 mm) less than the button size. On a one-inch- (25.4-mm-) diameter button, 0.0003 inch (0.008 mm) provides an excellent press fit. However, it is important to be watchful of buttons near edges. If a button hole in a hardened section is close to an edge, do not finish-grind that edge until the button is pressed into place.

As an example, consider the die section shown in Figure 5-4. This particular section is from a high-precision progressive die and is fitted into a pocket with other hardened sections. If it is completely ground before the button is pressed into place, a minute bulge will appear in surface A. This bulge, small as it is, will not permit re-entry of the section into the pocket unless it is forced and will also disturb hole location. Though the disturbance of hole location is generally of negligible importance, it could be serious if the hole were located in a shave station. Hence, the prudent path is to first finish the hole, then press in the button, and grind surface A.

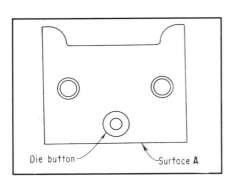

Figure 5-4. *Surfaces adjacent to die buttons such as surface A should not be finish-ground until the button is in place.*

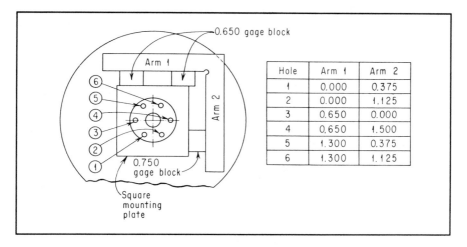

Figure 5-5. *Holes in circular details can also be jig-ground on an ID grinder.*

Jig Grinding Odd-shaped Sections

If the part to be jig-ground on the ID grinder is out of square, the preceding setup can be used by the addition of one more tool. The work is mounted on a square plate and the solid square is set to the plate instead of to the die section. This plate must be hardened and ground to perfect squareness and it should be well perforated with tapped holes of all sizes.

Figure 5-5 shows the setup and chart for grinding six equally spaced holes on a 1.500-inch (38.10-mm) circle in a round section. The section is mounted on the square plate, the chart is made, and the solid square is set to the plate.

Jig Grinding Angular Holes

Of all the jig-grinding operations possible on an ID grinder, angular holes present the greatest challenge, but the adept diemaker can learn to handle these as well as the others already discussed. Figure 5-6 shows the setup for grinding an interchangeable progressive die detail with a hole at 60 degrees to the bottom surface. The entire section is dimensioned from a construction hole, the 60-degree hole being 2.250 inches ± 0.001 inch (57.15 mm ± 0.025 mm) from the construction hole.

First, the base and edges of the section are ground all over but not to finish sizes. When this has been done, the job is mounted securely on a sine plate

Progressive Dies

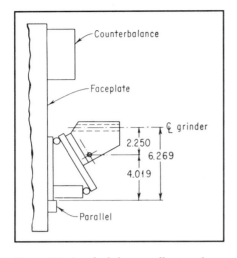

Figure 5-6. *Angular holes normally ground on a jig grinder can easily be ID-ground.*

at 30 degrees. The sine plate must have a locking device to prevent its opening once set. After the work is mounted on the sine plate, the sine plate is placed on a surface plate and against an angle plate. The distance from the construction hole to the end of the sine plate is then measured with gage blocks. This operation must be performed with absolute accuracy because the distance being measured represents a critical setup dimension.

The dimension in this example is 4.019 inches (102.08 mm). Knowing that the distance from the construction hole to the center line of the 60-degree hole is 2.250 inches (57.15 mm), by adding 2.250 inches to 4.019 inches (102.08 mm), the setup dimension of 6.269 inches (159.23 mm) is derived.

The next step is to mount a hardened parallel on the faceplate 6.269 inches (159.23 mm) from the center line. This dimension must be set with the greatest accuracy possible to avoid an accumulation of errors.

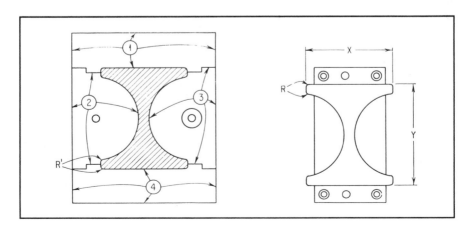

Figure 5-7. *Interlocking die sections and mating punch of a progressive draw-die blank stage.*

The sine plate is then mounted on the faceplate and the construction hole automatically falls 2.250 inches (57.15 mm) from the grinder center line, as it should.

It is now necessary to position the sine plate on the parallel to get centered transversely or sideways, a task which can best be accomplished by indicating the 60-degree hole. This hole can then be ground and the remaining surfaces of the section surface ground to this hole.

If tolerances are extremely close on such a job, it is wise to clean up the hole and then remove the section from the sine plate and check it. It can then be remounted on the sine plate and any minute errors corrected by the use of an indicator.

Grinding Round Blank Stages

Two other examples of jig-grinding work found frequently in progressive dies are the punch and the die section of the blank stage of a progressive draw die. These sections are shown in Figure 5-7 and are typical of the blank stages in all progressive cup dies.

Figure 5-8 shows the setup for grinding the punch, which has been mounted on a plate between two jig-bored construction holes. These construction holes correspond to the pilot holes in the die section. The plate is mounted on the faceplate of the ID grinder and one of the construction holes is indicated. With the construction hole on dead center, it is relatively simple to grind the radius to exact size by measuring between the work and the construction block. If tolerance is extremely fine, a planer gage (gage-block setup) can be made. When one side is finished, the process is repeated with the other construction hole at dead center. The workplate can then be removed from the faceplate, but dimensions X and Y are surface-ground to the construction holes before the punch is removed from the workplate.

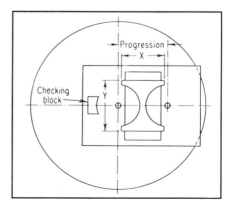

Figure 5-8. *Setup for grinding contour of hourglass punch.*

Progressive Dies

Radius R, which is never a critical dimension but one which must mate with a corresponding radius in the die section, can be put on with a surface grinder setup and in some cases with a hand grinder.

The die sections which mate with this punch present a slightly more complicated problem, but one which can easily be accommodated between the ID and surface grinders. The first step, after cleaning top and bottom, is to indicate and grind the die button hole and the pilot hole which will be referred to henceforth as the construction holes.

The flat surfaces are ground next and a flat is ground on the radius 0.001 inch (0.03 mm) greater than the desired dimension as shown in Figure 5-9.

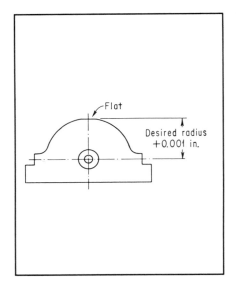
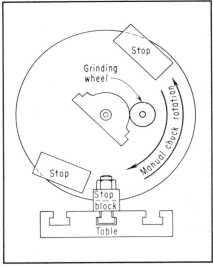

Figure 5-9. *Flat ground on circular die section as an aid to grinding of the contour.*

Figure 5-10. *Shops that have no jig-grinding equipment use this setup for grinding circular die sections.*

The section is then ready to go back to the ID grinder where the construction hole is indicated to dead center (Figure 5-10).

The next step is to disconnect either the belts or the power to the headstock. The faceplate can then be rotated back and forth manually as the grinding wheel is fed into the work. Remember that a flat was ground on this radius 0.001 inch (0.03 mm) over the desired dimension. When this flat "cleans up," the radius is on size. If, however, the tolerance on the radius is less than 0.001 inch, cleaning up the flat is not sufficiently accurate and the job must be checked with gage blocks before it is removed from the faceplate.

The remainder of the grind stock on the radius is ground on the surface grinder (as shown in Figure 5-33, page 000). This setup is discussed later in this chapter.

An ID Radius Dresser

In many progressive dies, such as those with circular draw sections, there are one or more sections with a ground hole ending in outside radii.

An effective way to dress these holes is by way of a shop-made radius dresser such as the one shown in Figure 5-11. The time invested in building this tool pays many dividends in time saved and increased accuracy. It can be used to grind either an inside or outside radius and is also useful for dressing either side of the wheel. To make this dresser, a slot is milled into the round block dead center to the pivot-pinhole. The reamed hole for the diamond is also centered in the bar so that the axis of the diamond intersects the axis of rotation.

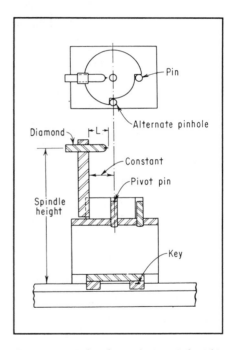

Figure 5-11. *Radius dresser for ID grinder. This tool is necessary for accurate radius work on draw stages.*

The distance from the bar to the axis of rotation is stamped into the block. This is the constant of the radius dresser and is used to locate the diamond for the correct radius. If L is equal to the constant, the tip of the diamond is on the axis of rotation and no radius can be dressed. If L is greater than the constant, an internal radius will be dressed. If L is less than the constant, an external radius will result.

As such, the dresser has two equations for finding the proper setting of L:

1. For external radii,

$$L = \text{constant minus desired radius}$$

2. For internal radii,

$$L = \text{constant plus desired radius}$$

Dressing an ID Wheel

On some ID jobs it is convenient to dress the front of the wheel; on others, the back. Because the axis of the spindle of the machine may be, and usually is, out of parallel to the axis of the table, a sound rule to follow is:

When dressing a radius or the flat part of a wheel, dress the side that is in contact with the work.

OD Grinding

Surface and ID grinders perform the bulk of the grinding operations involved in progressive die work, and the OD grinder, by comparison, gets very small play. There are, however, two simple OD jobs worth discussing. The first is the small die button found in the great majority of progressive dies. Many shops spend valuable time grinding both the ID and OD of these buttons, when actually only the outside diameter needs to be ground.

These buttons should be turned on a lathe to size plus grind stock. The ID should then be bored to size to as smooth a finish as possible.

The button should be made about 0.063 inch (1.6 mm) longer than required, and the hole centered as shown in Figure 5-12. The button should then be reversed in the chuck and the other end centered as well. After heat-treatment the button can be OD-ground between centers and OD and ID will be concentric. With this procedure, grinding time will be cut to less than half.

After the button is pressed into place, the excess stock containing the center can be ground off.

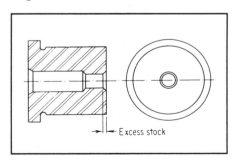

Figure 5-12. *Button-die design when commercial buttons cannot be used. Inside diameter need not be ground.*

Guide Pins

In many progressive die designs, the vendor must build the die set because certain required specifications are not available in commercial die sets. The guide pins can be purchased and are generally about 0.015 inch (0.38 mm) over nominal sizes. The guide-pin holes can then be drilled and reamed to a standard size, usually 2.0, 2.5, or 3.0 inches (50.0, 63.5, or 76.2 mm). The guide pins are then ground on the OD grinder to a 0.0015-inch to 0.0020-inch (0.038-mm to 0.051-mm) press fit. The "lead" is of the utmost importance and should be ground in two steps as shown in Figure 5-13. The first step should be ground 0.001 inch (0.03 mm) under hole size, permitting pin entry under manual pressure. The second step should be ground to the same dimension as the hole. When this area has been pressed in, the pin is in perfect alignment. This double lead configuration is recommended, not only for guide pins, but for all press-fit jobs.

A second recommendation is to break the edge of both leads with emery cloth. This additional precaution is a guarantee that the pin will not shear the hole.

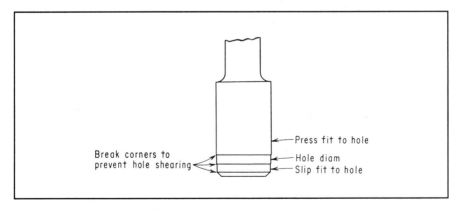

Figure 5-13. *Double lead for press-fit components.*

Grinder Operation

For an ordinary surface-grind job, approximately 95 percent of the diemakers operate the table too slowly and "infeed" the work too rapidly. In most cases they are hesitant about taking a heavy cut for fear of damaging the spindle bearings. Although a heavy cut combined with a heavy infeed shortens bearing life, a heavy cut combined with a light infeed does no damage whatever.

A veteran grinder does not hesitate to take 40 to 50 thousandths of an inch (1.0 to 1.3 mm) per cut, as opposed to the more cautious diemaker who makes innumerable passes at from 3 to 5 thousandths of an inch (0.08 to 0.13 mm) per cut.

The chuck of a surface grinder should be ground regularly with a coarse wheel to keep it parallel to its ways, and it should frequently be stoned to remove the burrs that are always being raised. When grinding close-tolerance die sections, the careful grinding specialist will stone the chuck each time a detail is put on it.

Warped Sections

Most flat sections return from heat-treating slightly warped because of the stresses induced by quenching. Such sections should first be stoned, and then placed on the chuck with the concave side down. The magnet is thrown on all the way and then reduced to about half strength. In this way, residual magnetism is induced in the workpiece. This residual magnetism, plus half the strength of the magnet, is sufficient to hold the work in place without pulling the concavity down to the chuck. It can then be cleaned up without distortion and turned over with the magnet at full power.

Test of Flatness

If absolute flatness is a requisite, as it generally is in progressive die sections, the diemaker can check accuracy by spinning the section on a surface plate. If it spins freely, it is not making full contact with the plate and, therefore, is not flat. Conversely, if it spins with difficulty, full surface contact is being made and the job is flat.

Grinding Thin Stock

One of the more difficult jobs encountered in surface grinding work is that of grinding thin stock without distortion due to heat. A very coarse-grained wheel should be selected, the table speed should be fast, and the infeed extremely fine. If the class of surface finish is not critical, and size is the only object of grinding, the following is helpful:

1. Dress the wheel.

2. Loosen and then retighten the lock nut holding the wheel.
3. Grind the work with a fine infeed.

Loosening and retightening the lock nut puts the wheel a thousandth of an inch (0.025 mm) or so off-center. Wheel contact with the work is lessened, heat is reduced, and distortion is minimized.

Radius Grinding

For grinding radii and tangent flats, common practice is to:

- Grind flats first on external radii, thereby boxing in the radii centers (Figure 5-14).
- Grind the radii first, on internal radii, then the flats (Figure 5-15).

Although these rules are general in nature, in most cases following them usually simplifies the grinding operation.

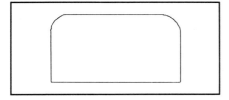

Figure 5-14. *On outside contours grind flats first.*

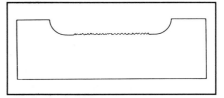

Figure 5-15. *On inside contours grind radii first.*

Full Outside Radius

Dress the wheel (Figure 5-16) so that it has not only a full radius but two small lands tangent to the radius. Grind straight down until the horizontal land starts to contact the work. Then grind in until the vertical land begins to contact the work. Marking the work with a soft lead pencil is helpful in that it provides a guide to the diemaker that he is within several thousandths of contacting the flats.

Full Inside Radius

Dress the wheel as shown in Figure 5-17 so that it has a full radius and two tangential lands. Grind the work until the radius and both lands have

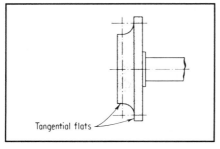 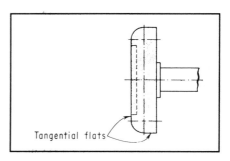

Figure 5-16. *Tangential flats necessary for outside radius grinding.*

Figure 5-17. *Tangential flats necessary for inside radius grinding.*

"cleaned up." Check the dimensions of each land and then grind down and over until the required dimensions have been obtained. The remainder of the inside surfaces can then be ground from these flats.

Partial Outside Radius

A frequent problem in progressive die work is the partial outside radius shown in Figure 5-18. To make the setup for this job, dimension Y of the radius dresser (Figure 5-19) must be determined as well as the constant C. Usually Y and C are both stamped on the dresser or in any case should be so stamped. The steps to take are:

1. If Y is greater than X, place the radius dresser directly against the back rail and dress a 90-degree radius into the wheel. The chuck now must not be moved in or out until the job is finished, because the radius in the wheel has a dimensional relationship with the back rail.

2. Position the workpiece so that the horizontal land on the wheel can be brought into contact with its upper surface. Note the reading on the handle and remove the work.

3. Place Y minus X in gage blocks between the work and the back rail.

4. Beginning at the contact setting noted, grind straight down a distance of D, as indicated on the up-and-down handle.

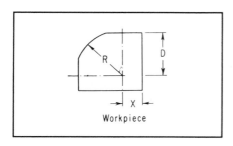

Figure 5-18. *Partial outside radii can be accurately ground, although the operation must be performed without the aid of tangential grinding flats.*

Grinding Operations

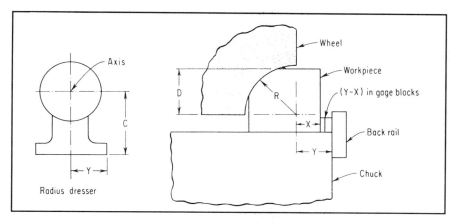

Figure 5-19. *Setup for dressing grinding wheel for partial outside radius.*

5. Referring again to step 1, if X is greater than Y, place X minus Y in gage blocks between the radius dresser and the back rail when dressing R. The process is now the same except that the workpiece will be placed directly against the back rail.

Two final rules regarding these steps are, first, not to memorize them, but rather analyze them and understand the reasoning behind them. Second, not to take a chance on back-rail squareness. Back rails should always be cleaned up at the start of such a job.

Partial Inside Radius

The preceding process can also be used to grind a partial inside radius. Another method that is useful in grinding partial radii, either inside or outside, is shown in Figure 5-20. In this setup a piece of cold-rolled stock is mounted on the die section a distance of X minus R from the end of the detail. Dress the full radius R on the wheel, then contact the cold-rolled block with the wheel. That done, grind straight down until dimension D has been obtained, then finish the remainder of the flat.

Small Outside Radii

When an edge is to be finished with a small outside radius, the diemaker may be tempted to send it to heat-treatment without roughing on any of the radius. This is not good practice because a square workpiece edge will ruin

101

Progressive Dies

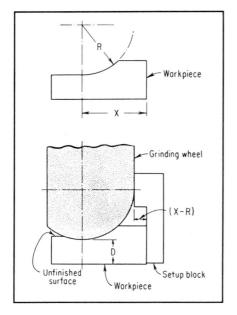

Figure 5-20. *Grinding setup for partial inside radius.*

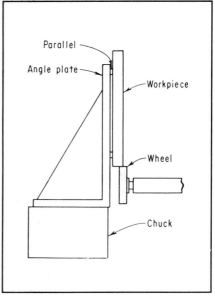

Figure 5-21. *Setup for grinding the ends of long sections.*

the contour of the radius, making it necessary to redress the wheel. Such edges should be filed on in the soft form or hand-ground in the hard to save the wheel contour.

Long Sections

Occasionally, diemakers encounter the problem of grinding the ends of extremely long sections. If such sections are only an inch or two (25 mm to 50 mm) longer than the range of the grinder, they can be hung over the end of the chuck and clamped to an angle plate. If the section is too long to be ground in this manner, it can be ground as shown in Figure 5-21. An angle plate of the proportions shown should be made and kept in the grinding room.

The diemaker should be cautioned, however, that when working with this setup, sparks fly in the reverse direction. To prevent serious eye damage, the grinder is advised to look at the job from the left side instead of the right.

Long Radii

The angle plate shown in Figure 5-21 is useful in making setups for the grinding of long inside and outside radii. A liner bushing is pressed into a bored hole in the angle plate and another liner is pressed into a flat plate, as shown in Figure 5-22. The two plates can then be joined with a shoulder screw, making an adequate setup for grinding long radii as shown.

Roll Dimensions

A grinding problem often encountered in progressive die work is illustrated in Figure 5-23. A dimension D is required between a square face and an angular face. The angular face is, of course, ground on a sine plate. The problem is to hold D to close tolerance.

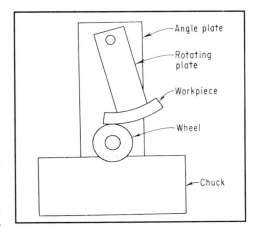

Figure 5-22. *Setup for swinging a long radius.*

This problem can be handled very easily, as shown in Figure 5-24. First, the detail should be finished all over except for the angular face, which should barely be cleaned up. Next, an angle plate should

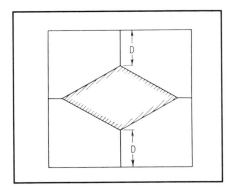

Figure 5-23. *Frequently encountered problem in progressive die manufacture. Dimension of a surface is taken between a square and an angular surface.*

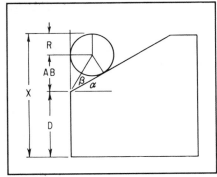

Figure 5-24. *Roll dimension X is established by the diemaker to hold D.*

be placed against the workpiece to hold the roll flush with the edge of the work. (Dowel pins make excellent rolls.)

Calculation of X, a dimension which can be measured with an indicator and gage-block setup, is an exercise in basic trigonometry. It equals D plus AB plus R. To make the calculation, it is first necessary to determine angle β. If the angle of inclination is denoted by α, β is equal to 90 degrees minus α divided by 2.

$$\beta = \frac{90° - \alpha}{2}$$

When β has been determined, AB can be determined by algebraic transposition of the following equation:

$$\tan \beta = \frac{R}{AB}$$

Transposing

$$AB = \frac{R}{\tan \beta}$$

If a formula is desired for the roll dimension setup, it will read:

$$X = D + \frac{R}{\tan \beta} + R$$

Conical Sections

Conical sections like those shown in Figure 5-25 are frequently used in progressive dies with interchangeable die sections. They present a variation of the problem mentioned in the preceding topic in that two rolls are necessary. In this case

$$X = D + 2\frac{R}{\tan \beta} + 2R$$

Grinding to an Internal Angle

The cross section of a very popular retainer of round sections and punches in many progressive dies is shown in Figure 5-26. The dimension to the angle is always given as shown by X, never by Y. The reason for this is that D_1 is clearance and is allowed a comparatively large tolerance. This

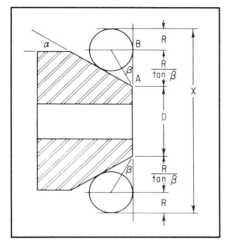 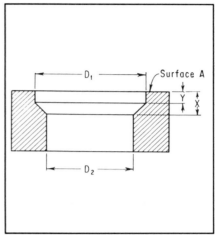

Figure 5-25. *Double-roll setup for holding a dimension between two angles.*

Figure 5-26. *Punch retainer used by some of the automotive concerns. The problem is to hold dimension X.*

tolerance permits an equal variation in Y; thus the section must be dimensioned at the other end of the angle as X. The angle is generally 45 degrees.

The best way to hold dimension X is to grind D_1, a nonworking diameter. With a stringent dimension assigned to D_1, the dimension Y can be calculated in terms of X.

The problem now is to hold Y. The angle should be ground in an ID grinder to a depth of about 0.010 inch (0.25 mm) greater than Y as measured by a depth gage, after which surface A can be ground exactly to dimension Y.

A handy shop-made depth gage comes into play in measuring dimensions such as Y (Figures 5-27a and b). A 0.359-inch (9.13-mm) hole is drilled in a piece of 0.75-inch (19.1-mm) tool steel (preferably oil-hardening gage stock) which is then hardened and surface-ground. The 0.359-inch hole is then ground and lapped to a 0.375-inch (9.53-mm) gage fit.

One end of a 0.375-inch (9.53-mm) diameter by 2.0-inch (51-mm) dowel pin is ground to a ball nose and the other end is ground flat. The length L of this pin is carefully measured as is the thickness T of the plate. These two dimensions are then etched on the side of the plate.

The pin is then inserted in the plate and pushed down against the wall until it rests upon the angle. The correct gage-block setting is placed on the

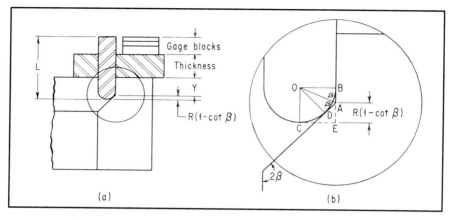

Figure 5-27. *Gage for checking internal angles. Gages of this type are not made commercially; they must be made in the shop.*

gage and the difference in height between the gage blocks and the pin, as measured by an indicator, will be the amount to grind off to obtain Y.

The mathematics of this gage are:

R = radius of pin
L = length of pin
β = 1/2 of angle DAB
$OB = OD = OC$ = radius of ball nose
Y = dimension being held
T = thickness of gage
? = gage-block dimension

From Figure 5-27a and b,

$$L = AE + Y + T + ?$$
$$AE = R - AB$$

But $$AB = R \cot \beta \left(\text{because } \frac{AB}{R} = \cot \beta \right)$$

Only one unknown remains; the gage-block thickness designated as ?.

Transposing the equation,

$$? = L - AE - Y - T$$
$$? = L - (AE + Y + T)$$
$$? = L - [(R - R \cot \beta) + Y + T]$$
$$? = L - [R(1 - \cot \beta) + Y + T]$$

Once the correct gage-block setting has been established, holding dimension Y becomes a simple surface-grinding job.

The Horizontal Whirligig

Another important shop-made tool is the horizontal whirligig shown in Figure 5-28. In shops specializing in progressive dies with interchangeable die sections, such a tool is a necessity. A case in point is the comparatively simple punch shown in Figure 5-29. This job called for two identical dies with three complete sets of details each. This means eight identical punches.

The punch itself is simple enough but the 0.125-inch (3.18-mm) radii present a problem. They must be held to a tight tolerance and be ground all the way to the shoulder. Each radius could be "roughed on" soft and

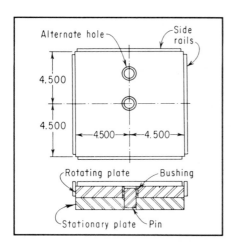
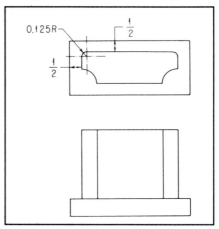

Figure 5-28. *Horizontal whirligig for radius grinding. Alternate hole is placed on pin when exceptionally large sections are "swung."*

Figure 5-29. *Typical punch design in progressive die work. The problem is to grind the 0.125-inch radius.*

finished with a high-speed head, but this job is easily done by the horizontal whirligig.

The whirligig consists of two square plates approximately 9.0 inches by 9.0 inches (229 mm by 229 mm) in size, one mounted on the other. The lower plate has a hole in its center which is ground to a 0.500-inch (12.7-mm) press fit. A 0.500-inch dowel pin is pressed into this hole. The upper plate has a 0.500-inch slip-fit hole in its center which enables it to rotate on the lower plate by means of the dowel pin (Figure 5-30). There are a number of tapped holes in the upper plate which are necessary for clamping and several in the edges which are necessary for screwing on side rails.

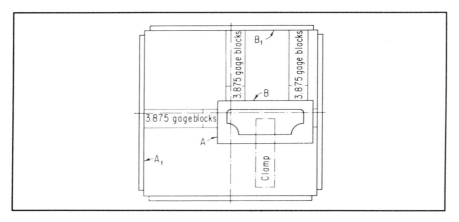

Figure 5-30. *Punch positioned on whirligig for radius grinding. Whirligig must be perforated with tapped holes for purposes of clamping.*

The two plates should be made of machine steel or boiler plate which has been carburized and hardened. After the plates have been ground, the distance of the hole (in the upper plate) from the edges should be measured to three decimal places. If the hole is centered 4.500 inches (114.3 mm) from all edges, the centers of the 0.125-inch (3.18-mm) radii are 0.500 inch (12.7 mm) plus 0.125 inch (3.18 mm) from the edges of the punch shoulders, Figure 5-29. In other words, each center is 0.625 inch (15.88 mm) from the two closest edges.

Because the purpose of the whirligig is to rotate a part on the axis of a desired radius, calculation of these dimensions is necessary. In this case the axis of the desired radius lies 0.625 inch (15.88 mm) from the edges of the punch. To place this axis in the axis of the whirligig, it is necessary to subtract 0.625 inch (15.88 mm) from 4.500 inches (114.3 mm) for a remainder of 3.875 inches (98.43 mm). By positioning the punch on the whirligig so that surface *A* (punch) is 3.875 inches from surface A_1 (whirligig) and surface *B* (punch) is 3.875 inches from B_1 (whirligig), the two axes are coincidental. The whirligig can then be placed on the chuck of a surface grinder and rotated through 90 degrees, the face of the wheel being used to grind the 0.125-inch (3.18-mm) radius.

Positioning the punch on the whirligig is accomplished by placing 3.875 inches (98.43 mm) in gage blocks between the side rails and the punch shoulders. The punch is then clamped to the whirligig with the gage blocks tightly in place.

To perform this operation, all dimensions must be exact and all surfaces must be clean and free of burrs. Manual rotation of the setup is tricky and requires a certain amount of practice. It is best to practice this kind of radius work on one or more pieces of scrap steel before attempting to grind an otherwise finished die section.

The Vertical Whirligig

Vertical whirligigs are also useful additions to grinding-room equipment. As in horizontal whirligig applications, the objective is to grind a radius by placing the axis of the whirligig. In the design shown in Figure 5-31, a centrally ground slot is used to limit side movement of a V block. The block can be moved up or down, but one of its coordinates is always

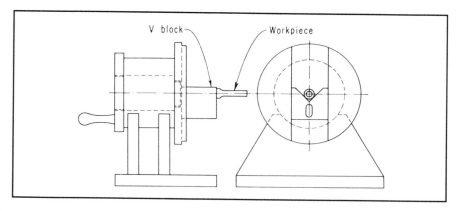

Figure 5-31. *Vertical whirligig for small punch grinding. Tools such as this are mandatory in progressive die work.*

coincident with one of the coordinates of the whirligig. If a round punch is to be ground in this tool, it is necessary only to move the V block up or down, using an indicator to find the other coordinate. In simpler language, it is necessary only to clamp the punch in the V block and "indicate" it.

This tool is particularly valuable for the grinding of punches and pilots which are not standard in size. Because many progressive die designs utilize nonstandard size holes which receive a long line of pilots, this tool is essential for efficient grinding operations. Needless to say, if punches and subsequent pilots are standard in size, it is more cost-efficient to buy them than to grind them down from straight blanks.

This tool is also useful in grinding the type of punch whose cross section is shown in Figure 5-32. This punch is first ground to 0.250 inch by 0.750 inch (6.35 mm by 19.05 mm) "central" after which it is placed in the V block and its shank indicated 0-0. In this way, the axis of the punch falls on the axis of the whirligig. After this has been done, it is necessary to rotate the punch until its flats are parallel to the Y coordinate of the whirligig. This can be done by indicating the edge of the V block or the keyway itself to 0-0, and then indicating the flat to 0-0.

When this has been done, the axis of the punch is on center and the flats are parallel to the V block. It is then necessary to move the V block so that

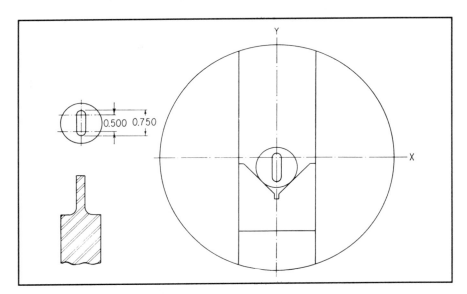

Figure 5-32. *Punches such as the one illustrated are easily ground on the vertical whirligig.*

the center of one of the desired radii falls on the axis of the whirligig. This means moving the V block 0.250 inch (6.35 mm) in either direction, a move that can be made with the aid of a planer-gage, gage-block setup. The whirligig is then rotated back and forth through 180 degrees beneath a grinding wheel. The ends of the 0.250 by 0.750-inch (6.35-mm by 19.05-mm) rectangle (the short flats) should be "blued up" as an aid in grinding; when the blue disappears, the radius is on size.

Upon completion of this radius, the V block must be moved back 0.500 inch (12.7 mm) after which a similar radius is ground on the other small flat. Again, punches such as this are available commercially in a wide variety of sizes. Whirligig setups like this one are expensive and should be used only on odd sizes.

Grinding S Curves

So-called S curves are constantly recurring phenomena in progressive die work. In Chapter 6, which is devoted to blank development, a formula is evolved for determining the amount of metal in such a curve. At this point, the focus is on the method used to dress this curve into a grinding wheel. But in referring back to the blank stage illustrated in Figure 5-9, note that R_3 is left unground because it must be blended into R_1.

The setup used to finish-grind R_3 and R_1 is illustrated in Figure 5-33. Like R_3, dimension D_1 has been ground and it is only necessary to pick up this flat

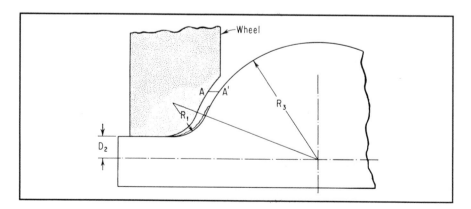

Figure 5-33. *Method used to finish-grind S curves. Desired contour is dressed into wheel. When wheel picks up finished portion of R_3, radius R_1 is complete.*

Progressive Dies

and advance the work into the wheel until R_3 on the detail starts to clean up. At this point the die section is finished.

Actually, there is no problem in grinding this S curve — the problem is in dressing the curve itself into the wheel. This particularly difficult dressing setup requires a top-quality radius dresser, one who not only dresses a perfect radius but has an accurate protractor-type head as well. This is important, for S-curve dressing involves dressing through a specified angle.

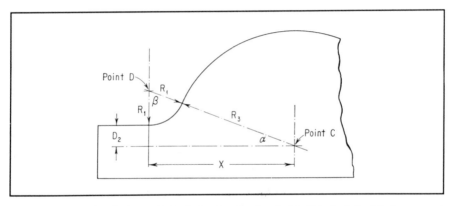

Figure 5-34. *Grinding the S curve is simple, but dressing the wheel is difficult. Axis of the dresser must move from point C to point D.*

To dress the curve in Figure 5-34:

1. Angle *a* must be determined. This is necessary for two reasons: first, it will be used in calculating X; second, it is the complement of the angle of arc to which the dresser will be set for dressing R_1.

2. Dimension X must be determined as well. This is the distance the dresser will be moved to dress R_1. Calculation of X and *a* is not difficult because the value of one leg is known and the hypotenuse of a right triangle is easily calculable. Leg D_3 is known because it equals D_2 (an arbitrary dimension) added to R_1, a desired radius. The hypotenuse of the triangle is equal to the sum of the two radii. Therefore

$$\sin \alpha = \frac{D_2 + R_1}{R_1 + R_3}$$

Substitution of known quantities in this equation and reference to a trigonometry table will provide a value for α. Using α, it is possible to determine X as follows:

$$\frac{X}{R_3 + R_1} = \cos \alpha$$

$$X = (R_3 + R_1) \cos \alpha$$

With this information, the radius can be dressed by following the procedure below:

1. Set the diamond in the dresser to R_3.
2. Place X in gage blocks between the dresser and the back rail.
3. With the protractor locked at 0 degree, dress the wheel parallel to the chuck.
4. After moving the dresser clear of the wheel, lower the spindle a distance equivalent to R_3 minus D_2.
5. Dress radius R_3 into the wheel. (Caution: The chuck may not be moved in or out again once R_3 is dressed.)
6. Move the dresser away from the wheel (longitudinally), and change the setting of the diamond from R_3 to R_1.
7. Set the protractor head to swing through the arc b. (This equals 90 degrees minus α.)
8. Remove the gage blocks (dimension X) from between the dresser and the back rail. Place the dresser firmly against the back rail.
9. Lower the grinding wheel the distance ($D_2 + R_1$) while moving the dresser through arc b.

The procedure outlined for dressing the S curve should not be memorized but it should be studied until the sequence is thoroughly understood.

Grinding Carbide Dies

The Banko patent and the improved techniques in electric discharge machining are important factors in the emergence of tungsten carbide as a die material. Another equally important development is the improvement in grinding equipment and techniques. Without such improvements the maintenance of progressive carbide dies would be impossible.

Unlike most tool-steel progressives, which can be sharpened on almost any type of surface grinder, a carbide die requires the very best in grinding equipment. This includes a machine large enough to carry the comparatively heavy weight of the lower die. (Normally, the entire die is placed on the surface grinder, whereas in conventional sheet-metal progressive dies, the blanking and piercing details are removed from the die for sharpening.)

Grinders. Grinders that mount a peripheral-type wheel are preferable to those with vertical spindles for grinding carbide dies. The vertical type, using cup wheels, covers a greater portion of the die than the horizontal (peripheral) type, generating higher heat levels and impairing visibility. Regardless of the type of grinder used for sharpening carbide dies, the grinder should have a sturdy spindle which runs true on good bearings. All sources of vibration within the machine should be located and eliminated. External sources of vibration must be isolated from the machine by dampeners, because vibration from any source has the same effect on surface finish as an unbalanced wheel.

The coolant is more important in the grinding of carbide than in any other material. Its flow should be uninterrupted and heavily directed at the point of contact. While maintaining a uniform temperature in the workpiece, it also washes away the grit produced by grinding. This is important in reducing wheel "load-up" with its attendant glazing and chatter. In selecting coolants it is important that the die grinder avoid those containing sulfur. Sulfur will chemically attack both the cobalt used to sinter the carbide and the resins used as bonding agents in the wheel.

Wheels. A resinoid bonded diamond wheel of proper grit should be used for grinding carbide dies. A specially reinforced bakelite core will do much to add to the stability of the wheel. Green grit wheels should be avoided because they present too many hazards to the die. An ideal combination of wheels for grinding carbide would consist of a 100-grit wheel for roughing and a 180-grit for finishing. However, the use of a 150-grit wheel for both roughing and finishing is practical. Whether the combination of wheels or the all-purpose wheel is used, all wheels should have the same general specifications other than grit size. These would include a diamond concentration of 100, an R bond hardness, a resinoid bond for wet grinding, and a 0.125-inch (3.18-mm) depth of diamond section. Peripheral speed of the grinding wheel should be approximately 4200 surface feet per minute (sfpm) (1260 surface meters per minute [smpm]) with table speeds of approximately 85 fpm (25.5 mpm) for roughing cuts. Speed should be increased above 85 fpm for finishing. Cross-feeds of 0.25 inch to 2.7 inch (6.3 mm to 69 mm) are correct for roughing cuts. For finishing cuts the cross-feed should be reduced to 0.03 inch (0.8 mm).

Wheel Runout. After the diemaker has selected the proper machine and wheel for grinding a carbide die, he or she should inspect both the spindle and adapter for runout. This can be done with a dial indicator mounted on a surface gage or height gage. Any burrs which may interfere with wheel fit should be stoned. When the wheel is mounted on the adapter, the diemaker should be certain that the hole and adapter mate in such a manner that the wheel can come to rest against the shoulder.

Diamond Wheel Dressing. For initial and all subsequent dressings of the wheel a brake-controlled dresser is recommended. This dresser should have a silicon-carbide wheel of the following specifications: 37C60Q5V Norton or equivalent (dimensions: 3-inch [76-mm] diameter; 1-inch [25-mm] face; 0.5-inch [13-mm] bore).

The dresser should be placed at some convenient place on the grinder table with its spindle parallel to the spindle of the grinder; its wheel should travel in line with the direction of travel of the diamond wheel. When the brake-controlled dresser is properly set, it will travel at approximately 1500 sfpm (450 smpm) when driven by contact with the diamond wheel traveling at normal speed.

Before the dressing wheel is brought in contact with the diamond wheel, it must be set in motion. This is important; if the dresser is not spinning, the diamond wheel will grind a flat on it, causing permanent damage. A small wood stick can be used to set the dresser wheel in motion. Some lamination diemakers add a knurled knob to the dresser spindle. This knob provides increased efficiency as well as improved safety.

The diamond wheel should be passed over the dressing wheel by means of the cross-feed. All dressing is done without the use of coolant. Downfeeds not exceeding 0.001 inch (0.025 mm) should be taken at each pass with the diamond wheel clearing the dressing wheel each time. After a downfeed of 0.005 inch (0.13 mm) has been made, the diamond wheel should be stopped and examined for results. This action should be repeated if necessary until the diamond wheel is dressed true and smooth. This accomplished, the color of the wheel will be the same over its entire face. The purpose of this operation is to dress away the wheel bond, thereby truing the wheel and exposing the diamond particles. Cleaning sticks supplied by wheel manufacturers should be used frequently to keep the wheel opened.

Sharpening. When a carbide die is ready for sharpening, it should be cleaned of all oil, slugs, and foreign material. All burrs should be removed from both the chuck and die shoe to ensure that the die will lie flat.

Cutting edges should be inspected to determine the amount of stock that must be removed. If excessive wear or chipping is a localized condition, it may be expedient to raise or replace that particular section to avoid grinding the entire surface of the die.

When the die has been placed on the chuck, the stops should be set so that the work will pass at least 1.0 inch (25 mm) beyond the wheel. The cross-feed and speed should be set as noted previously, and the grinding can begin. There are, of course, no sparks when carbide is ground with a diamond wheel. In this form of grinding, the diemaker listens for a light hissing sound which indicates wheel contact with the work.

After initial contact has been made, the wheel should be raised 0.002 inch (0.05 mm) before it is again brought in contact with the work. It should then be down-fed in increments of 0.0002 inch (0.005 mm) for roughing cuts and in increments of 0.0001 inch (0.003 mm) for finishing. It is also important that the wheel be permitted to make at least two free passes over the work between feeds.

To establish the required depth of cut, some grinder hands "plunge grind" at the point showing greatest wear. They then cross grind until the path of the plunge cut is removed. If the total amount of the grind is to exceed 0.010 inch (0.25 mm), an aluminum oxide wheel should be used to remove as much of the shoe and die material as possible. This should be done to save the wheel, because steel will cause a diamond wheel to break down much faster than will carbide.

Grinding of the upper die is performed in a similar manner except that small perforating punches must be supported. This support can be given in one of two ways: use of the stripper and use of a strip from a previous run. Using a stripper presupposes very close support of the punches by the stripper, which is fitted in place without its springs. Resting on parallels, it proves most effective as a means of supporting small punches during grinding.

Using a strip from a previous run is a much less desirable method. The strip can be fitted over the punches and do an adequate job; however, the use of a strip instead of a stripper is dangerous in that it is easy to chip or fracture small punches when the strip is put on or removed.

Whichever method is used for die sharpening, it is necessary that punch exposure be kept at a minimum. The brittleness of carbide makes it an almost delicate material.

6

Blank Development

Cold Development

Among the serious pitfalls in designing a progressive die for a close-tolerance part is that of relying on charts or formulas to dimension a blank stage. Charts and formulas, although helpful, should be used for reference only; final blank development must be made in the die shop.

U.S. auto makers design some of the finest progressive dies in the world, but seldom do they dimension any stage of a die, nor do they establish the progression of a die in decimals. Needless to say, this is not an economy measure with these companies, but it is wise policy. When a diemaker completes development of a blank stage for one of these dies, it will match almost perfectly with the designer's preconceived blank. Although these manufacturers do allocate the necessary time and money on development, by omitting dimensions they place the responsibility (and authority) for the finished part on the diemaker, which is where it belongs.

This does not imply a contempt for mathematical processes; rather, it punctuates the point that if a part has close limit tolerances, charts and formulas should be used as guides only. It is equally true, however, that designers must have a sound knowledge of the theory of blank development if they are to be successful in progressive die work. This is necessary because:

- The progression of a die is always dependent upon the blank size. Without an accurately predetermined progression, the designer will find it impossible to lay out and design a progressive die with any degree of confidence.
- Lacking knowledge of the blank, designers will find it impossible to determine the approximate sizes of details in the blank stage. This can result in details that are either too small, weak, and crowded or too large, which is a waste of expensive tool steel and labor.

Nearly all progressive dies can be classified in one of four broad categories as concern blank development:

Progressive Dies

1. The formed part whose approximate development can be determined mathematically;
2. The symmetrically drawn part whose approximate development can be determined by formula, addition of geometric components, or centrobaric calculation;
3. The unsymmetrical part, either drawn or formed, whose approximate development must be determined experimentally;
4. The flat blank which is pierced or notched progressively.

Bend Allowance. Figure 6-1 details three of the most useful formulas in tool and die engineering.

In some cutoff progressive dies, a part is formed into a curved S shape whose length must be developed before work proceeds on the die. In these S-curve progressive dies, the part development yields easily to mathematics, as shown in Figure 6-2.

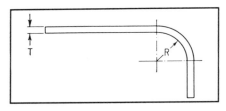

Figure 6-1. To calculate the length of stock in a 90° bend: If R is less than T, length = T/4 + R x 1.5708. If R is one to two times T, length = T/3 + R x 1.5708. If R is greater than two times T, length = T/2 + R x 1.5708.

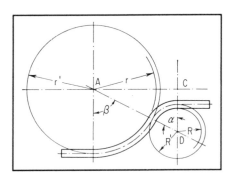

Figure 6-2. Mathematics of an S curve. Length of curved portion of part equals angle/360° (2π) (R' = r').

All calculations are based on an imaginary line running through the center of the part. By adding half of metal thickness to R, R_1 can be obtained and, likewise, half the metal thickness added to r provides r_1.

R_1, added to r_1, is the hypotenuse of a right triangle whose legs AC and CD can be deduced from part-print dimensions.

Values for angles α and β which are equal, can now be formulated.

$$\sin \alpha = \sin \beta = \frac{AC}{R_1 + r_1}$$

The complete length of the small circle equals $2\pi R_1$, and the complete length of the large circle equals $2\pi r_1$.

Of interest is that portion of each circle that serves to make the S curve; that portion is to the total length as the angle is to a full 360 degrees. Thus, the entire length of the S curve can be calculated by calculating these two proportional parts. This now provides a formula for determining the length of metal in an S curve.

$$\text{Length} = \frac{\beta}{360}(2\pi R_1) + \frac{\alpha}{360}(2\pi r_1)$$

$$\alpha = \beta$$

$$\text{Length} = \frac{\alpha}{360}(2\pi)(R_1 + r_1)$$

Diemakers can also get an extremely close development by accurately drawing the S curve 10 times size on a piece of sheet metal and then "stepping off" the centerline with a pair of dividers. This method will derive an answer within a few thousandths of an inch (mm) of the mathematically derived solution.

Draw Formulas

Between Figures 6-3 and 6-34 inclusive are a number of different shapes often encountered in single-operation and progressive draw dies. By using their accompanying formulas diemakers can greatly reduce the time involved in predetermining blank sizes. The formulas also have a secondary, problem-solving function. Often in the development of progressive draw dies the size of the finished part is known but the size of certain preliminary stages is not.

To design and build a progressive die to make the cup shown in Figure 6-35, it would be impossible to draw the part in one stage. This is because, first, the punch and die radii are too small, and second, the part is drawn at an angle less than 90 degrees. (It is much easier to set up plastic flow in metal at a 90 degree angle than it is at sloping angles.) To attempt to draw this panel would be futile because it would consistently fracture.

In cases such as this, a "bubble" of metal is drawn in a preliminary draw stage. This bubble, of course, conforms to the mechanical necessities of a draw stage. It is then possible to force the bubble into the desired shape with a second forming stage.

The size of this bubble must be accurately predetermined because if it is too small, fracture will occur in the second form. Or even worse, if it is too large, there will be too much metal to compress.

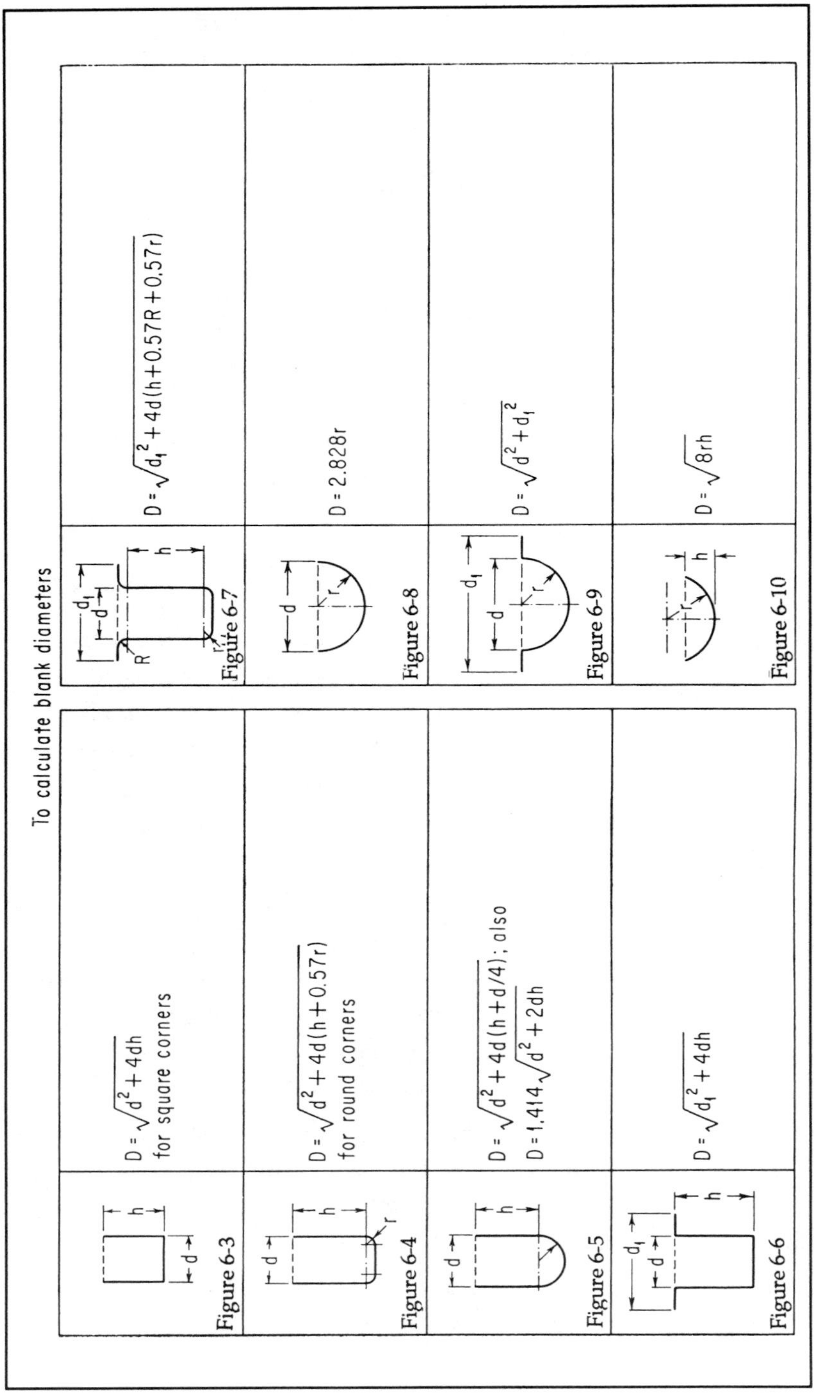

Progressive Dies

To calculate blank diameters

Figure 6-3	$D = \sqrt{d^2 + 4dh}$ for square corners
Figure 6-4	$D = \sqrt{d^2 + 4d(h + 0.57r)}$ for round corners
Figure 6-5	$D = \sqrt{d^2 + 4d(h + d/4)}$; also $D = 1.414\sqrt{d^2 + 2dh}$
Figure 6-6	$D = \sqrt{d_1^2 + 4dh}$
Figure 6-7	$D = \sqrt{d_1^2 + 4d(h + 0.57R + 0.57r)}$
Figure 6-8	$D = 2.828r$
Figure 6-9	$D = \sqrt{d^2 + d_1^2}$
Figure 6-10	$D = \sqrt{8rh}$

Figures 6-3 through 6-10.

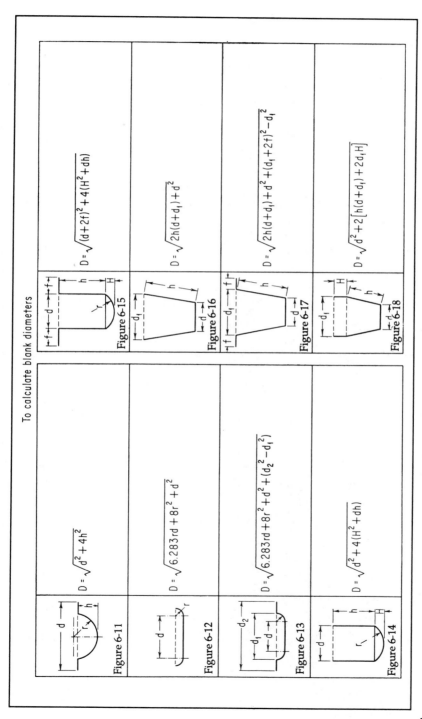

Figures 6-11 through 6-18.

Progressive Dies

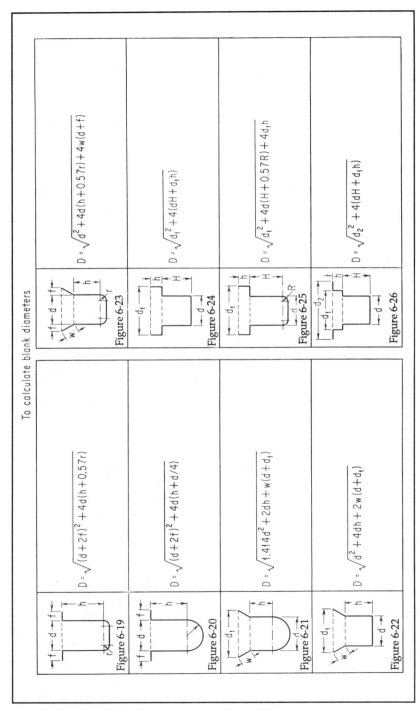

Figures 6-19 through 6-26.

Blank Development

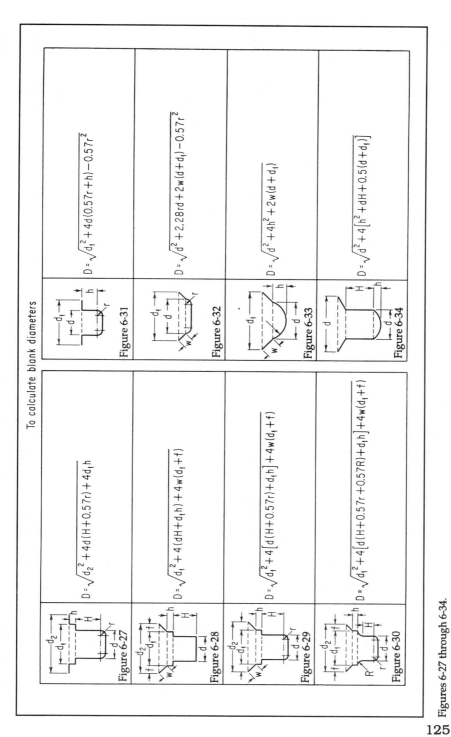

To calculate blank diameters

Figure 6-27	$D = \sqrt{d_2^2 + 4d(H + 0.57r) + 4d_1 h}$
Figure 6-28	$D = \sqrt{d_1^2 + 4(dH + d_1 h) + 4w(d_1 + f)}$
Figure 6-29	$D = \sqrt{d_2^2 + 4[d(H + 0.57r) + d_1 h] + 4w(d_1 + f)}$
Figure 6-30	$D = \sqrt{d_2^2 + 4[d(H + 0.57r + 0.57R) + d_1 h] + 4w(d_1 + f)}$
Figure 6-31	$D = \sqrt{d_1^2 + 4d(0.57r + h) - 0.57r^2}$
Figure 6-32	$D = \sqrt{d^2 + 2.28rd + 2w(d + d_1) - 0.57r^2}$
Figure 6-33	$D = \sqrt{d^2 + 4h^2 + 2w(d + d_1)}$
Figure 6-34	$D = \sqrt{d^2 + 4[h^2 + dH + 0.5(d + d_1)]}$

Figures 6-27 through 6-34.

Progressive Dies

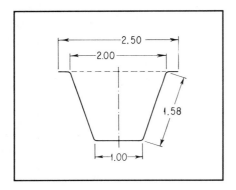

Figure 6-35. *A final form stage. The problem is to develop a preliminary form stage of equal area.*

Ideally, the preliminary bubble should have the exact surface area of the final form stage, but ideal or perfect conditions remain stubbornly elusive. However, by making the bubble slightly smaller in area than the second or final form, a slight amount of metal stretch results, which is acceptable, providing the panel does not fracture or thin down too much. Whenever such fracture or thinning down does occur, consider:

- Whether enough metal has been pushed into the first form;
- Whether the second form is trying to perform a mechanically impossible operation, in which case it is necessary to alter the shape of the first form;
- Whether an appropriate number of preliminary form stages have been incorporated in the die.

Naturally, if the diemaker is forced to consider the third point, failure almost always will follow. Dies have been designed that form an exotic shape without sufficient preforming. In such a case, experience, combined with careful study of strips of progressively drawn panels, will provide the direction necessary to determine the appropriate amount of preforming needed.

In the example, surface area of the finished part should be equal, in theory, to that of the bubble, which in turn should equal that of the blank. Since it is known that the area remains constant, it is not necessary to know its value in square inches (or square centimeters). All that is necessary is to calculate D, the diameter of the blank required for the final draw, and then work back to the bubble to determine the size of the bubble that can be drawn from a blank of diameter D.

To illustrate this procedure, refer again to the example. The formula for area in this case, which is illustrated in Figure 6-17, is

$$D = \sqrt{2h(d + d_1) + d^2 + (d_1 + 2f)^2 - d_1^2}$$

From the part drawing, assign the following values:

$$d = 1 \text{ in. } (25 \text{ mm})$$
$$d_1 = 2 \text{ in. } (51 \text{ mm})$$
$$f = 0.25 \text{ in. } (6.4 \text{ mm})$$

in which case $h = 1.58$.
Substituting these values in the formula,

$$D = \sqrt{2 \times 1.58 \times 3 + 1 + (2 + 0.5)^2 - 4}$$
$$= \sqrt{9.48 + 1 + 6.25 - 4}$$
$$= \sqrt{12.73}$$
$$= 3.567$$

The diameter of stock required for this blank is 3.567 inches (90.6 mm).

With this value of D it is now possible to determine the size of a bubble, i.e., preliminary form stage, which can be drawn from this amount of stock. A possible shape for such a stage is shown in Figure 6-20, the formula of which is

$$D = \sqrt{(d + 2f)^2 + 4d\left(h + \frac{d}{4}\right)}$$

where D must remain 3.567

$$d = 2$$
$$f = 0.25$$
$$h = ?$$

Substituting these values in the formula,

$$3.567 = \sqrt{(2 + 0.5)^2 + 8(h + 0.5)}$$

Squaring both sides of the equation,

$$12.73 = 10.25 + 8h$$
$$2.48 = 8h$$
$$0.31 = h$$

These calculations now make it possible to design the preliminary form stage, shown in Figure 6-36.

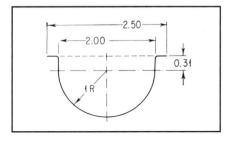

Figure 6-36. *Preliminary form for Figure 6-35.*

Calculation of Area

Many geometric shapes encountered in progressive draw die work fit no known formula, and it falls on the designer to calculate area in a step-by-step manner and then put this area into the blank. The known formulas for calculating area are:

1. Area contained in a circle of diameter D is $0.7854D^2$ (Figure 6-37).

$$\text{Area} = 0.7854D^2$$

2. Area of a sphere of radius R is $4\pi R^2$ (Figure 6-38).

$$\text{Area} = 4\pi R^2$$

3. Area of a zone of a sphere is equal to the product of a great circle and the altitude of the zone (Figure 6-39).

$$\text{Area} = \text{altitude} \times \pi \times \text{diameter}$$

A zone of a sphere is that portion included between two parallel planes. The altitude of a zone is the perpendicular drawn between the two planes. A great circle of a sphere is the line of intersection of the sphere and a plane passed through its center. (Put another way, a circle whose radius is the radius of the sphere itself.)

4. Area of a cylinder with radius R is equal to the product of the circumference and the height (Figure 6-40).

$$\text{Area} = 2\pi R \times h$$

5. Area of a pyramid is equal to half the product of its slant height and the perimeter of its base (Figure 6-41).

$$\text{Area} = 1/2 l(p + p_1)$$

Blank Development

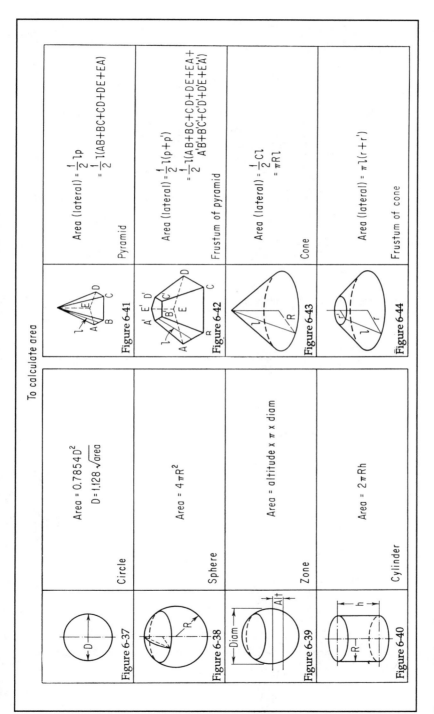

To calculate area		
 Figure 6-37	Area = $0.7854 D^2$ $D = 1.128 \sqrt{area}$	Circle
 Figure 6-38	Area = $4\pi R^2$	Sphere
 Figure 6-39	Area = altitude $\times \pi \times$ diam	Zone
 Figure 6-40	Area = $2\pi Rh$	Cylinder
 Figure 6-41	Area (lateral) = $\frac{1}{2}lp$ $= \frac{1}{2}l(AB+BC+CD+DE+EA)$	Pyramid
 Figure 6-42	Area (lateral) = $\frac{1}{2}l(p+p')$ $= \frac{1}{2}l(AB+BC+CD+DE+EA+$ $A'B'+B'C'+C'D'+D'E'+E'A')$	Frustum of pyramid
 Figure 6-43	Area (lateral) = $\frac{1}{2}Cl$ $= \pi Rl$	Cone
 Figure 6-44	Area (lateral) = $\pi l(r+r')$	Frustum of cone

Figures 6-37 through 6-44.

6. Area of the frustum of a pyramid is equal to the product of half its slant height and the sum of the perimeters of its bases (Figure 6-42).

$$\text{Area} = 1/2 l(p + p_1)$$

7. Area of a cone is equal to half the product of its slant height and the circumference of its base (Figure 6-43).

$$\text{Area} = 1/2 cl$$

8. Area of the frustum of a cone is the product of the slant height and the sum of the radii of the bases multiplied by π (Figure 6-44).

$$\text{Area} = \pi l(r + r_1)$$

In the example shown in Figure 6-45, the panel illustrated is a composite of four of the geometrical figures just mentioned.

1. Area of a cylinder
2. Area of the frustum of a cone
3. Area of a larger circle minus area of a smaller circle
4. Area of a hemisphere

To calculate, in theory, the area of the blank, compute and add the four areas.

1. Area $= d_2 \times h$
2. Area $= \pi l \left(\dfrac{d_1 + d_2}{2} \right)$
3. Area $= 0.7854 d_1^2 - 0.7854 d^2 = 0.7854(d_1^2 - d^2)$
4. Area $= \dfrac{4(d/2)^2}{2} = 2 \dfrac{d^2}{4} = \dfrac{d^2}{2}$

This total is the area of the blank. It is now arithmetically simple to calculate D, the blank diameter.

$$D = 1.128 \times \text{total area}$$

Centrobaric Calculation

If shapes are symmetrical about a common axis there is still another method, called centrobaric calculation, which can be used to complete blank area, as detailed in the following steps.

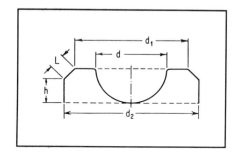

Figure 6-45. *Panel outline representing a combination of four geometrical shapes.*

1. Draw (to as large a scale as practical) a cross section of the part about its axis as shown in Figure 6-46.
2. Locate the center of gravity of the line AB. The perpendicular erected to the center of gravity from the axis is radius R.
3. Determine the length of line AB. At 10 times scale it can be accurately stepped off with dividers.
4. Multiply the length of this line by the circumference of a circle whose radius is R. This gives us the area between A and B.
5. Calculate the area of BB^1.
6. Add the two areas to obtain the total blank area.
7. Find D, the blank diameter, by using the conversion formula D = 1.128 area.

The centrobaric method lends itself well to the geometric shapes shown in Figures 6-3 through 6-44 and is useful as a double-check on these computations. Designers need only separate the cross section of the panel into its various components, find the center of gravity of each component, determine R (the distance from the axis to the center of gravity), and multiply R x 2 x length of component. Totaling the areas gives the blank area, and the conversion formula gives D.

Unsymmetrical Blank Development

Formulas and methods presented here are basic to all sheet-metal die work, but certain progressive dies are designed to produce unsymmetrical

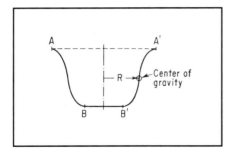

Figure 6-46. *Locating center of gravity simplifies area calculation.*

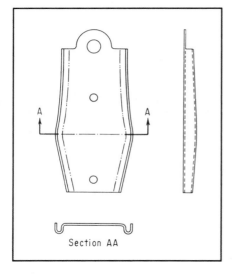

Figure 6-47. *Although this is a relatively simple panel, its blank must be determined experimentally.*

parts whose blank development defies routine shop mathematics. Figure 6-47 depicts a classic example of such a part. To solve the mystery of this part's blank, the two forming stages were mounted in scrap die shoes and placed side by side in the same press, as shown in Figure 6-48. The first experimental blank was laid out with a combination square, compass, and French curve, dimensions of reference points being noted on paper.

The estimated blank was run through the two forming stages and errors in excess of 0.125 inch (3.18 mm) were noted. The second blank was carefully laid out with a height gage and sawed and filed to the corrected dimensions. It was then traced on paper and the new dimensions were penciled in lightly. The second blank was then hit and still found to be in error. Corrections were made on the lightly penciled dimensions and the process repeated three or four more times until a blank had been developed that would satisfy all part-print dimensions.

There are more powerful mathematics to predetermine this blank without so much experimental work, but such mathematics lie above and beyond the confines of the die shop.

Having developed the blank, the diemaker is confronted with the task of locating it in the strip. To do this, it is necessary to keep in mind that two or more reference points in the blank must be kept — points that are easily identified in both the preliminary and final form stages. This is usually

Blank Development

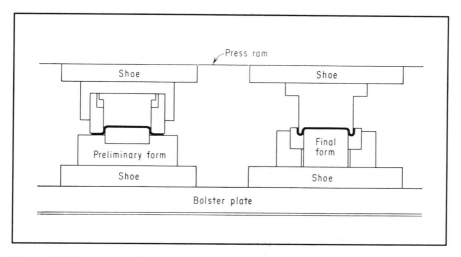

Figure 6-48. *Press setup for developing the blank for the panel shown in Figure 6-47.*

made easy by pierced holes in the part and easier still if these holes are used for piloting.

As shown in Figure 6-49, the first form stage forms the legs complete at a 45 degree angle and the second form stage forms the legs down in their final position. The hole in the center of the part is used for piloting; all the diemaker need develop is the angle and location of the trim punch.

The location of notch N can be taken from the part print, but the exact location of the trim line in the second stage is not known. The first step, then, is to turn up three small round plugs and press them into pilot holes in the die.

Next, through several pieces of stock of the correct thickness, jig-bore three holes, in line and a progression apart, as depicted in Figure 6-49. Third, working from these holes, lay out what appears to be the development.

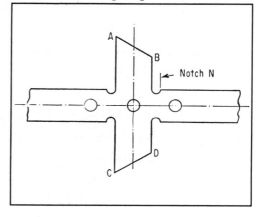

Figure 6-49. *This blank, like the blank in Figure 6-48, was developed experimentally within the die itself.*

133

Progressive Dies

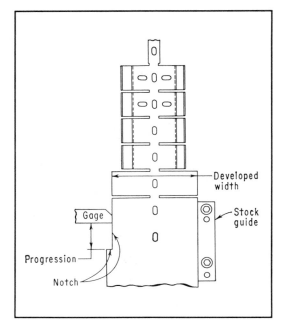

Figure 6-50. *The notch serves a dual purpose: it narrows the stock to correct width, and, when advanced, it places the strip on location.*

Points A, B, C, and D are laid out with a height gage and filed as closely as possible to the layout line. This part is now placed over the plugs in the die at the first form stage and hit. The part is then moved up one stage and hit again.

At this point, the part is checked and measured for errors at A, B, C, and D. It is simple to make the necessary corrections on a second part and locate the final position of these points. This simplified, and valid example of development work is basic in shop practice and can be used in making highly intricate parts.

Found in many progressive dies is a notch placed in the edge of the stock, as pointed out in Figure 6-50. This notch has a twofold purpose: it notches the stock to the proper width and, when the stock is advanced, it contacts a running gage to register the strip on the progression.

This part is particularly easy to develop since all that is required by way of development is proper overall stock width. Half of this problem is solved by use of an adjustable stock guide. On this type of die, it is possible to grind all of the die sections and fit them into the pocket. Stock width can then be cold-developed with extra grind stock left on the die section at the notching stage. The guide can be set in approximately the right position after a notch is filed in the cold development. The strip can then be advanced through the die and the finished part inspected for size.

After correct stock width has been determined, the stock guide is adjusted and doweled, and the notch die section is ground to corrected size. It is then possible to mount the notching punch and set the running gage, after which the die is ready for its final tryout run. Care must be taken in setting the running gage, because any error in its location is reflected in the part, multiplied by the number of stations in the die.

In developing a blank experimentally, common sense should prevail in selecting the method to be used. For example, diemakers must decide whether to develop the part by building the die on the way back to the blank stage, as in Figure 6-49, or to mount the form stages in separate shoes, as shown in Figure 6-48. The die for the part shown in Figure 6-47 might well have been built complete back to the blank stage. However, being an extremely large progressive die, the development work involved would have tied up a large press for the better part of a day at a time when it was in demand for tryout work; hence, the better decision to mount the form stages in scrap shoes and use a relatively smaller press.

In some cases, it is impossible to build the die back to the blank stage because the die is one blank (or more correctly, pierce stage) after another up to the form and cutoff stages. In this case, necessity dictates that the final form stage be built first, the part developed in a separate die set, and the progression established before completion of the die. However, if it is at all

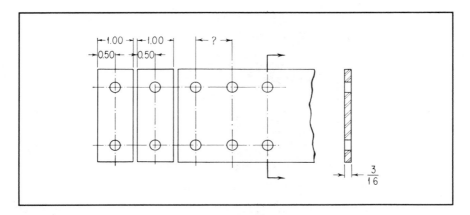

Figure 6-51. *An extremely simple layout. As shown, the panels are one inch wide with centered hole locations. The question: How far apart are the holes in the strip?*

135

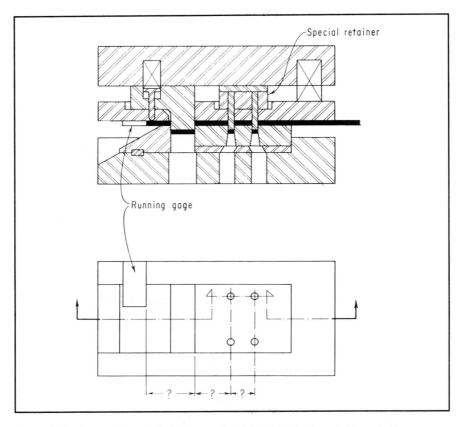

Figure 6-52. *Cross section and plan of progressive die for the strip shown in Figure 6-51.*

practical, the diemaker should develop the blank within the die itself in the interest of time and material economy.

Double-blank Development

There is one particularly persistent problem that arises in progressive die work, that of dimensioning in double-blank development.

To illustrate, picture the following example: a progressive die is needed to make brackets, two per press stroke, as shown in Figure 6-51. A cross section and plan view appear in Figure 6-52.

As shown in the cross section of the elevation, a special retainer must be made to hold the four piercing punches because they are too close to permit the use of separate retainers. The question, "How far apart are the punches in the special retainer?" seems ridiculously simple — one inch, apparently. But is it?

An examination of a cross section of the strip shown in Figure 6-53 reveals that, on 0.180-inch (4.57 mm) stock the clearance (one side of die – punch + 2) is 6 percent of 0.180, which is 0.011 inch (0.28 mm).

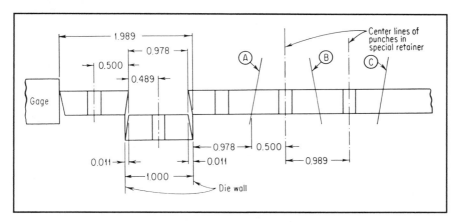

Figure 6-53. *This sectional layout shows the effects of breakage on progressive die dimensions.*

When this cross section is dimensioned and the critical factor of breakage is accounted for, the diemaker finds that to make two 1-inch (25.4-mm) blanks at a time, the distance between punch centers must be 0.989 inch (25.12 mm), or to express it in formula:

Distance between centers = blank width − die clearance

The die opening is 1.000 inch (25.4 mm) and the blanking punch is 1.000 inch (25.4 mm) − 0.02 = 0.978 inch (24.84 mm).

The distance from the die wall to the running gage is 1.989 inches (50.52 mm). In this there is nothing mystical or difficult. The designer and diemaker need only remember that the edge of a blank tapers according to the die clearance, and if a cross section of the strip is laid out as shown in Figure 6-53, the dimensions can be picked off and transferred to the die drawing, as illustrated in Figure 6-54.

The determining factor here is laying out and dimensioning the strip as shown. If this is not done, it will be impossible to get parts identical in width with holes in the center. If the running gage is set so that the blanks are of

Progressive Dies

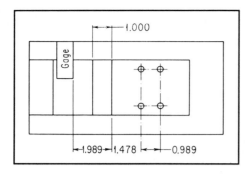

Figure 6-54. *Dimensioned plan view of the die. Critical hole spacing is not 1.000 inch, but 0.989 inch.*

equal width, the holes will be off center; if the gage is moved around in an effort to obtain centered holes, the difference in blank width increases.

To further complicate an essentially simple problem, the firm purchasing the die very often supplies the special retainer to the jobber. In these cases experience shows that the problem of metal breakage has rarely been considered; the punches are separated — almost invariably — by the exact width of a blank. The diemaker in such instances should examine the tolerance allowed this part. If, for example, the tolerance cited in Figure 6-51 is 0.020 inch (0.51 mm), the diemaker, by transposing the formula, will find the correct blank width to be 1.011 inches (25.68 mm).

Blank width = distance between centers + clearance

Blank width = 1.000 inch (25.4 mm) + 0.011 inch (0.28 mm) = 1.011 inches (25.68 mm)

If the tolerance is only 0.010 inch (0.25 mm), the diemaker and the buyer's representative will have to agree to either grant larger tolerance to the part or furnish a new special retainer.

To illustrate a different phase of the same problem, picture the building of a progressive die to make the part shown in Figure 6-55, two per press stroke.

The notches at the ends of the part are made identical in size (0.187 inch (4.75 mm)) by the product designer to facilitate the die work involved. What size punch should be used to notch the strip? At first glance, it would appear that a punch ground to 0.374 inch (9.50 mm) wide is required, since the rule is that in notching stock, the notch will equal the punch size.

Examining a cross section of the strip at section AA shows that by dimensioning the cross section as shown, the notch should be 0.187 + 0.187 - 0.0075, or 0.3665 inch (9.31 mm). Dropping the last decimal brings the notching punch to 0.366 inch (9.3 mm).

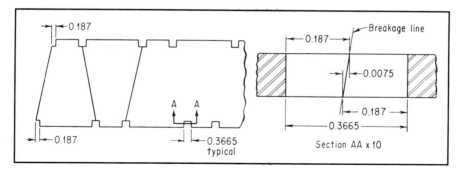

Figure 6-55. *Effects of breakage must be considered when close-tolerance notches fall on a parting line.*

Argument could be made that this is a lot of work for such a small amount of stock; but it should be remembered that although a part may have large tolerance, it is usually necessary that all parts be uniform in size. If, for instance, this particular part were notched with a 0.374-inch (9.50-mm) punch, the parts coming out of the die would show glaring discrepancies.

Cutoff Dies

In most cases, when a simple part such as is shown in Figure 6-51 is required, it is better to design and build a die that stamps out two at a time rather than just one. Obviously, production is doubled at little additional cost; but more importantly, if the die makes only one part at a time, it functions as a cutoff die, and cutoff dies always give parts a "rolled" edge. This rolled edge may or may not be acceptable, depending on the tolerance assigned the part. Needless to say, the die designer should examine the part print carefully for tolerance before he designs a cutoff die.

Figure 6-56 illustrates a cross section of a typical progressive cutoff die. As the drawing shows, the panel is unsupported as it cuts off; this is

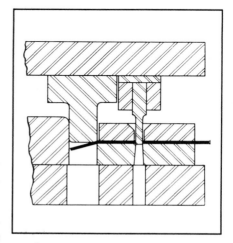

Figure 6-56. *Sectional view of a typical cutoff die. Rolled edges and out-of-tolerance panels always result from dies of this design.*

what causes the rolled edge. Whenever a die is built like this, the temptation is to grind an acute angle on the cutoff punch. This will not help. The only solution is to add a spring pad to support the part at cutoff and an air blowoff to remove it. However, this additional work makes the die nearly as expensive as it would have been had it been designed for two parts per press stroke.

Basic Types

Progressive dies like any other type of tooling can be very simple or very complex. How difficult they are to design and build depends almost entirely on the parts they produce. They can be simple two-stage dies that pierce holes, advance, and cut off; or they can be complicated tools that carry a panel 20 or more stages before all operations are completed.

The fundamental principle of progressive work is that operations are performed on stock in sequence within a single die. For this to happen, stock must advance from one station to another. Whatever distance it advances is called the progression. The progression imposes certain mathematical limitations on progressive dies — limitations not found in other classes of dies. All stations must be a progression apart, and yet the details or sections must butt against each other. This implies that the most inconsequential filler block must be ground to an exact dimension that is generally unknown and must be calculated. This one factor probably does more to intimidate the diemaker than any other.

The second factor that seems to provide a mental obstacle to many diemakers and designers is general lack of understanding of cutting operations. That the *back* of a punch cuts the *front* of the trailing panel while the *front* of the punch cuts the *back* of a leading panel is confusing to many. In progressive operations the part is not blanked — the scrap is blanked. This, of course, is just the reverse of single-operation dies.

While the mathematics of progressive dies are treated generally throughout this book and specifically in Chapters 13 and 16, a review of the fundamental theory of progressive operations is warranted here.

Classification

Broadly speaking, there are four families of progressive dies:
1. Progressive blank dies,
2. Progressive cutoff and form dies,

3. Progressive draw dies,
4. Cut-and-carry progressive dies.

Among these, a certain amount of overlap exists in that many dies can fit into two different categories. Because of this overlap, some dies are difficult to classify, but this is not critical. The breakdown into types is for purposes of study only.

Progressive Blank Dies

As the name implies, progressive blank dies produce unformed parts, which usually have pierced holes. For many years, industry favored compound dies for production of these parts. However, the superior production ability of progressive dies, when compared to that of compound dies, makes them increasingly acceptable for making blanks.

A progressive die is illustrated in Figure 7-1. It is probably the most elementary type of progressive die possible. After a hole is pierced, the strip advances and is cut off. Simultaneously another hole is pierced. Note that the cutoff punch is supported by a heel block which also acts as a gage. In this operation, the operator pushes the stock against the heel block, which is spaced one progression away from the die block. The punch dimension then equals one progression minus die clearance.

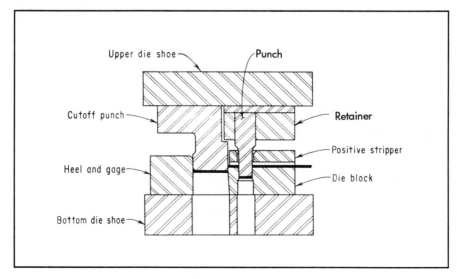

Figure 7-1. *The most simple progressive die design — a pierce and cutoff die.*

The cutoff punch is cleared away to provide room for the punch retainer. Because of this crowded condition, such dies usually incorporate what is called an idle stage between the pierce and cutoff stages. Idle stages provide room for various details. However, when an idle stage follows the first operation, it is necessary to use a preliminary gage to pick up the stock for the second hit. At the third hit the stock has reached the heel and gage block. Hence, the preliminary gage is used for only one hit per length of stock. To eliminate the necessity of this infrequently used gage, the designer has altered the shape of both retainer and cutoff punch. This practice is acceptable as long as the alterations do nothing to weaken any of the details.

A point of interest in this design is the timing of the two punches. The piercing punch is longer than the cutoff punch. Although it does not have to be longer — it could be shorter — it is important that some difference in length exist to ease tonnage requirements in cutting operations. In larger dies with multiple piercing punches, it is especially important that punch lengths be staggered.

Notch and Pilot Stages

A slight alteration in part design can cause a substantial change in die design, as shown in Figure 7-2. Two corners of the panel have been cut off on 45-degree angles and a third corner has received a rectangular notch. In addition, one edge of the blank is trimmed to hold a tight decimal dimension. Such a part calls for two additional stations — one to notch and trim, the other to pilot.

The pilot guarantees the progression. As such, the pilot must be spaced a distance from the punch equal to the progression. For this reason pilot location is critical. Any error in its setting will cause the strip to be off location. Although the remainder of the die is built to exacting dimensional tolerances, the accuracy of the part is equal to the accuracy of pilot location — no more, no less.

There are several important considerations in designing pilots. First, they must be long enough to be the first part of the upper die to contact the strip. When progressive dies with spring strippers are designed, pilots must extend beyond the stripper. This is because the pilots must register the strip before the spring pressure behind the stripper locks the strip in an incorrect position.

Progressive Dies

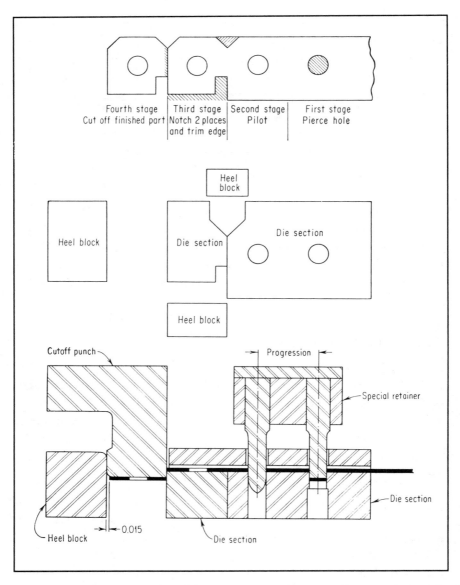

Figure 7-2. *A slightly more complex strip design (top) and an ideal layout of die sections to create this strip. Sectional view of die is shown at bottom.*

Second, pilots should have bullet noses, not rounded hemispheric noses. A bullet nose has greater efficiency in registering the strip. Even though a strip is off location by as much as a complete hole diameter, a pointed pilot will return it to location. Moreover, a pilot is often forced to pierce a hole in

144

the stock. This happens whenever the punch that pierces the pilot hole breaks and the press is not stopped immediately. In such cases a bullet-nosed pilot will pierce the strip much easier than a round-nosed pilot will, and with considerably less chance of breakage. Because pilots will cut slugs from the strip, clearance holes at pilot locations should be drilled through the shoe for slug relief.

It should be emphasized that gages are secondary to pilots in locating the strip. In progressive dies where piloting is impossible because of a lack of space for pilot holes, gages must be relied upon to locate the strip. In all other cases, gages are used for preliminary locating only as illustrated in the cross-sectional view of Figure 7-2. As in the cross-sectional view of Figure 7-1, the heel block serves as an end gage, but its dimension from the die block is equal to a progression plus 0.015 inch (0.38 mm). This is important in progressive die construction; end gages must provide a feed greater than the progression. Although the 0.015-inch dimension need not be held — 0.010 inch or 0.020 inch (0.25 mm or 0.51 mm) would serve as well — 0.015 inch is the generally accepted constant.

The relationship of gages and pilots can be established by the rule: It is the function of the gage to locate the strip closely enough for the pilot to register the strip in final location.

Two-per-stroke Die

Although notches were added to the basic part in the preceding example, the operations remain fundamental. The panel in the strip can be seen from start to finish. In Figure 7-3 the same part is produced but the operations involved are beginning to assume the complexity of normal progressive design.

Although the part remains exactly as shown in Figure 7-2, this die is designed to produce two parts per press stroke instead of one. Perhaps the best way to study this strip is from the last stage back to the beginning. First, it should be noted that the progression has doubled; it is now equal to two part lengths. In the fourth stage and latter part of the third stage, the part outlines are easily seen. The trailing part is punched out of the strip (note the die plan), automatically forming the leading part.

In the first part of the third stage and central part of the second stage, two triangular punches create the 45-degree corners. The 45-degree corners of

Progressive Dies

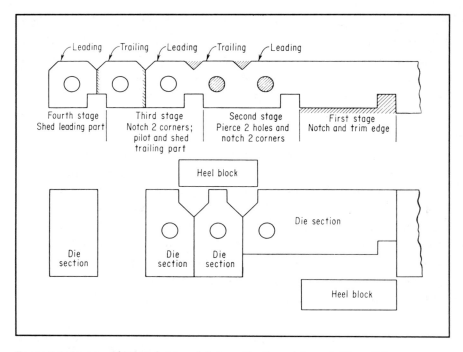

Figure 7-3. *A more sophisticated strip and die layout for the part shown in Figure 7-2.*

the trailing panel are created entirely in one hit. However, the leading panel receives its corner outline in two hits.

The second stage pierces both center holes. The first stage trims the lower edge and creates the rectangular outline of both parts — all with one punch.

Determining Stages

Just where the stages of a progressive die begin and end is sometimes vague, especially in dies such as this one. It must be remembered, however, that no matter where one starts to mark off lengths, each length should be equal to the progression of the die. In this example, the trailing edge of the rectangular notch punch is selected as a starting point. Progressions are marked off from this point with the resulting stages beginning and ending on a line going through the part. Many designers would prefer to mark off the stages with lines coincident with the edges of the parts. For example, the two finished panels might be selected as the fourth stage and preliminary stages marked off from their edges in multiples of the progression. In practice, though, it is rather unimportant; regardless of where one starts marking stages, the mathematics of the die remain the same.

Cutoff-and-Form Dies

The three dies discussed to this point were designed to create flat parts. However, the great majority of progressive dies involve forming operations; therefore we shall begin with the most elementary type of progressive form die possible, that of Figure 7-4.

This die is a three-stage progressive. The strip is pierced in two places, advanced through an idle stage, and then cut off and formed. Detail 2 is a form punch, and detail 7 serves as both cutoff punch and forming die. In the die shop it is referred to simply as the cutoff punch.

The die area of the cutoff punch barely covers the formed part. If it does not cover the part completely, only the covered portion will be spanked. There will also be a sharp line of demarcation between the spanked and

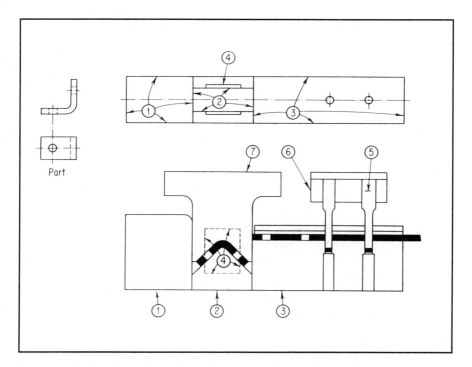

Figure 7-4. *Typical pierce, idle, cutoff, and form layout for the part illustrated.*

unspanked areas — an undesirable condition. Therefore when this die is sharpened, it will be necessary to regrind the form area in the punch.

The gage indicated by detail 4, two of which are required, is a small piece of heavy sheet metal which prevents the part from turning in the interval after it is cut off and before it is formed. In many cases these gages are unnecessary, but it is a wise precaution to drill and tap holes through this punch. If, in tryout, gages are found necessary, the holes are available. If gages are unnecessary, the holes do no harm.

Springback, the bane of the diemaker's existence, is easily corrected in a die such as this because the angles can be easily reground. However, if the radius in the cutoff punch is shortened so that it does not spank the panel, overbend could result instead of springback. This, of course, provides an excellent correction factor for springback and overbend alike. In starting the die, the diemaker should plan to use the theoretically exact radius and the required angles. If in tryout springback results, the upper radius can be *eased-off*. If this fails to provide sufficient correction, the angles must be changed.

As in all cutoff and form dies, the form punch should be slightly below die level. Die level is the height of the die block, detail 3.

This die design is good only for panels having legs of equal length. Should one leg be longer than the other, it will be impossible to control the part after it is cut off. In such cases a two-per-stroke design should be used.

Cutoff and Flange

Strip design of a progressive cutoff and flange die and a cross-sectional view of the die itself are shown in Figure 7-5. In the first stage of this die the part contour is completely finished, the central tab being used as a bridge to carry the panel through the remainder of the die.

The second stage is of necessity an idle stage; the third stage cuts the part from the strip and forms it immediately. In dies such as this, a pressure pad (detail 7) is necessary to support the panel and prevent it from slipping during the flanging operation. In this example, the flanging operation covers a relatively small area and exerts little pressure. In other dies, where flanging areas are larger, slippage becomes a major problem.

One good solution to the problem of slippage is to insert one or more cat's-paws in the pad and/or the forming punch, as shown in Figure 7-6. The holding power of a cat's-paw is tremendous, especially when two of

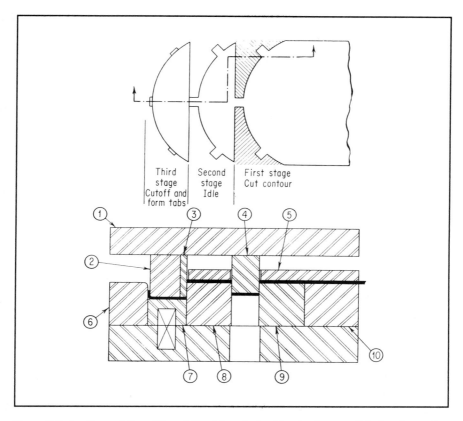

Figure 7-5. *Another variation of the cutoff and form design. Forming is accomplished against spring pressure.*

them are placed in opposition as shown. The astute diemaker will make provision for cat's-paws by drilling and reaming holes in sections to be hardened, if the die design indicates that slippage may become a problem. If there is no slippage, the holes can be plugged with cold-rolled steel and ground flush with the surface. The main shortfall of the cat's-paws is that they mark the parts. Parts used as decorative hardware must have unmarred surfaces which argues against the use of these holding devices. Therefore, it is especially important for designers to consider the importance of part appearance when designing progressive dies. It may be necessary to alter an entire design to remove potential slippage.

Detail 3 of Figure 7-5 is an insert in the forming punch serving also as a cutoff device which can be sharpened and shimmed back to the level of the punch. It provides indefinite life to the forming punch, for without it, it

Progressive Dies

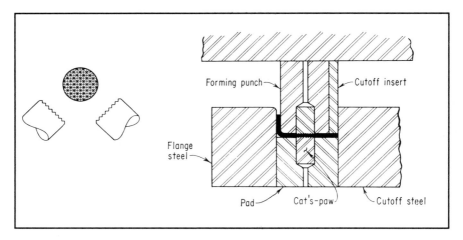

Figure 7-6. *Design and use of cat's-paws. These are necessary to prevent slippage of the panel.*

would be necessary to grind the entire punch surface. This would also necessitate reworking the radii at the flanging areas. Eventually the entire detail would have to be replaced after repeated sharpenings had exhausted punch life.

Design of this insert is recommended as shown in Figure 7-7. Although the shoulder design is not absolutely necessary, it is to be recommended. The elongated counterbored screw hole and hardened backup plate are mandatory.

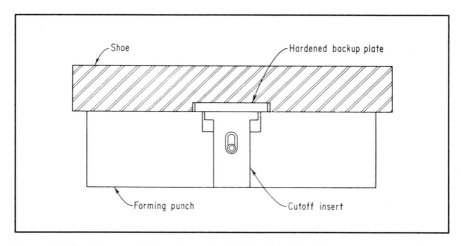

Figure 7-7. *Cutoff insert used in forming punches. Without this device die maintenance personnel must grind the entire punch surface and regrind the radii.*

Springback

It is impossible to predict the amount a piece of metal will spring back after a flanging or forming operation. Although there are formulas which some diemakers use to calculate springback, they are not worth noting. There are simply too many different analyses of sheet metal for any universal springback formula to be successful.

In a die such as the one shown in Figure 7-5, springback may or may not be a problem, depending upon the analysis of the stock and the tolerance given the part. If springback returns flanges to an angle outside of tolerance, the following corrective approach can be used on operations of this type.

> Before tryout, do not dowel flanging steels such as in detail 6. Set this detail to metal thickness, screwing it down as tightly as possible. Put the die in the tryout press and run through a few panels, observing the springback characteristics of the tabs or flanges as the case may be. Remove the forming punch (detail 2 in this example), and grind a negative angle of 1 degree in the flange-forming area. After the forming punch has been returned to the die, reset the flanging steel (detail 6) at from 0.001 inch to 0.002 inch (0.03 mm to 0.05 mm) less than metal thickness. After a panel is hit, springback may well have disappeared. It is even possible that overbend will result. Should overbend result, it can be corrected by easing-off the tightness at the flanging area. In this way the diemaker can use two variables — tightness and negative angle — to control the angles of flanges. It is entirely possible to overbend a tab by as much as 3 degrees when such a condition is desirable.

As soon as the correct tab angle has been established, however, the diemaker should pull the die from the press and dowel the flanging steel. If the die is continued in operation without dowels, the flanging steel will eventually move, causing the tabs to deviate from the desired location.

Progressive Draw Dies — A Case History

Because the press operation called drawing is the most formidable of all press operations, it follows that progressive draw dies are among the most

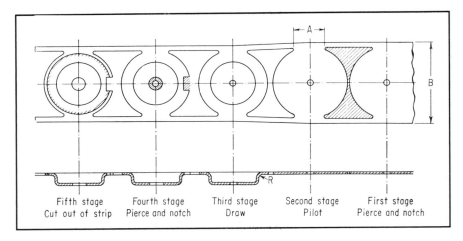

Figure 7-8. *Strip design for a progressive draw die. Hourglass cut between first and second stage typifies this design.*

difficult of all progressives. In Chapter 1 the basic nature of the draw operation and some of its attendant problems are discussed. Chapter 6 deals with the formulas involved in draw work. This chapter will touch only upon the two basic types of progressive draw dies in an effort to show the potential of these tools.

In Figure 7-8 a plan and cross-sectional view of a strip from a simple five-stage progressive draw die is shown. This strip is typical of the draw-die group, but it is noteworthy that it contains basic errors which made the die something less than a resounding success. These errors were not occasioned by designer's mistakes, but by unrealistic limitations imposed by die engineering, product, and process engineering departments.

At first glance the sequence of operations appears to be not only sound but ideal. The first stage pierces a pilot hole and blanks the hourglass contour, which is typical of all progressive dies that stamp round drawn shells. The second stage is idle except for piloting, and this too is typical. (Stages following an hourglass cut are invariably idle.) The third stage is the all-important draw stage.

At the fourth stage, a hole is pierced at the pilot-hole location and a notch is made in the flange. The fifth and final stage cuts the drawn panel from the strip.

It was found that this die worked well in the vendor's tryout press (at 11 hits per minute) but failed completely at the purchaser's plant when set up in a dieing machine. Although shallow, the draw stage would not perform under high-production requirements of over 60 hits per minute. All cups fractured when approximately two-thirds deep.

The die was returned to the vendor (accompanied by the purchaser's chief tool engineer) and again was set up in the tryout press — where it delivered a perfect part.

An explanation for the unusual behavior of this die is readily seen in the design of the strip. The first basic error is in the size of R, which is a metal-thickness radius. R should be equal to four times metal thickness, if the draw is to be successful. Normally, the designer would draw this shell over a correctly sized draw radius and then shorten the radius to part-print size with a restrike stage. His failure to do so in this case was because of a lack of die area. The process engineer specified the size of press the die was to be run in before the die was designed. In effect he specified a five-stage die for a panel requiring six.

The second basic error in this design is in the width of dimension A. This dimension should have been made as small as possible to reduce resistance to plastic flow of metal — the draw action. As it stands, this panel meets an enormous resistance to plastic flow because of the stiffness of the strip. However, to shorten A, it would be necessary to widen B so that the hourglass punch could more nearly encircle the panel. This condition is also a result of faulty process engineering, for although there is sufficient metal to draw the shell, the metal must be *stolen* from the carrier. To avoid this condition, designers should calculate the size of the necessary blank and then add metal for bridge and carrier, as shown in Figure 7-9.

The result of this lack of coordination between process engineering and die engineering was that the die was turned over to a stamping company which ran it in a slow press on a contract basis.

A cross section of this die is illustrated in Chapter 2, Figure 2-4. However, in that view the notch punch of the fourth stage has been omitted.

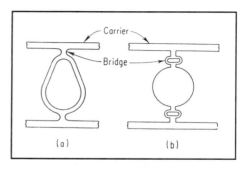

Figure 7-9. *Two types of bridges used in progressive draw dies.*

This example is typical of progressive draw dies that draw metal down into the lower die. More complex examples are shown in subsequent chapters, but suffice it to say that the great majority of progressive draws follow this pattern. The first stage pierces pilot holes and blanks the hourglass. Second stages are idle except for piloting operations. First draws begin at the third stage with subsequent stages being used for redrawing, restriking, embossing, etc.

Bridges and Carriers

In Figure 7-9a the correct design of a bridge is shown. Since it has been reduced to minimum practical size, metal can be drawn without undue resistance from bridge or carrier.

Another bridge design, although one more wasteful of metal, is shown in Figure 7-9b. By creating a circular bridge of relatively narrow dimension, the designer enables the drawn panel to be pulled in without disturbing overall strip width. The design is also an effective solution to the lifter problem referred to in Chapter 2, Figure 2-10.

Progressive Draw Dies — A Second Type

The type of progressive die just discussed forms metal down and, as such, has definite limitations. As pointed out in Chapters 1 and 2, it is necessary to place the binder in the stripper. Because the stripper is spring-actuated, draw pressure is uneven, starting low and ending high. While this condition is acceptable in shallow draws such as shown in the preceding example, it is an insurmountable difficulty when deeper draws are made. In such cases it is necessary to put the form punches in the lower die. A pressure pad or binder can then be built around the punches (Figure 7-10). This pad should be actuated by air pressure to obtain correct and uniform pressure throughout the draw.

As shown in Figure 7-10, draw dies which draw *up* are more versatile than those that draw *down*, but they are generally more costly to build. If draws are not deep, it is probably better to design the dies down with binding action being provided by the stripper.

Cut-and-carry Progressive Dies

This type of die is by far the most popular of all the progressives, although it might be argued that progressive draw dies are also cut-and-carry dies.

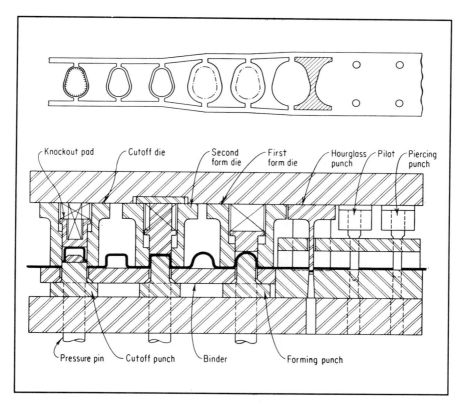

Figure 7-10. *Inverted type of progressive draw die. When draws are relatively deep, this design should be used.*

Figure 7-11 shows a strip from an elementary cut-and-carry progressive die as well as the plan view of the four most important die sections. If this part, which is comparatively simple, were to be made in a line-die setup, at least three separate dies would be required: one to blank the developed panel, one to bend it, and a third to pierce the three round holes.

If hole locations did not have to be held to tight tolerances, it would be possible to make this part in two dies: a compound blank and pierce die and a form die. Even with only two dies, labor costs would be more than doubled on a per-part basis — hence the value of a progressive die for this operation.

The die illustrated notches the outline of the part outline in the first stage, leaving only a small bridge to keep the part in the strip. The second stage pierces the center hole, which serves as a pilot hole in the third stage.

155

Progressive Dies

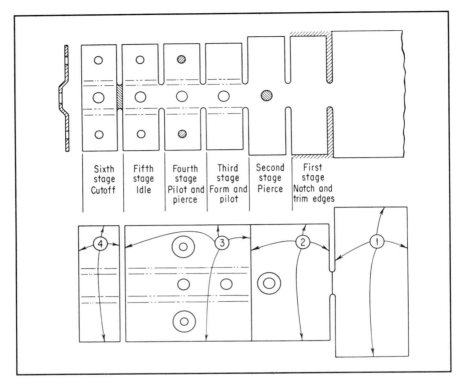

Figure 7-11. *Strip and die plan of a basic cut-and-carry progressive die.*

The third stage not only pilots but forms the panel into the finished shape. The fourth stage pierces the two outside holes, holding as close a tolerance as is desired. The part is now finished and all that remains is to separate it from the strip. However, this separation cannot as yet be effected because of a crowded condition among the *upstairs* punches. To provide space, an idle stage is incorporated as the fifth stage of the die.

In the sixth stage a punch cuts out the bridge, separating the finished panel from the strip. An air blowoff is then used to blow the part out of the die, as illustrated in Figure 7-12.

It should be noted that this strip is not perfect, nor is it above criticism. Many designers would criticize it on three important counts. First, the center hole used for piloting is not pierced until the second stage, which means that the strip will not be piloted until it reaches the third stage. Good design practice calls for piloting to begin as soon as possible, in this case, the second stage. A second feature which might well have been changed is the use of two punches for trimming edges. In a die of this type, it is necessary to trim only one edge of the stock.

A third criticism is the use of punches that notch and trim simultaneously, as shown in the first stage. Such punches are difficult to make, grind, and maintain because of the thin projection. Die size permitting, it is much better to separate these operations as shown in Figure 7-13.

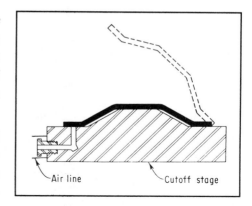

Figure 7-12. *Air blowoff design utilized in cutoff stages.*

After the enumerated changes have been made, the sequence of operations would read:

1. Pierce round hole and notch between panels,
2. Pilot,
3. Trim one edge,
4. Form complete,
5. Pierce outside holes,
6. Idle stage,
7. Cut off part from strip.

A Two-per-stroke Layout

One of the great advantages of progressive dies is that many parts can be produced on a two- or four-per-stroke basis. Figure 7-13 shows the layout of such a strip; it can be observed that such a part is actually easier to make on a two-per-stroke basis.

Starting at the first stage, the die pierces and notches and the second stage pilots. This is the revision that was suggested for the die in Figure 7-11, but a special retainer to hold the pierce punch and pilot as well as the notch punches is almost a necessity.

The third stage trims the edge of the stock to the part development, another revision that was suggested for Figure 7-11.

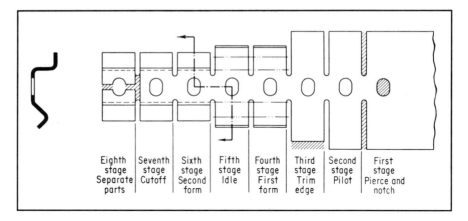

Figure 7-13. *Two-per-stroke design showing the form at 45-degree stage.*

The fourth stage forms the legs complete at a 45-degree angle as shown in the strip cross section.

The fifth stage is idle to avoid an overcrowded condition in the upstairs punches. The sixth stage is the second form wherein the legs are formed down at 90 degrees. The seventh stage sees the bridge cut out and the eighth stage separates the two pieces from each other. A single T-shaped punch separates the two parts while simultaneously cutting them from the strip.

To return to the fourth stage, this type of forming action is rapidly gaining in popularity, particularly in automotive die manufacture. The advantage of not forming the parts completely in one stage is *metal stretch*. The part can be formed completely in the sixth stage, but putting four 90-degree bends in a piece of sheet metal simultaneously invariably stretches it.

Metal stretch in itself implies a bad condition. On some occasions, it is absolutely unavoidable, but where it can be eliminated or minimized it should be; it runs contrary to good mechanical practice and it makes it impossible to hold close tolerances. In this particular die, the overall size is established in the third stage. If metal stretch is a factor in obtaining this development, it follows that a change to a stock of a different analysis will produce a part of different length. However, a preform stage, such as the one shown in this strip, will form the outside angles without metal stretch regardless of the stock analysis. The final form stage merely pushes the legs down from 45 degrees to 90 degrees. Here again, metal stretch does not enter the picture.

A 45-degree preform stage is also an excellent solution to the problem of springback. The outside angle of the part can be corrected to a perfect 90 degrees in this stage. To correct the springback of the final stage, refer to the discussion of Figure 7-5 in this chapter.

Criticism can be made of the T-shaped punch of the seventh and eighth stages because it is difficult to build and maintain. In this respect it resembles the notch and trim punch of Figure 7-11.

Right- and Left-hand Parts

Just as progressive dies lend themselves quite well to two-per-stroke designs, they are equally adaptable to production of right- and left-hand parts. In many of these designs, the die layout is essentially that of a two-per-stroke die. A contour punch is shown in the sixth stage of Figure 7-14; the fact that the holes deviate from the part center line causes two symmetrically opposite panels instead of two identical panels to be made.

This die follows the conventional cut-and-carry layouts described in this chapter; therefore no review of the sequence of operations should be necessary. However, the punch outline of the fifth and sixth stages should be studied carefully. This punch not only cuts the part out of the strip, but notches a desired contour as well. Note how the rectangular portion of the notch *overlaps the progression*, a design principle used whenever possible in progressive die work.

In the plan of the die sections, note the manner in which the sections are split to facilitate the building of this portion of the die. It is not unusual to find designs for such a station to be made from one solid block of steel. Though not an impossibility, such a design greatly increases the cost of the die. The hole must be entirely finished before heat treatment, a risky procedure, particularly when thin stock is to be run in the die. Thin stock means small die clearance, in which case it is difficult to make an internal contour, to say nothing of holding it through heat treatment.

Splitting the sections as shown allows them to be ground after heat treatment to tolerances of 0.0001 inch (0.003 mm) if necessary.

It should be added, however, that this type of split should not be made unless the sections are pocketed in the lower die.

If these sections are mounted on the surface of the lower die shoe, sections 5 and 6 should be combined as one piece as should sections 3 and 4. In this

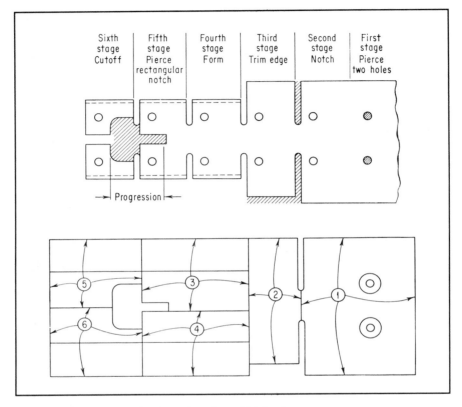

Figure 7-14. *Strip and die plan layout for right- and left-hand parts.*

way, side pressures are effectively contained. The only remaining opening pressure then lies between the two solid details and this pressure can be contained by heavy-duty keys at each end of the dieblocks (Figure 7-15).

In those progressive dies using pockets to contain die sections, the additional sectionalizing shown in Figure 7-14 is optimum. Fabrication is simplified and slug pressure does not enter as a design factor.

Double-strip Cut-and-carry Dies

An increasingly popular type of progressive die design is that in which a double row of parts is made. A strip from such a die is shown in Figure 7-16. Such progressives are for extremely high-production runs and are well worthwhile when production is of a permanent nature. Millions of parts

Basic Types

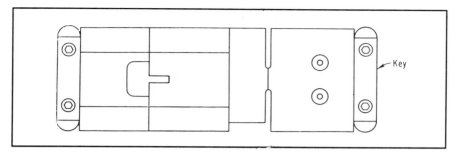

Figure 7-15. *A variation on the die plan shown in Figure 7-14.*

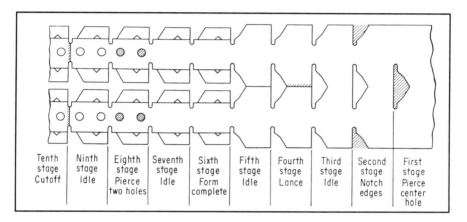

Figure 7-16. *Another kind of two-per-stroke progressive layout.*

have been made on this die; therefore the increased cost of the second row was easily amortized.

Starting at the first stage, a single hole is pierced on the stock center line. This hole creates the outline of each side of two inside flanges. The second stage notches the outline of the exterior flanges so that the part outline is completed in the third stage. (Because the third stage is idle, this may sound confusing. Actually, nothing happens in the third stage, but the panel emerges from this stage complete because of the notched edge created in the second stage.)

The third and fifth stages are idle because of the lance action in the fourth stage. This particular lance required a great deal of space, and so it was deemed advisable to separate it from the notching punches of the second stage and the forming punches of the sixth stage by two idle stations.

The lance, of course, is necessary to separate the panels before forming action begins. The sixth stage forms the parts complete, putting triangular stiffeners on each side of the panel. Like the lance, the forming punches require a great deal of room; therefore the form operation is also isolated between two idle stages. The eighth stage pierces two holes in each part, the ninth is idle, and the tenth removes two completed parts from the strip.

Double-strip designs such as this lend themselves quite well to production of right- and left-hand parts and can be utilized in some cases, such as the part designs of Figures 7-13 and 7-14, to make four parts per press stroke.

8

Strip and Stamping Design

Strip Design

To say that a well constructed die begins at the drafting board would be to overwork an obvious cliche. Not quite so obvious is that responsibility for success or failure of the die rests much more heavily on the engineering department than on the die shop.

If a die design is basically sound, the journeyman mechanic can be relied upon to produce a satisfactory tool, but if the design is fundamentally faulted, the finest mechanic available cannot possibly salvage it. Against the backdrop of this axiom, it would be well to define the designer's primary responsibilities in order of their importance. These are:

1. To design a strip showing a logical, workable sequence of operations,
2. To establish an approximate progression, accurate to within 6.0 inches (152.4 mm),
3. To provide adequate piloting and/or gaging,
4. To design the die around the strip resulting from the first three steps.

Strip design and the importance of this phase of die design cannot be overemphasized. The die is constructed from the die drawing; the die drawing is made around the strip; and the strip represents a sequence of logical, workable operations, which is to say a sequence of ideas.

If this sequence of operations is in error, the error will surely emerge in the tryout press; it is therefore incumbent upon the designer to make certain that the strip is 100 percent sound. Other errors, even basic mechanical errors, in the design can be corrected, but if the strip sequence is unworkable, the die is scrap, which may or may not be salvaged by redesigning.

Unfortunately, a method of determining proper sequence cannot be formulated. The strip must be a product of the designer's experience, sense of logic, and knowledge. The experience can be gained only by working; sense of logic must be developed.

A Simple Cut-and-carry Strip. Figure 8-1 shows a basic cut-and-carry strip. Having designed this strip in a logical, workable sequence, the designer will have little difficulty in designing the die, nor should the die present any major difficulties to the diemaker.

The diemaker will have to develop the trimmed contour in the second stage. This is an excellent example of a die that can be used to develop its own part: all stages except the second are built and mounted in the die. Experimental blanks are then hit in the third stage to obtain the trim line development of the second stage.

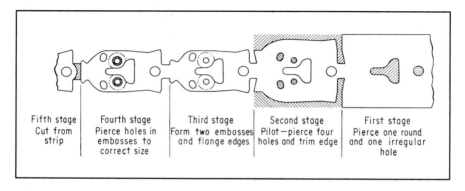

Figure 8-1. *Typical five-stage strip design.*

Because of the close spacing of the holes in this strip, the designer has put two of the holes in the first stage and the remaining four in the second. A special retainer is required to hold the punches of the second stage because the proximity of the punches does not permit the use of individual retainers. The round hole pierced in the first stage is used for piloting the strip at the second and third stages.

Note that two of the holes pierced in the second stage are at the center of the embosses formed in the third stage. In the fourth stage the holes are repierced to the desired size. Obviously, these holes could not be pierced in the second stage because they would open when the embosses are formed. It would be possible to leave the two holes out of the second stage entirely, but they do serve a useful purpose.

It is easier to form a good emboss if a hole is first pierced at its center, because this allows metal to "flow" out from that center. If there is no hole, the emboss must be made by stretching metal. In this particular case, if the die were to stretch metal, this action would disturb the size and shape of the irregular hole in the center of the panel. The use of a preliminary hole allows the emboss to be formed "out" from the preliminary hole. In this manner, distortion of contour of the central hole is kept to a minimum. Whenever an emboss is to be pierced with a center hole — as most are — it is wise to provide a small preliminary hole. Even if there is no close contour for the emboss to disturb, the preliminary hole is still preferred. Without it, metal must stretch; with it, metal can flow.

Although this strip design is quite satisfactory, it does not have a perfect sequence, nor could it have. Trouble occurs here and in every die, progressive or single-operation, when holes are placed close to bend lines. In this example, the oblong holes pierced in the second stage are badly distorted by the flanging action in the third stage. The only alternative would be to pierce the holes after the flanging operation has taken place. This option, however, is actually the greater of two evils because of the thinness of the resulting die wall, the fact that the locus of the flange radius cuts through the holes, and the difficulties in sharpening the section. Thus, the error in this design is the location of holes too close to bend lines.

This strip could have been further improved by the use of four separate trim punches, instead of two, to define the developed contour. End contours between the panels could have been notched in the first stage and side contours only in a second stage. This would have been more in accordance with good design practice which does not allow a punch to have a long extended member.

A Progressive Notch and Form Layout. Figure 8-2 exemplifies good layout from both economical and mechanical points of view. Economically, the designer has made maximum use of the commercial punches and die buttons currently on the market.

In the first stage, a round pilot hole is pierced and the sides are notched with two V-shaped punches. These notching punches are simple and represent a small investment in material and labor.

In the third, fifth, and sixth stages the developed outline of the panel is completed entirely by the use of commercial punches and die buttons. As the blanks enter the idle seventh stage, they are completely defined, and the eighth stage forms them into finished panels. The ninth stage separates them from the strip and pushes them through the escape holes in the shoe.

Progressive Dies

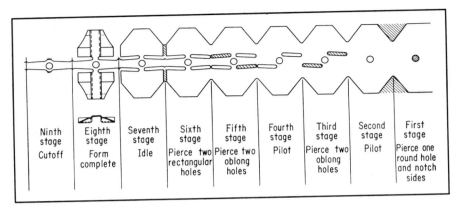

Figure 8-2. *An example of good strip design.*

From a mechanical perspective, too, this strip design is optimum; one would be hard pressed to find a better, faster way to build a die to produce these parts in volume.

This die, unlike that in the preceding example, is one in which the forming stage must be built first. This stage is then mounted in a small die set to develop the blank outline and establish the progression in decimals.

The progression is equal to the developed width plus the width of the piercing punches of the sixth stage. Normally, the development of a part such as this is quite simple, requiring no more than half a dozen experimental blanks. However, locating the blank in the strip and dimensioning the die layout is a lengthy mathematical operation with great potential for error.

Unfolding the Panel. As the complexity of the part increases, its strip is correspondingly more difficult to design, due largely to more problems in panel processing and blank development. The part print in Figure 8-3 illustrates this point. (Dimensions have been omitted since procedure only is of concern.) Observe that:

- There are two bend lines,
- There are two flat surfaces,
- There is a third surface following an irregular curve,
- Two holes are pierced in the panel.

From these observations, the following conclusions can be drawn:

- The finished part, as shown in the part print, will be either the last or the next to the last stage;
- While the flat surfaces represent no particularly difficult problems, the curved surface must be formed before it is thrown into perpendicular;

- The part, when unfolded and put entirely into one plane, will represent one stage;
- An initial stage will be required to pierce the two round holes;
- There will be a "carrier" between panels;
- A logical preliminary sequence of operations would be
    a. Pierce two round holes,
    b. Blank out the part outline,
    c. Form the curved surface,
    d. Form the panel at the two bend lines.

This sequence represents four stages, but the veteran designer need only glance at the part print to ascertain that additional stages will be needed. A sequence of operations must be not only logical but workable, and this *does* have a bearing on the novitiate's approach to strip design. The new designer must determine logical sequence intellectually, then prove workability by drawing the strip according to this predetermined sequence.

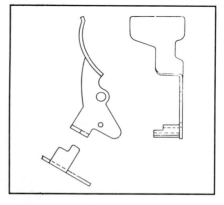

Figure 8-3. *Complex automotive panel. Problem: to design a progressive die strip for this part.*

One way to do this is by way of a cardboard or plastic template made to the cold-developed blank. By sketching or tracing a number of blank outlines on an arbitrary center line, the designer can get an accurate picture of the cutting stations required. This procedure will also give a good idea of space requirements for the forming operations.

However, this procedure applied to the example of Figure 8-3 — the finished strip of which is shown in Figure 8-4 — reveals that completion of the part outline in one stage is impractical: it would require two very complicated punches. These punches would have not only heavy and thin cross sections, but would be difficult and costly to build and replace. Hence, the designer must create this blank outline in two stages by using four separate punches.

Starting with the blank outline (which ultimately becomes stage four) and drawing two blanking operations to the right defines the beginning of

Progressive Dies

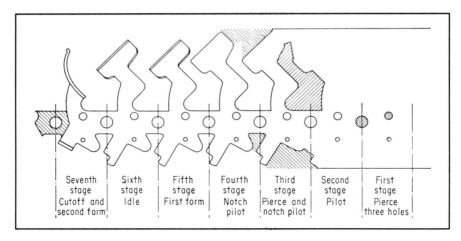

Figure 8-4. *Strip for part shown in Figure 8-3. The basic problem here is to create curvature in a flanged segment.*

a strip. Continuing to the right, the design calls for two pierced holes, as well as a pilot hole. It would be possible to pilot in one of the holes in the part but, because there is a carrier between panels, additional space is available which can be used for a larger pilot. A larger pilot is preferable, since it ensures greater accuracy in strip alignment. Owing to space requirements, it is necessary to place an idle stage between the blank station and the pierce station. Accurate piloting is one of the three primary responsibilities of the designer.

With the strip to the right of the developed blank completed, the design effort now moves to the left. The final stage will be represented by the finished panel, and at some point between the developed blank and the final form will be a forming stage which puts curvature into one surface. Experimenting with the cardboard template, the designer can locate the curvature immediately following the developed blank stage. However, to throw the curved surface into a vertical position, an idle stage following the first form must be incorporated. The designer is now ready to draw the strip as shown in Figure 8-4. Through freehand sketching and use of the template, the following logical workable sequence of operations results.

1. First stage: Pierce three holes;
2. Second stage: Pilot (otherwise idle);
3. Third stage: Pilot, pierce, and notch;
4. Fourth stage: Notch and pilot;

5. Fifth stage: First form;

6. Sixth stage: Pilot (otherwise idle);

7. Seventh stage: Cutoff and final form.

A Two-per-stroke Strip. A type of strip that appears with regularity in the automotive world is shown in Figure 8-5. The two flanges are spot-welded to the body or to some other piece of hardware, the round hole used for a fastening bolt. A multitude of strip designs exist for such a panel, and the layout shown in Figure 8-5 embodies their best features.

First in importance is that this strip is a two-per-stroke layout: production is doubled at very little additional die cost.

Second, the two forming operations ensure consistently uniform parts. The unwitting die designer could well take the developed blank (represented by the fourth stage) and finish-form it in one operation. Such a forming operation invariably stretches metal, and when metal is stretched, tolerances cannot be held because of variations in stock thickness and hardness. These two variables can cause a great deal of difference in panel size if the part is moved from stage four to stage six without preforming it in stage five. In contemplating a forming stage, the designer at this point has to determine whether the panel will be stretched by a forming stage. If, indeed, the stretch will be a troublesome factor, then one or more additional stages is indicated to complete the forming. But if a certain length of die is specified, and that length precludes the use of additional stages, limitations will have to be accommodated. This strip layout, like that of Figure 8-2,

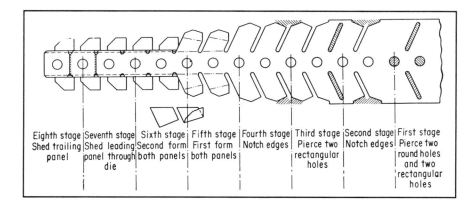

Figure 8-5. *Eight-stage strip showing 15 parts in varying stages of completion.*

makes good use of commercial punches and punch blanks to define the developed blank outline. The die designer has thus reduced the cost of the die. Good die design mandates use of these whenever possible.

With this strip, the designer starts with the two developed panels (which is stage four) and designs to the right to find the best way to establish the desired contour. Here again a cardboard or plastic template of both blanks could be made and traced on a center line.

In completing the strip to the left, the designer works from the "known" — a 45-degree first form, a finish form, and a cutoff stage. These can be sketched freehand following the template layout, which will give a fairly good idea of the strip to be designed.

Because the flanges of the finished panels overlap, it would appear that such a strip is an impossibility. Such overlapping is possible, but the forming punches cannot cover the entire flange, nor is it necessary that they do. Because the flanges are preformed in the fifth stage, it is necessary only that the legs be folded down from 45 degrees to 90 degrees. Therefore, if the finish forming punch covers the bend line, the mechanical action in this stage is acceptable. In building such a die, the diemaker should bear in mind that:

1. The forming punches of the sixth stage will have the shape of the finished flange. This angle means that the forward part of the punch will contact the strip first, and forming to 90 degrees will be accomplished with a shearing action. If this punch is set too tightly, the shearing form will distort the panel (Figure 8-6).
2. Usually the flanges are square, making springback a problem. If a negative 2-degree angle is ground in the die steel, springback can be neutralized (Figure 8-7).
3. The most difficult sections in such a die are the first-form die steels.

Strips for Frames. Often, the die designer encounters comparatively fragile part designs such as the one in Figure 8-8. Frames of this type are used extensively in automotive work and in decorative hardware. Designing a die to make this part can be a simple task if the designer is mindful that the part must be completely finished in the strip before the center portion is blanked out. Any other approach to such a part is impossible because the center is needed to lend strength to the panel during the various operations performed. Therefore, when looking at such a part print, the designer

Strip and Stamping Design

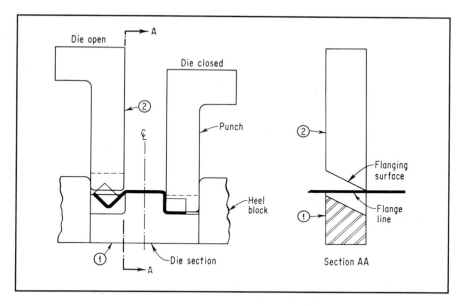

Figure 8-6. *Angular forming stages for strip illustrated in Figure 8-5.*

immediately knows that the last two operations in the strip sequence will be to blank out the center contour and then cut off. Unfolding this part (with its center intact) will give the blank development stage.

Piloting will be required, but this is easily accomplished by putting the pilot hole in what eventually will be the scrap.

Following the developed blank stage will be a form stage. The countersunk screw holes can be put in one stage, but two stages are preferable.

From this analysis of the part print the designer can say with a fair degree of certainty that the sequence will be:

1. Pierce pilot hole and notch contour,
2. Form panel,
3. Pierce the two round holes,
4. Countersink the holes,
5. Blank out the inside contour,
6. Cut off.

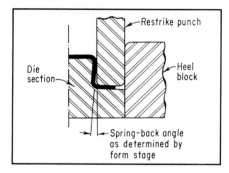

Figure 8-7. *Restrike stage for strip illustrated in Figure 8-5.*

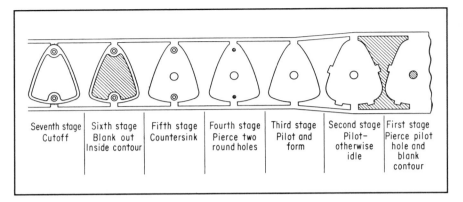

Figure 8-8. *Typical strip layout for frames used as decorative hardware. Part is complete when center portion is removed.*

The second stage of the die must be idle, though it will certainly be used for piloting. With the idle stage incorporated, the sequence proves workable and logical.

The designer should note, however, that parts like this are most often made of light-gage metal, and as such a problem arises in that the strip itself is fragile and will tend to buckle. The designer should keep carriers and bridges as wide as practical and use bar-type stock lifters rather than the round-post type. Post-type lifters have regularly caused problems in dies running fragile strips.

Extended Members. Another fundamental type is shown in Figure 8-9. This part presents no particularly difficult problems to the designer or diemaker and the strip as shown should be self-explanatory.

The stock for this part is light-gage steel, so that it is possible to form the emboss without first piercing a hole to let the metal flow. The second stage pierces a 0.31-inch- (7.87-mm-) diameter hole. Between the second and third stages much of the developed contour is defined. One pierced contour punch finishes the side and one end of each of the four lugs. Two more comparatively simple punches between the third and fourth stages completely finish the contour.

This strip demonstrates the inherent advantage of progressive dies over single-operation work: in this case, the designer has achieved a fragile outline without the use of correspondingly fragile punches and die sections. The pitfall to avoid is the temptation to disregard the natural advantages of

Strip and Stamping Design

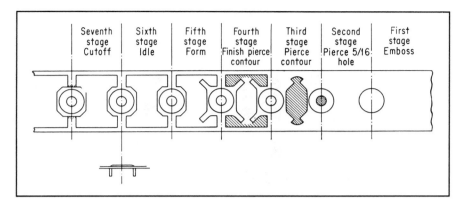

Figure 8-9. *Development of a delicate outline by relatively heavy punches.*

the progressive die in cases such as this and, given this part, complete the entire contour with a complicated fragile punch between the second and third stage, as is shown in Figure 8-10. Granted, a station could be omitted in this way, but cost of such a die would increase greatly as would punch and die mortality rate. Elimination of stages in a die through careful design should be the goal of all designers, but only if it can be accomplished without overloading the other stages. It is a sound rule of good die design to always complete a fragile extension such as these lugs in two or even three stages.

The stock used for this strip is 0.020-inch (0.51-mm) material, but when designing such a strip for heavier stock, for instance 0.060 inch (1.52 mm) and above, three contour piercing stages become necessary, as shown in Figure 8-11. This is because the lugs, as they enter the fourth stage, are unsupported and, consequently, will be unable to resist the pressure exerted by the finish-piercing punch. They will deform and the edges will have a sheared effect. When the strip is designed, the only serious drawback is that a special retainer will

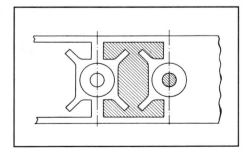

Figure 8-10. *Correct only in theory is the blank stage shown above. Although it could be built, the maintenance would be prohibitively expensive.*

173

Progressive Dies

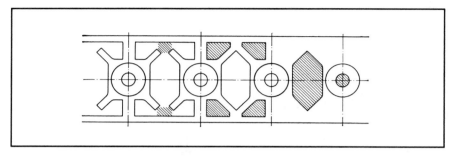

Figure 8-11. *The blank stage is an improvement on either of the examples shown in Figures 8-9 and 8-10. Its superiority results from careful balancing of blanking pressures.*

be needed — and they are costly. However, if the die is designed for a long run, this extra cost is easily amortized and the buyer has a much better tool.

An Electric Connection Die Strip. Figure 8-12 shows the completed blank development at the end of the fourth stage. Creating this outline is the most difficult part of the strip, because to the left there are only two forming stages after which the part is finished and ready to be cut off.

To the right of the completed blank are four stages, three of which are used to define the blank outline. This strip could be improved only by extending the pierced area of the third stage up around the radius of the tab in the manner shown in Figure 8-13. Such an alteration would make it easier to grind the notching punch of the fourth stage. Whenever a 180-degree radius must be cut in two stages, the designer should put 90 degrees on each punch. Obviously, less than 90 degrees on one punch, as in stage three (Figure 8-12), means more than 90 degrees on the other punch, as in stage

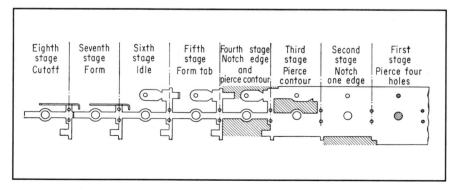

Figure 8-12. *Strip design for electric switch component. This is a good design but it can be improved.*

Strip and Stamping Design

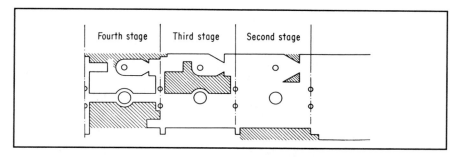

Figure 8-13. *Design improvements possible in the blanking stages of the example shown in Figure 8-12.*

four. Grinding a radius over 90 degrees on a punch may be difficult or even impossible depending on the amount of interference there is to entry of the grinding wheel. In this case, there is indeed a triangular projection (Figure 8-14) which would interfere with grinding. As a result, the punch would have to be jig-ground. Therefore, in any progressive die, a 180-degree radius should be cut in two stages. A punch should never cover both sides of a radius in one stage because such designs are generally fragile and permit no correction for grinding errors (see Figure 8-15). As an example, the work done by the upper punches in the third and fourth stages could be done by one punch in the third stage. Such a punch would create the entire upper outline of the blank, but it would

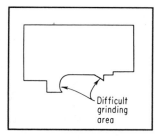

Figure 8-14. *Radii extending over 90° present difficult problems in grinding. Had the radius illustrated stopped at 90°, this punch could have been surface-ground.*

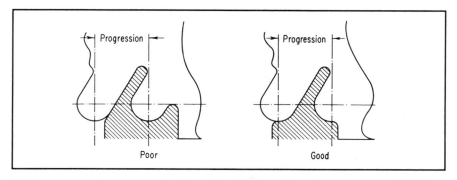

Figure 8-15. *The design at the left is mechanically acceptable but difficult to build. Punches should never go around a radius in this manner.*

cover both sides of the tab radius and, as such, would represent an impossible grinding operation. And, as illustrated in Figure 8-10, the die sections would be extremely fragile.

A further possible refinement of this strip design would be to place the two triangular notches (in the tabs) in the second stage (Figure 8-13), thus permitting straight punch and die sections in the fourth stage.

General Considerations. As stated, there are few inviolate rules governing strip design. Successful designers are those who have accumulated a great deal of practical hands-on experience. In many engineering offices, designers who demonstrate marked talent focus exclusively on strip design. Others create progressive dies around the strips these designers develop.

Despite the elusiveness of hard and fast rules covering all die design tasks, certain guidelines prevail that prove helpful to the designer.

- Learn to unfold a part accurately. This provides both a precise blank definition and an accurate progression.
- Starting with the flat blank, work first to the right to obtain blank and pierce stages and then to the left to obtain forming stages.
- Make piloting as thorough as possible, and where possible, pilot in the scrap. Often this permits use of heavier pilots.
- Use templates of standard retainers to determine pierced-hole stages. Even if the expense of special retainers is warranted, individual retainers are still preferable.
- Avoid use of complicated punches in defining part outlines. When blank outlines are difficult, utilize additional blank stages.
- When possible, avoid metal stretch in double bends by use of the 45-degree preform stage.
- Visualize the completed strip. If it is fragile, specify bar-type instead of post-type lifters.
- Where possible, make any embosses first, then define the contour. An emboss near a finished contour is certain to draw metal.
- If the part print shows a pierced hole in the center of a round emboss, pierce a smaller hole at this point before the emboss is formed. In this way, metal can flow out from center. Later in the sequence, a hole of correct size can be pierced.
- In designing draw stages, follow the four-times-metal-thickness rule for draw radii. If corner radii are smaller than this, shorten them with a restrike stage.

- Always consider restrike stages in progressive draws and progressive forms. Parts showing stringent tolerancing almost always require restrike stages.
- Drawn panels with sizable holes can often be more effectively drawn if the strip design incorporates a preliminary pierced hole.
- Avoid pierced holes near flange lines. If at all possible, pierce the holes after flanging has taken place. To pierce them before runs the risk of hole distortion.
- On progressive draws ensure that maximum strength remains in bridges and carriers.
- Before designing the die, make certain that part disposal and scrap disposal have been accounted for.
- Ensure that grinding considerations are accounted for when designing stations. A mechanically perfect strip design could in reality be one that uses punches and dies that are impossible to make.
- Work toward a goal of maximum production efficiency. It may be possible, at a slight increase in time and money, to design a two- or four-part-per-stroke strip that is much more efficient than the conventional one-per-stroke strip.
- Work for simplicity of design. Avoid eccentric or overelaborate designs.

Stamping Design

Design of stampings to be run in progressive dies is closely allied to strip design. Without an understanding of the principles of progressive opera-

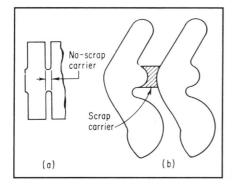

Figure 8-16. *Two designs for linking parts in a strip. In (a) there is an easily parted area. In (b) a scrap slug is used.*

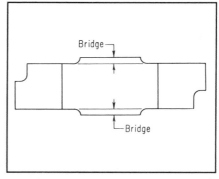

Figure 8-17. *The no-scrap design is ideal in panels with straight sides.*

tions, it is highly unlikely that a product designer will be able to design a stamping that can be created in a progressive sequence. Therefore, certain "musts" of stamping design need to be adhered to.

Bridges. Because stampings are linked in the strip, a carrying device between parts is necessary in many cases. In some instances, such as in progressive cutoff dies, one panel is cut immediately adjacent to another and no carrier is required. If forming action takes place — and it generally does — the stamping must have an easily parted linkage with adjacent stampings. Bridges, the term given those parts of a strip that unite two adjacent panels, can take one of two basic forms: an easily parted area (Figure 8-16a) or a scrap slug (Figure 8-16b) which is blanked out to free the finished part. When two panels have symmetrical edges, the no-scrap bridge is practical. In all other cases, a scrap slug must be incorporated. However, use of the no-scrap bridge requires that a certain portion of the panel protrude. As shown in Figure 8-17, this design can be used quite often in formed panels with straight edges.

Double Panels. Because many panels are made on a two-panel-per-station layout or on a right- and left-hand layout, a half-bridge should be used. As an example, the panel depicted in Figure 8-18 is a "natural" for a two-panel-per-station progressive die. The designer therefore has added dimension A in two places. This addition enables the panel to be carried two at a time.

In operation, these panels would be lanced in the third station from the end of the die. The subsequent station would be idle and the final station would cut both panels from the strip.

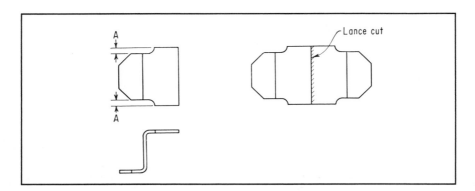

Figure 8-18. *This part lends itself quite well to a two-per-stroke design.*

Strip and Stamping Design

Slug Bridges. The panel shown in Figure 8-19a is typical of a part requiring a slugged-out bridge. However, the panel must be notched to allow bending action to take place at the flange. Therefore edges A, B, C, and D must be cut in a preliminary stage of the die. After the part is completed, the slug is cut out in the final stage of the die, which means ac must be coincident with A and C. Likewise, B and D must coincide with bd.

This is a particularly problem-prone condition, because it is difficult to maintain perfect alignment of these edges, especially after the die "seasons" in production. As a result, fine, almost microscopic slivers of steel are constantly being formed. These are hazardous to operators because of the air blowoffs typically used in progressive dies. They also manage to work back to other areas of the die, fouling the working members.

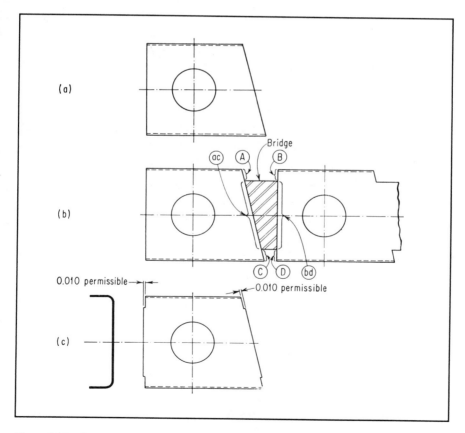

Figure 8-19. *When straight line cuts are made, the 0.010-in. permissive dimension should be specified.*

A proven solution is the use of a 0.010-inch (0.25-mm) step at the edges where the slug is removed. In this case, the stamping designer merely specifies "0.010 permissible," as shown in Figure 8-19c. This allows the diemaker to narrow down his cutoff punch to a level sufficient to avoid slivers and a generally bad parting condition.

Straight Line In Blank. Often an emboss or drawn area is placed in a position in the stamping where it draws metal from an edge. This may or may not be critical, depending on the function of the edge. For instance, the panel illustrated in Figure 8-20 has a drawn area which has pulled metal from the edge. This natural condition should be anticipated by the stamping designer and considered in the light of part function.

For many parts, it is absolutely necessary that the edge be straight — as in the case of an exposed panel. This consideration is critical to stamping design. Straight edges in panels can be achieved only by developing the edge to compensate for metal pull-in. Inasmuch as blank development with resulting irregular blank contours is costly, it should be avoided whenever possible. If it is possible, the stamping designer should make the notation, "Straight line in blank."

Radii. In general, the smaller a radius, the more likely it is to cause complications or additional expense in fabricating the die. In the less well equipped die shops, a 0.030-inch (0.762-mm) radius is regarded as a "broken edge" and any radius less than 0.030 inch a total impossibility. This is not to say that product design should be tailored to meet the vendor's standards; it merely points up that small radii are expensive. If they are

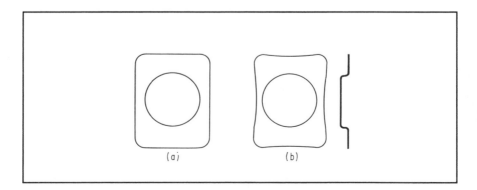

Figure 8-20. (a) The panel as it was originally designed; (b) the pull-in resulting from draw action.

necessary, the stamping designer has no choice; he or she should, however, make certain that larger radii are truly unacceptable.

This is particularly true when specifying radii for drawn panels. Since a draw radius should equal four times metal thickness, designers should try to adapt their parts to this formula. Often the radii are much shorter than this, making a restrike stage mandatory.

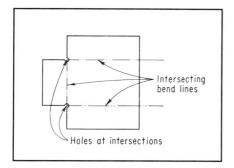

Figure 8-21. *Holes should be specified at the intersections of bend lines.*

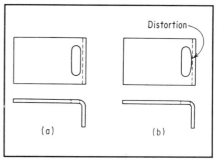

Figure 8-22. *(a) Part design as the product designer conceived it; (b) the result of this design.*

When metal is flanged, it should not be pulled over a radius of less than metal thickness. To do so seriously weakens the panel.

Intersecting Bend Lines. When two flanges of a stamping are bent, as shown in Figure 8-21, metal generally tends to gather in the corners, creating an undesirable condition. A good solution to this problem is to design the blank with holes at the corners. This type of design allows the flanges to form without any metal dislocation and subsequent irregular surfaces.

Die design for this stamping is quite easy. Two holes are pierced in the first stage, after which notching punches are cut to the center of the hole in a subsequent stage.

This design also works well in flat panels notched to a square corner for assembly. If a notched area of a stamping is fitted to a square-cornered component, this clearance is mandatory.

Holes Near Bend Lines. A design to avoid is one calling for a hole near a bend line. A hole at a bend line is difficult to obtain with a progressive die since bend lines set up stress in the metal. Therefore, good design practice

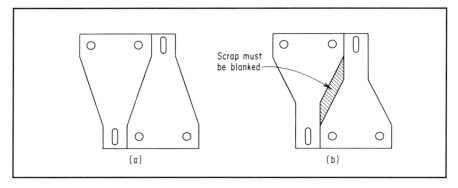

Figure 8-23. *(a) Perfect flip-flop parts designed for a no-scrap die; (b) imperfect flip-flops necessitate a blanking stage.*

argues against locating holes at points where bend stress will distort the holes (Figure 8-22).

Symmetrical Edges. Flat parts to be produced on progressive dies should have symmetrical edges wherever possible. This enables them to be made without scrap, as shown in Figure 8-23*a*. If edges are asymmetrical (Figure 8-23*b*), a scrap slug must be blanked out between them. Loss of material is not important (it takes the same amount of metal to produce two stampings by either layout), but the addition of a sizable punch to blank the scrap is costly. Here again, the designer can become frustrated by part function and prevented from designing perfect flip-flop parts, but by staying constantly aware of this problem, he or she can vary designs to eliminate scrap.

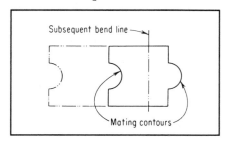

Figure 8-24. *Blank with mating contours. This design enables the die designer to specify a no-scrap die.*

Interlocking Parts. In addition to designing parts with symmetrical edges, designers should be aware of the possibilities of interlocking parts. For example, the part illustrated in Figure 8-24 can be made no-scrap by mating the ends of the stampings. Again, this is not always practical, but it is an often neglected opportunity to simplify strip and die design.

Conventional Progressive Dies

An examination of some of the more common progressive dies is a valuable aid to die designers and diemakers who are not acquainted with this specialized type of press tool. Although none of these dies is particularly difficult for experienced designers and diemakers, all fit into the category generally referred to as Class A dies.

Extrusion and Double Flange

Extruding is not frequently used in progressive die work largely because product designers seldom specify extruded holes.

Development of the pierced hole for an extrusion is generally a necessity — it cannot be accurately predetermined if the height of the extrusion is established as a decimal dimension. A cross-sectional view of an extruding stage of a progressive die strip is seen in Figure 9-1. A binder is necessary in this operation because, although stock lifters are used to raise the strip from die level to feed level, the strip will tend to "hang up" at the extruding punch.

The tabs that are formed down in the sixth stage of this die are partially defined in the first stage. This definition is necessary because inner edges of the openings adjacent to the tab are in the extruded area. However, it would be unwise to create the entire tab outline before extrusion takes place, because to do so would place unrestrained metal at each side of an area subjected to upward extruding pressure. This would result in an uneven wavy edge on the extruded boss. Completion of the tab outline in the third stage circumvents this problem. In any case, this U-type notch should be completed in two stages. Its completion in one stage requires too delicate and difficult a piercing punch to be practical.

Many strips incorporate flanges; and though formation of flanges may appear quite simple, many designers still employ cams for this operation, which are costly. In such designs, the panel at the fifth stage advances to the equivalent of the seventh stage. At this point the major flange is formed

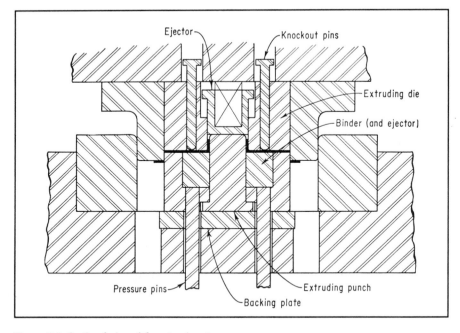

Figure 9-1. *Sectional view of the extrusion stage.*

down. Following an idle stage, two cams form the minor flanges around a die block. Although this cam operation is effective, its use creates additional cost in labor and material. Generally, wider shoes are required, and this additional width may necessitate a larger press. Cam action in a progressive die should be resorted to only when other options have been exhausted.

Forming an Emboss

Where embosses are specified, the critical consideration from the designer's perspective is the manner in which the emboss is formed. The die should be designed such that metal is restrained from flowing in from the outside. Good design practice would specify two punches in the second stage to define the exterior contour of the embossed area as well as the leading edge of the trailing panel. Forming punches in the third stage would then create the emboss. After this operation, it is possible for the remainder of the panel outline to be created in the fourth stage.

The only development work required in a die of this design lies in the counter punches of the second stage. Because metal will pull in an indeterminate amount, the diemaker must build the die backward to the second stage. Experimental blanks placed in the third stage will show how the metal will move and will give a layout of the large pierced holes in the second stage.

The outline of the central hole in the panel will not be pierced until the emboss has been completed. If it were punched before the third stage, it, too, would be subject to distortion from the embossing punches. While the ends of the tabs are blanked in the first stage, this hole is far enough away from the emboss to resist distortion.

Form and Cutoff

The ingenuity of product designers can be a significant factor in the efficiency of a die's design. In a typical case, just by interlocking the part design, scrap was reduced to a minimum. Such designs not only reduce the quantity of the scrap but also shorten the required working area of the die and simplify the job of building the tool itself.

Normally, operations that combine forming and piercing require the use of air pressure against the forming pad. (In many cases spring pressure suffices but air is always preferable.) Typically, the forming operation is completed against the air-supported pad which descends only after the flange is formed. When the cutting edge of the combination form and blank punch reaches die level, it starts to cut the finished part out of the strip. When the pad hits "home," the combination punch "spanks" the part, removing the radius created by the forming action (Figure 9-2).

Timing a die such as this is necessary because of a lance-form action in the third stage. The lance must bottom simultaneously with the form-pierce punch of the last stage. This can best be accomplished by finishing and setting the form-pierce punch and then grinding the stop blocks to correct height. The lance punches can then be ground to size by trial and error. The angle is ground, the lance is inserted, and the die is closed. By taking a gage-block measurement between the stop block and the upper shoe, the diemaker can readily determine the amount of metal to grind off, remembering that the derived measurement is a function of the sine of the acute angle of the lance. Further timing is required to cause the pierce punches to contact the strip after the lance has started to work and before the form-pierce punch of the last stage hits home. It is also necessary to time the notch-trim punches

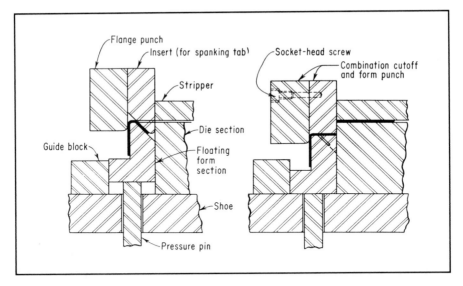

Figure 9-2. *Sectional views showing part formation (left) and part cutoff (right).*

to hit after the pierce punches reach die level. Timing of these punches to hit individually eases the shock of operation by distributing the load through a greater portion of the press cycle. It is also helpful if the diemaker adds a metal-thickness shear to his trim punches.

Cutoff and Form Acute Angle

Many different parts requiring formation of acute angles (angles created by flanging beyond 90 degrees) can easily be finish-formed in the final stage of a progressive die. Cam action can be used in such a die, but, again, it is generally good policy to incorporate cams only as a last resort. If the angle is only 1 or 2 degrees past vertical, the method discussed in Chapter 7 (Figure 7-5) is effective, although it makes no provision for spanking the flange.

A simple and effective way to create this angular flange is illustrated in Figure 9-3. Advancing through the eighth stage, the primary flange is formed at a 90-degree angle. A cutoff punch, serving also as a forming punch, shears off the panel as it advances into the ninth stage. The geometry of the part is such that the panel is unable to escape the punch and no shift

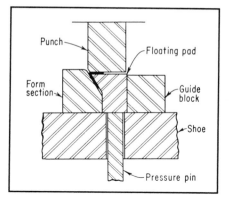
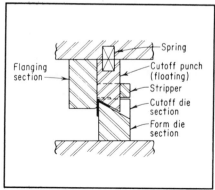

Figure 9-3. *Floating punch assembly. This design is excellent when flange has curvature.*

Figure 9-4. *Simple cutoff and form stage. Flanging section must be set "loose" to avoid part distortion.*

in the bend line is possible. Spanked against a hardened pad, the part is forced to conform to the angle of the cutoff punch. Inaccuracies in the angle of the part due to springback or overbend are easily overcome by correcting the angle of punch and pad. Performance of this cutoff and form stage can be further improved by spring or air pressure beneath the pad. Another method of completing this flange is by way of a complex punch assembly, Figure 9-4. This assembly is also an optimum device for putting contour in flanges, an operation that might otherwise require a secondary operation. In operation, the spring-loaded floating punch cuts the preformed part from the strip. Continuing its descent, the floating punch depresses the pad which can be actuated by either air or spring pressure. Once the pad has bottomed, the floating punch can descend no farther and the flanging steel, which continues to descend, effectively forms the flange at the proper angle.

A necessary precaution in the use of a floating punch is provision for adequate support to prevent any lateral motion. If the punch is able to move sideways the slightest amount, it will shear the die as well as its own cutting edge. Keeper blocks, hardened and ground and "set on the tight side," are usually sufficient support for the punch.

Floating punches are most effective on light-gage metal (0.032 inch [0.81 mm] or less) because metal thicker than this generally requires more spring pressure above the punch than is practical. It is essential, of course, that the part be cut off and the pad depressed on bottom before the floating punch begins to retreat within its assembly.

Unsupported Forming

The strip layout of Figure 9-5 presents two examples of unsupported forming. The curling operation is one. The second, in which the two curled flanges of the panel are formed past 90 degrees, is highly unusual although it follows well-known laws of mechanics.

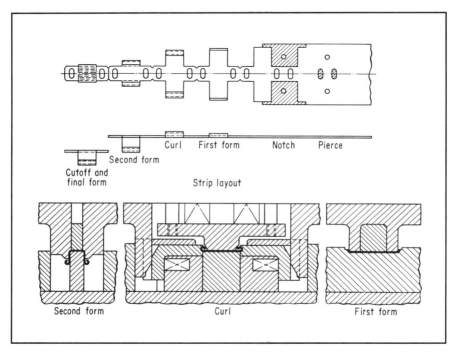

Figure 9-5. *Preliminary forming stages showing strip layout and die design of an unusual curling die.*

The curling operation follows the procedure outlined in Chapter 3 in that the designer has used the third stage to put initial curvature into the flange. In this preforming function, almost 90 degrees of what is approximately a 300-degree curl is completed. Thus, as the strip enters the fourth stage, a good lead has been established for the curling action of the advancing cams. This is a standard progressive die operation, although it can be difficult to perform on small diameter curls.

After the ends of the flanges have been curled, a 90-degree forming operation takes place on each flange. This, too, is a conventional operation, designed to prepare the flanges for the final form stage. The forming punches extend around the radii, but although this feature is desirable (it spanks and sets the radii), it is omitted in many designs in lieu of a straight wall. Straight-wall forming punches are less expensive to build and present no problems in timing the die. (Timing a die, though not difficult, is time-consuming when more than one station is involved. In this design, any correction in the curl stage calls for a corresponding correction in the flange- and first-form stages.)

The final station of this die provides a superb example of unsupported progressive forming. It should be of interest to designers of single-operation dies, since parts such as this are generally formed over mandrels or forming sections by cam action. Although cams are among the most useful devices at the designer's disposal, they should be passed over whenever possible in favor of straight forming.

As shown in Figure 9-6, the action is basic. Forced down by the cutoff punch, which separates the panel from the strip, the part is forced to conform to the angular die walls. As the punch rises after completion of its downward stroke, an ejector raises the part into an air jet which removes it from the die. The springs used in such ejectors must be comparatively light or they will cause distortion of the panel.

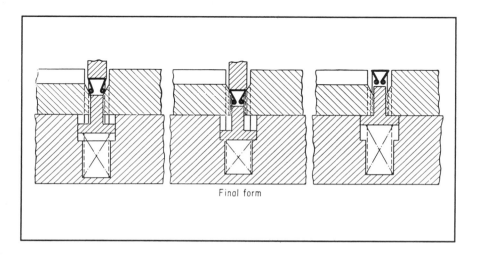

Figure 9-6. *Final form stage showing unsupported formula action.*

An Electrical Component Progressive Die

Pierce, cutoff, and form dies present no insurmountable difficulties in design or fabrication, but are noteworthy for two features. The first is the intricacy of the part outline and the steps needed to establish this contour; the second is the tab and the manner in which it is formed. A strip such as this is particularly difficult to divide into stages, for no matter where one starts, there is an overlap of piercing or notching operations. A series of piercing and notching operations reading into the fourth stage almost completely defines the outline before the part is advanced into the fifth stage.

Definition of the outline of the tab follows well established practice. The long side is cut in the first stage and the short side in the second. In this way the tab receives maximum possible support in each of the piercing operations. At the third stage, the notching punch — which defines the ends of the legs — is extended around to cut off the end of the tab. It would be possible to complete this outline in one stage but in the interest of simplifying punches and die-section design, it is wise to refrain from doing so.

In the fifth stage, the end of the tab receives a 60-degree curl by way of a conventional forming operation. In the sixth stage an unusual operation takes place in that the tab is again flanged, this time by a floating punch (Figure 9-7). Although this could be accomplished by a stationary punch, panel distortion would be caused by the punch in rising as the dies open.

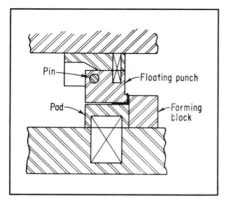

Figure 9-7. *Floating punch design required to clear radius at end of tab.*

To effect the forming operation, the punch is designed to revolve around a dowel pin. Light spring-action forces the punch down until, in descending, it contacts the pad. It then moves into place and forms the tab as the dies close.

As the die opens, the punch remains in this position until the pad reaches its maximum height. At this point the spring pressure forces the punch down, allowing it to swing out of the way of the hook. A sequence similar to this is performed by cam action described in Chapter 3 (see Figure 3-3).

The remainder of the action in this die is conventional. Between the seventh and eighth stages, a notching punch completes the definition of the part outline. In the ninth stage the part is cut off and formed by a typical cutoff-forming punch. The cycle is completed by an air blast which removes the finished part from the die.

Tilting a Panel

Placing an off-center hole in a progressively drawn panel is normally accomplished by a cam action pierce stage or by use of a secondary operation. But the use of cam-pierce operation is often impractical and secondary operations involve another die, another press, and an additional operator. These considerations combine to increase part expense significantly.

It is possible in some cases to perform the piercing operation within the progressive die itself — without the use of cams. Such an operation is accomplished in the strip illustrated in Figure 9-8 by tilting the drawn panel and piercing the hole, returning the panel to its normal position, and finally, cutting it out of the strip.

The tilting operation is performed in the tenth stage of this die. As the dies begin to close, the stripper plate grips and holds the strip in position. As the dies close further, a locating punch emerging through the stripper begins to tilt the panel out of its normal horizontal plane. When the locating punch has emerged its full distance through the stripper, it has rotated the panel to the exact position required for piercing the hole. Further downward travel of the die causes a piercing punch to emerge from the locating punch. This punch pierces the desired hole in the panel at the bottom of the press stroke.

As the dies begin to open, the piercing punch withdraws through the locating punch to strip the panel. As the dies open further, the main stripper plate advances past the locating punch, stripping the metal from the remainder of the punches and pilots.

The locating punch is mounted on guide pins and is spring-actuated. Its comparatively limited travel provides a telescoping action of the die at this stage. In relation to the upper shoe, the main stripper travels alone at first.

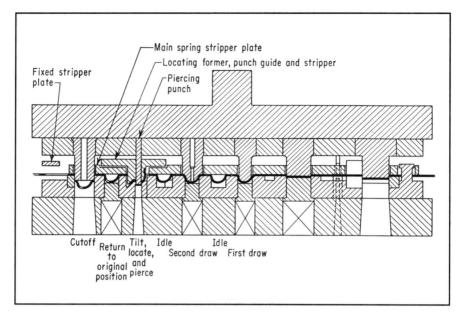

Figure 9-8. *Piercing off-center hole by tilting the panel.*

Then it and the locating punch travel together to allow the piercing punch to come through. As the dies open, this relative action is reversed.

After the tilting and piercing has been completed in the tenth stage, the tilted panel is advanced to the eleventh stage. In this stage the main stripper returns the tilted panel to a horizontal plane. In the twelfth stage the finished panel is cut from the strip and ejected through the die.

In addition to the unusual tipping action at the tenth stage, this die carries another noteworthy design feature. With regard to the manner in which the drawing action is contained, most progressive draw dies are designed in a way that causes the strip to narrow down at the draw stages. As a consequence, proper piloting and stock-guide location is a problem. This die quite effectively minimizes this difficulty by creating a thin web to support the blank. The web is made by first notching the outer contour of the strip, then pilot holes are pierced in the second stage, which is followed by an idle station. In the fourth stage the blank outline is lanced, leaving the blank suspended by four webs. Subsequent draw stages pull in these webs, but overall strip width remains constant. This permits piloting in the scrap throughout the length of the die.

Different Parts from One Die

Often similar parts can be stamped from one die by adding or omitting certain punches, as illustrated by the strip design in Figure 9-9. In this example four different parts are incorporated in the strip although the four parts are not made simultaneously. As is often the case, this die is also designed to produce similar right- and left-hand parts. Perhaps more correctly, this die produces two similar pairs of right- and left-hand panels. It is important to note, however, that this is a composite strip — no strip is produced on this die in this manner because the die makes only one pair of panels per stroke.

Differences in the parts are readily seen in the eighth and ninth stages. At the eighth stage the cutoff punch which separates the panels establishes the inner contour lines almost parallel to the die center line. At the ninth stage the cutoff punch establishes a much different contour in that it is parallel to the die center line for a short distance only before angling to the outside. The other basic difference in the two pairs is the location of the holes. All holes for panels AL and AR are produced in the second stage of the die. As the strip advances through the die, this pair of panels is completed at the sixth stage except for the inner contour lines. The two punches $P3$ and $P4$ indicated in the seventh stage are not placed in the die, so for panels AL and AR the seventh stage is idle. Advancing through the idle seventh stage, AL and AR are cut off in the eighth stage. As far as these two panels are concerned, the die is an eight-stage progressive.

To run the other pair of panels, BL and BR, it is necessary to remove the punches $P1$ and $P2$ from the second stage. The cutoff punch indicated in the eighth stage is also removed and punches $P3$ and $P4$ indicated in the seventh stage are inserted. (In practice, two punches will be removed from the second stage and inserted in the seventh.) As panels BL and BR advance through the die, they are completed in the seventh stage. The eighth stage, which is the cutoff for AL and AR, now becomes an idle stage and a pair of similar but entirely different panels are cut off in the ninth stage. Using the same die for an additional two parts has added only one stage.

Closed Forming. Closed forming is one of the more difficult operations to perform progressively. It is not impossible, however, and in many instances where final part formation is accomplished as a secondary operation, it might well have been included as part of the progressive sequence.

Progressive Dies

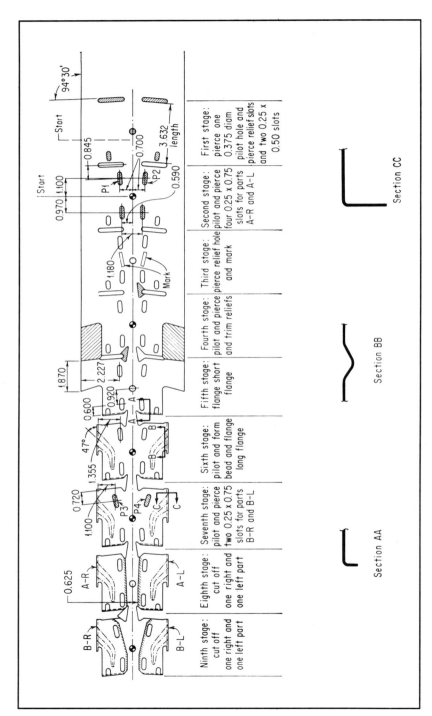

Figure 9-9. *Conventional but still unusual, four different parts are designed into the strip of this two-in-one progressive die.*

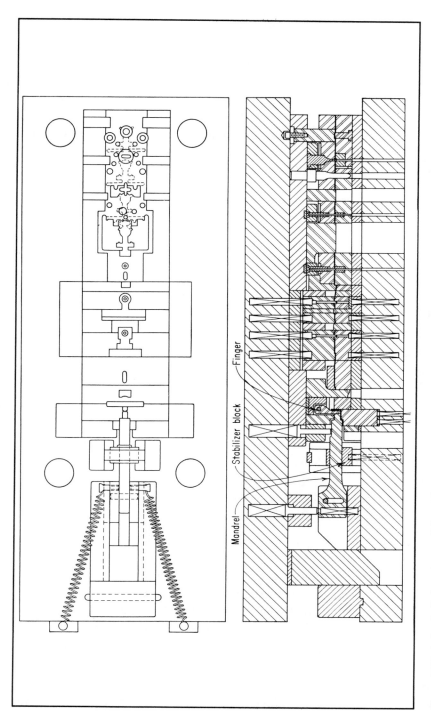

Figure 9-10. A chassis die for a cigarette lighter. The outstanding feature of this die is the manner in which the mandrel is activated.

Progressive Dies

Figures 9-10 and 9-11 show a nine-stage die used to make cigarette lighter chassis, a die using two-dimensional cam movements. The purpose of this type of cam is to accommodate a mandrel around which the part is formed. Activation of the mandrel (i.e., insert it into a form stage at or above feed level, lower it to die level, form, and then strip the part) is initiated well above the feed level. This allows the part to be cut off the strip in the final stage without interference. Spring pressure at two points below the cam mandrel supports it in its upward position, one spring being located directly beneath a stabilizing device.

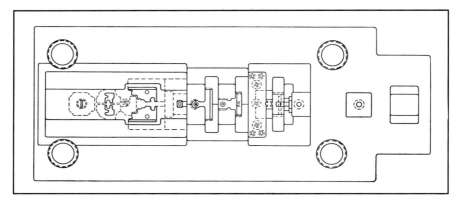

Figure 9-11. *Punch of the cigarette lighter chassis die.*

As shown in the plan and cross-sectional views, the stabilizer block is free to move in a vertical direction only. The mandrel itself is a free fit in the stabilizer, which leaves the mandrel free to move vertically and horizontally but not radially. When the die is in open position, spring pressures hold the mandrel in an up-and-out position.

As the die closes, the driver forces the mandrel into the form stage at a height above feed level. As die closure continues, spring-loaded pins in the upper shoe force the mandrel down to feed level where it picks up the part. As the final phase of die closure begins, a cutoff punch separates the part from the strip and the end of the part is wiped up. The upper forming punch is now in action and is forming the part around the mandrel. As die closure continues, the mandrel, under pressure from the punch, forces the pad to bottom and the part is spanked and set. This is the point arrived at in the cross-sectional view. Now note the details in the ninth stage. First, it is noteworthy that the mandrel does not perform the wiping action on the end piece. This is accomplished by a finger supported by the mandrel. The mandrel itself is backed up by the driver, which, in turn, is backed up by a keyed heel block. This finger strips the completed part from the mandrel. Because it is in a position where work must be performed, this finger is also required to serve as a forming punch.

The spring-loaded pins that bear on the mandrel serve to balance the pressure on this component, thereby reducing wear on the stabilizer block. In addition, these are the pressure devices used to lower the mandrel to feed level before the form punch begins to act. The springs actuating these pins must be stronger than the lower springs which return the mandrel to "open" height.

The designer of this die has made good use of commercial punch supporters to stabilize his perforating punches. Units of this type are generally preferred to solid punches of large shank diameter and they are less expensive and more easily replaced.

10

Progressive Transfer Dies

Description

Transfer dies retain almost all the characteristics of conventional progressive dies. Both are labor-saving devices, combining two or more sequential operations to produce a finished stamping, and employ automatic feeding mechanisms to move the stamping from operation to operation. Finally, both progressive and transfer dies are designed to run at higher operating speeds than their manually fed counterparts.

However, unlike progressive dies, stampings produced on transfer dies are independent of each other station-to-station (see Figures 10-1 and 10-2). The carrier strip used on progressive dies is not employed by transfer dies. Instead, transfer dies incorporate a shuttle device for progression.

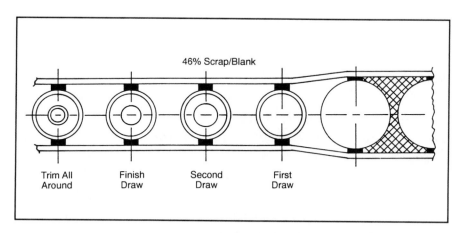

Figure 10-1. *Conventional strip layout of blanks for a progressive die. (Livernois Automation Co.)*

The Growth of Transfer Dies

The shuttle device used for transferring parts from station to station was originally an integral portion of the press, from which the name "transfer

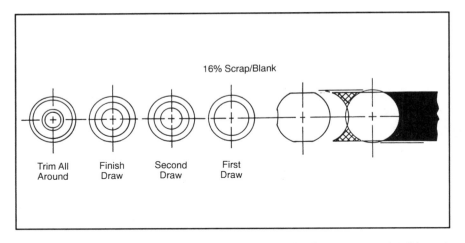

Figure 10-2. *The transfer method of laying out the same part as that shown in Figure 10-1. (Livernois Automation Co.)*

press" was derived. This posed two major obstacles to the development of transfer dies.

First, since the transfer mechanism was integral to the press, the press became a special, single-purpose machine, usually restricted to producing only one stamping. Second, because the investment in the transfer press was far greater than an equivalent size of a conventional press, transfer dies were built only for stampings which had unusually high production volume requirements.

In the early 1960s, these limitations were overcome. By divorcing the physical transfer mechanism from the press and allowing the transfer to become an automatic device independent of the press motion, transfer dies no longer had to be installed in a dedicated piece of equipment.

This divorce resulted in transfer mechanisms and dies that were jointly mounted directly to a master die set (Figure 10-3). Termed the portable transfer die, the previous qualifications that had retarded the growth of transfer die technology were eliminated.

The transfer dies and mechanisms were installed in the same manner as a conventional die set. When the production requirements were satisfied for a particular transferred stamping, the mechanism was removed along with the die. This allowed presses to be used for other applications, including regular progressive dies and single-operation work.

Progressive Transfer Dies

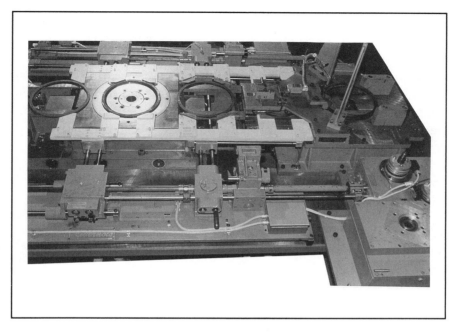

Figure 10-3. *A butting type of transfer die where each station is butted against the previous station. This is a light-duty transfer die which was designed for a 400-ton press with a press stroke of 10 inches, a speed of 40 spm and a finger movement of 2.5 inches. (Livernois Automation Co.)*

Transfer Mechanisms

The purpose of a transfer mechanism is to move parts station-to-station through the stamping progression. Upon completion of the working portion of the press cycle, the transfer mechanism moves either in or up, makes contact with the progression of independent parts, moves the parts forward one station, and releases them. The transfer mechanism then moves away from the parts, allowing clearance for the next press cycle.

In transfer dies, the conventional carrier strip, or ribbon, is replaced by transfer rails. Affixed to these rails are fingers shaped to conform to the part contours. The fingers are the devices to which the motion of the transfer mechanism is applied to move the stampings. This feature, however, is the root of one of the few problems of transfer dies. Since the parts being moved are not restrained by the strip, inertia can cause the part to mislocate in subsequent stations.

Put another way, bodies in motion tend to stay in motion. With a conventional progressive die, the motion is arrested by the carrier ribbon.

201

Progressive Dies

In the case of transfer dies, the motion is arrested only by the transfer fingers. This tends to limit the speed (strokes per minute) at which transfer dies can be run.

One other limitation of transfer dies presents itself in cases when the ram of the press is used to actuate the transfer mechanism. Since some portion of the press stroke is assigned to transfer motion, transfer dies typically require presses with longer strokes than those needed for similar operations with progressive dies (see Figures 10-4 and 10-5).

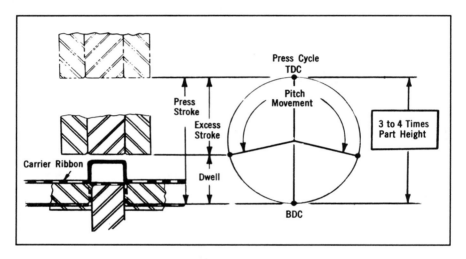

Figure 10-4. *Because the carrier strip is the vehicle of movement, a shorter stroke press can be used with progressive dies. (Livernois Automation Co.)*

Transfer Dies versus Progressive Dies.

The two forms of progressive operations are not mutually exclusive or even competitive. Each has its own place in progressive stamping.

It is virtually impossible to find a press operation where both transfer dies and progressive dies are equally applicable. The decision to produce a part on a progressive transfer die or a conventional progressive die is dependent on the design of the part and its production requirements.

It is fairly accurate to say that the principal difference in the cost of a portable transfer die system, as compared to a progressive die system, is the cost of the transfer mechanism. When part design or volume justify a transfer die, the advantages far outweigh the cost of the mechanism.

Progressive Transfer Dies

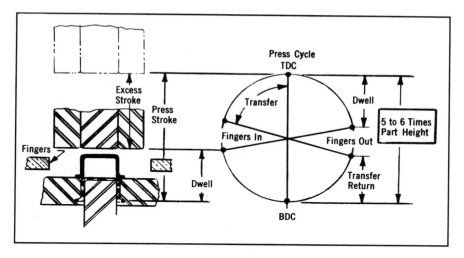

Figure 10-5. *A portion of the press stroke on transfer dies is dedicated to actuating the transfer mechanism, thus necessitating a press with a longer stroke. (Livernois Automation Co.)*

When doubt arises as to which method is best for a given stamping, a simple analytical formula can be applied. This formula, which takes into account the economic feasibility of a particular application, is:

$$X = \frac{T - M}{m - t}$$

where:

T = Total tool cost of transfer dies (including mechanism)

M = Total tool cost of alternate die (in this case, progressive)

m = Manufacturing cost per part with alternate die

t = Manufacturing cost per part with transfer dies

In this equation, X is the volume of parts that will result in a "breakeven" quantity for the choice between transfer dies or progressive dies. If a particular part volume exceeds X, then transfer dies are indicated. If the volume required is less than X, an alternative selection is better.

Given that the transfer mechanism is the primary difference in the cost between progressive and transfer dies, approximate calculations for T can be considered as $(M + U)$, where U is the cost of the transfer mechanism.

An example of this computation, related to a part selected for transfer dies, is shown in Figure 10-6.

Manual Dies	Press Cost per Hour	Parts per Hour	Manufacturing Cost per Part	Tool Cost	
Draw	$15.00	1,000	$0.015	$ 3,000	$x = \dfrac{T - M}{m - t}$
Trim	15.00	1,000	0.015	3,000	$= \dfrac{20{,}500 - 13{,}000}{0.062 - 0.006}$
Wipe Down	15.00	900	0.016	3,500	
Bulge	15.00	900	0.016	3,500	$= \dfrac{7500}{0.056}$
Total Manual			$0.062	$13,000	
Total Transfer	$15.00	2,500	$0.006	$20,500	$= 133{,}928$ parts

Figure 10-6. *Computing breakeven for a part manufactured by a transfer die. (Livernois Automation Co.)*

Theory of Progressive Transfer Dies

Transfer dies rely on "dwell" to properly time the motion of the mechanism to the transfer die. Dwell is defined as the period during punch press stroke when moving parts of the transfer mechanism are dormant on one or more of the axes of motion.

The most common type of transfer, termed shuttle, is a timed horizontal motion, followed by a timed longitudinal motion. In transfer die technology, the horizontal movement is called finger movement. The longitudinal motion, which is the delivery of parts from one working station to another, is called pitch, in reference to the distance the part must be transferred.

Dwell is applied to the finger motion and pitch motion in a certain sequence, depending on the position of the crankshaft of the press. During some portions of the press stroke, dwell is applied to both axes simultaneously. During other portions of the press stroke, one axis of the transfer is in motion, while the other axis remains in dwell. This is represented graphically in Figure 10-7.

This diagram is crucial to the successful design and building of transfer dies. First in importance is that the period of dwell be minimized and the transfer advance be at the maximum, in relation to the time it takes to complete one press cycle. In other words, the distance of transfer pitch should be minimized to allow for slowest possible transfer motion. Commonly, an idle station is incorporated in transfer dies to minimize this distance (see Figures 10-8 and 10-9).

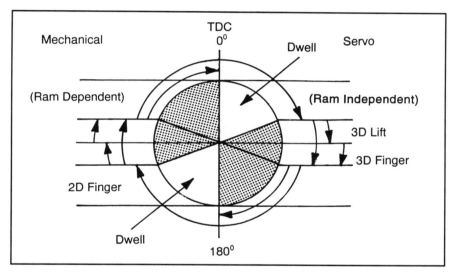

Figure 10-7. *Dwell is designed into the transfer cycle to properly sequence the finger motion and pitch motion. (Livernois Automation Co.)*

This leads to another important distinction between transfer dies and progressive dies. In transfer dies, the pitch distance is typically much greater than on an equivalent progressive die. Therefore, the physical strength of the die steels used in their construction does not have to be sacrificed to maintain progression, as is commonly the case on progressive dies.

Also noteworthy is that all the independent parts in a transfer die line must reach a common height at a common time. This point is the time when the fingers capture the part at approximately 120 degrees of the press crankshaft rotation. This is similar to a progressive die in that the ribbon on a progressive die must be returned to a common level before it is fed forward.

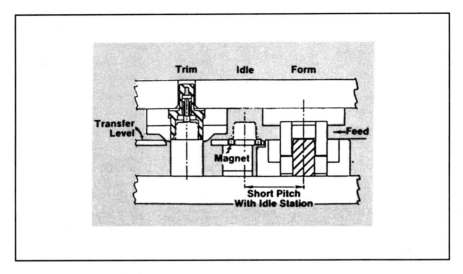

Figure 10-8. *Typical design of a transfer die with idle station. (Livernois Automation Co.)*

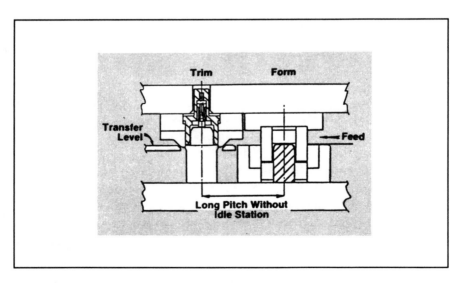

Figure 10-9. *Typical design of a transfer die without idle station. (Livernois Automation Co.)*

Another similarity of transfer dies to progressive dies is the location of the part relative to the die steels. In both, this positioning must be accomplished by some means other than the movement-inducing mechanism. Whether using a stock feeder or a transfer unit, the accuracy of part positioning relative to the die section is not left to the mechanism. Instead, pilots are incorporated to ensure proper guidance to the die. In conventional progressive dies, the female pilot holes are typically in the carrier strip. In the case of transfer dies, the piloting occurs on the individual stampings.

Differences Between Transfer Dies and Progressive Dies

Fundamental differences also exist between the two types of dies. One is the carrier strip found on progressive dies that is not needed on transfer dies. Its presence or absence has a direct bearing on the working height the part can have in a progressive die.

Because the carrier ribbon of a progressive die should never be flexed or bent, the working level of the part in each station must be very close to that of the adjacent stations. With transfer dies, this limitation does not apply.

The strip layout in Figure 10-10 shows the intrinsic capability of the transfer die. Note that in the third, fourth, and fifth station the part work line is significantly lower than the original feed level of the strip. This allows parts to be produced that require more die travel than is possible with a carrier ribbon.

Another significant carrier ribbon-based difference is flexibility in part metal gage weight. Parts being produced on a progressive die must always be attached to the carrier ribbon. In cases where a flange or form is required

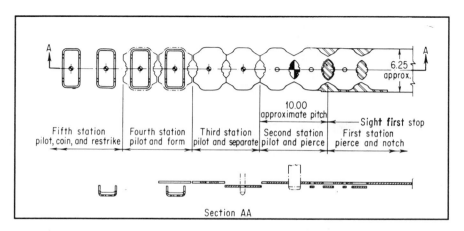

Figure 10-10. *Strip layout for a transfer operation. (Livernois Automation Co.)*

at the juncture of the carrier ribbon and the part, an almost impossible situation arises.

In light-gage metal, a pinch-trim can be used in a progressive die, but this type of operation does not work well with heavier metal. This sometimes forces a change in the product design to allow for an inclusion in the area that is being trimmed.

With transfer dies, this problem is significantly reduced, because the part can be separated from the carrier ribbon, and then moved into place by the transfer mechanism. In the plan view of Figure 10-11, note that a flange is produced all around the part, with no requirement for the die to have a special trimming operation.

Another significant difference between the two die types is the means by which the part is ejected from the upper die. Some conventional progressive die designs use the carrier ribbon to eject parts. While not particularly reliable, this method is nevertheless still prevalent in the industry.

In the case of transfer dies, there is no ribbon. Yet it is still extremely important to ensure not only a 100 percent efficient stripping method, but

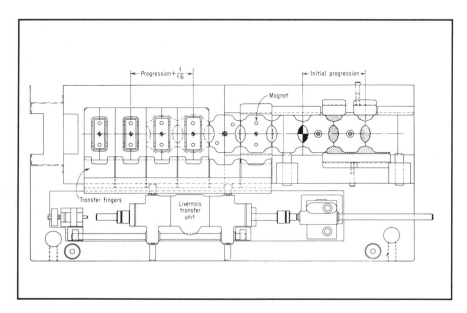

Figure 10-11. *Individual movement of parts in a transfer die reduces the need for design changes when heavier gage metal is specified. (Livernois Automation Co.)*

also one that maintains control sufficient to keep the part stable once it has been ejected.

Two means are incorporated to provide this stripping motion. One, found primarily on dedicated transfer presses, is a positive mechanical knockout. In this method, the action of the ram is used to move the part (in the die) against a fixed post that contacts the part. As the ram continues to move upward, the part is forced from the die because the position of the ram increases relative to the post.

The other style is a self-contained high-pressure nitrogen system. With portable transfer dies, this is the preferred method, since the ejector is removed in its entirety with the die set.

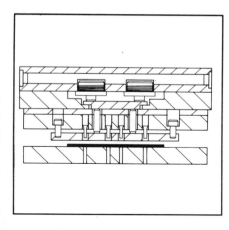

Figure 10-12. *Strip punch ejection systems apply high force with no preload, allowing the strip to be held securely. The punches can be stripped without breaking. (Livernois Automation Co.)*

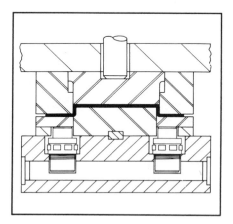

Figure 10-13. *In conventional draw operations, an adjustable force eliminates the lengthy trial-and-error process using die springs. Die springs give too much or too little pressure at the wrong time and are difficult to change. (Livernois Automation Co.)*

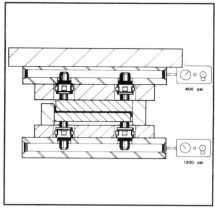

Figure 10-14. *Nitrogen die cylinders are ideal for forming up and down operations. By adjusting pressure, balance can be maintained to produce consistent quality. (Livernois Automation Co.)*

Typical nitrogen ejection systems are shown in Figures 10-12 through 10-14. It is an interesting aside that nitrogen systems are now commonly used on conventional progressive dies because of their increased pressures compared to regular die springs.

Still another difference between transfer and progressive dies is time of development. With markets becoming global and competition becoming even more intense, it has in turn become important that stamping users, stampers, and die shops reduce product cycle time.

Die designers, as a practical matter, design difficult stations in dies first. Since these stations typically determine the overall operating parameters and sizes of the die, it is logical to proceed with the design of these stations, then complete the design of the balance of the stations to match.

Because transfer dies are stand-alone dies that can be operated without a transfer unit, the designer or diemaker can elect to produce tooling for the most difficult station before proceeding with the entire die sequence. In this way, the proposed process can be verified without having to produce a completed die set, thus significantly cutting die development time.

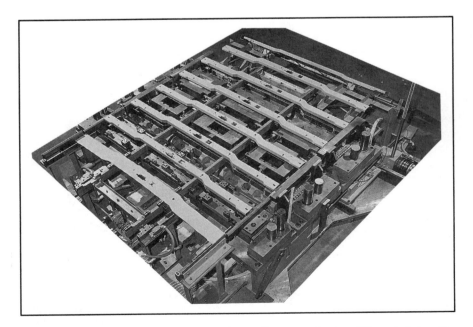

Figure 10-15. *Walking beam transfer unit provides significant cost advantage. (Livernois Automation Co.)*

Another advantage of this method is that it enables the diemaker to try out discrete transfer die stations on an individual basis. Final die development and tryout can take place in tryout presses smaller than those required to take an entire system. In this manner, construction or design can be "debugged" without having to produce an extremely large number of stampings or tie up a production stamping press.

It is necessary, however, that the transfer mechanism and transfer dies eventually be tried out in their entirety. This ensures that proper clearances are built into the dies, that fingers match the part profile, and that timing is correct between the dies and the mechanism.

The ability to try out individual dies typically reduces the amount of development work and, in worst cases, rework that must be completed before a stamping can be sent to market.

Die Differences Created by the Transfer Mechanism

One of the most common types of transfer is the lift and carry style, known as the walking beam. In this style of transfer, the fingers move vertically to meet the part, instead of horizontally.

In parts that have a length to width ratio of six to one or greater, this is the preferred method of progression. This style of part, when produced on either shuttle style transfers or progressive dies, tends to sag dramatically in the middle, causing feed problems. The walking beam style transfer eliminates this problem by supporting the part during transfer.

A walking beam transfer unit is shown in Figure 10-15. It is important to note that in this style of transfer, the lift to bring the part to the pass line is provided by the transfer unit, not by the dies. A significant reduction in tool price is gained with this type of transfer since the die does not usually require lifter pads.

Three-dimensional transfers are another common variation of portable transfers. These mechanisms move fingers inward horizontally, then lift them vertically, and finally move longitudinally to deliver the part to the next work station. In effect, this device combines a shuttle with a lift-and-carry (or walking) beam.

This type of transfer is most frequently used when parts must be nested on die steels for location or when there is no surface on the part to provide a stable base for sliding the part.

A typical three-dimensional transfer unit is shown in Figure 10-16. In this application, the parts are picked up by grippers, allowing the transfer rails to be attached to one side of the part.

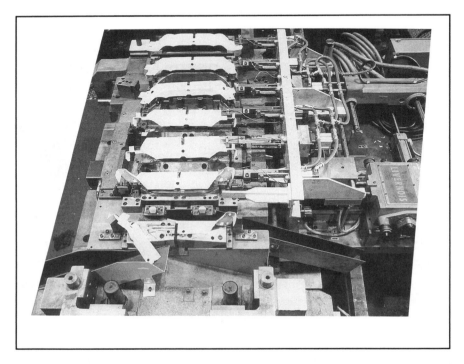

Figure 10-16. *Three-dimensional transfer units function essentially as combination shuttle and walking beam transfer units. (Livernois Automation Co.)*

The final transfer unit combines additional axes of motion, including "wrist" motion (part tip) and "elbow" motion (part turnover). Use of this type of transfer mechanism has increased dramatically in recent years.

One of its advantages is the ability to turn the part so that the strike plane of the die can be maintained perpendicular to the workpiece. Conventional progressive dies have capability to do this on a limited basis, but mechanical limitations prevent a tip of much more than 30 degrees.

Using the tip capabilities of a transfer can eliminate items such as aerial cam or out-of-perpendicular strike planes. Die up time is significantly improved, while initial cost of the die is reduced.

Progressive Transfer Dies

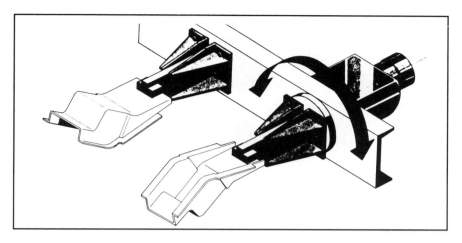

Figure 10-17. *A fourth axis adds "wrist" action to the transfer unit, allowing 180-degree turning of the workpiece. (Livernois Automation Co.)*

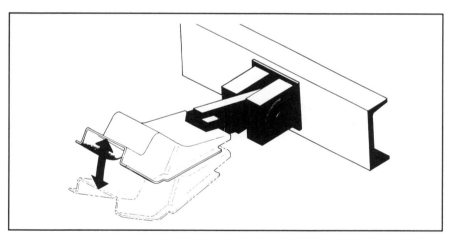

Figure 10-18.*Even greater workpiece orientation is available with a five-axis transfer unit. (Livernois Automation Co.)*

Additional axes of transfer motion are shown in Figures 10-17 and 10-18.

In certain cases, it is impossible to produce some parts using conventional progressive dies. Exemplary are drawn pans which have mounting holes, such as those found on automotive transmissions. Because the flange of the drawn shell is a seal surface, the burr from the pierced hole must be opposite the direction of the draw. However, by inverting the part 180 degrees, which allows the part to be produced in line with design intent, this seemingly perplexing problem is eliminated.

Application Decides Selection

However one regards the relation between transfer dies and progressive dies, it is clear that each has its own place in the stamping environment. By evaluating the advantages and limitations between the two, the die designer can make an informed decision on which type of die to use, based on the application.

11

Carbide Progressive Dies

Cemented Tungsten Carbide

Tungsten carbides cemented with cobalt have replaced steels for many tools used in metalforming operations, primarily because of their high abrasion (wear) resistance and compressive strengths. These materials are normally used for long production runs in which their higher initial cost can be economically justified as the result of longer tool life, reduced downtime, and decreased cost per part produced. Some stamping operations, such as the production of small holes in hard and tough materials, can only be done with tungsten carbide punches. Precise tolerances are maintained for long periods, thus improving product quality and reducing rejects.

The high elastic modulus (stiffness under bending loads) of cemented tungsten carbides permits their use for punches with length-to-diameter ratios exceeding 4:1. These materials also reduce the severity of galling, which is a problem common with punches and dies made from tool steels.

A possible limitation to the use of tungsten carbides, in addition to their higher initial cost, is the greater difficulty in grinding them, with resultant problems in finishing difficult shapes. Tungsten carbides are more brittle (nonductile) and have some physical properties lower than tool steels. Depending upon manufacturing techniques, their cobalt content, and the design of the punches and dies, however, it is feasible to make complex-shaped, high-strength, cemented carbide tooling for forming and blanking operations.

Applications. Cemented tungsten carbide dies are being used extensively for drawing wire, bars, and tubes; extruding steels and nonferrous alloys; cold and hot heading dies; swaging hammers and mandrels; and powder compacting punches and dies. Tungsten carbide is also used to make dies to draw sheet metal parts. Other important applications include punching, coining, sizing, and ironing tools for beverage and food cans and the production of a variety of laminations and other stamped metal parts.

Typical products produced by tungsten-carbide forming tools include automotive parts such as piston pins, bearing cups for universal joints, spark plug shells, bearing races for front-wheel drives, air pump rotors, transmission gear blanks, and valves. In the construction and farm equipment industries, tungsten carbide is used to form hitch pins, track link bushings, hydraulic hose fittings, diesel piston pins, and a variety of gear blanks. Tungsten carbide is also used in manufacturing fasteners, drawn and ironed beverage cans, cartridge cases, wrench sockets, bicycle drive cups, motor laminations, electronic terminals, and many other stamped parts.

Production of Cemented Carbides. Tungsten carbides are produced by a powder metallurgy (PM) process. Properties of the cemented tungsten carbides are determined by the compositions of the materials, the size of the particles used, the production techniques, and the metallurgical structure of the materials.

Although sintering of cemented carbides results in a density of virtually 100 percent, porosity can occur from several different causes. One cause is excessive carbon content. Such a defect cannot be cured in the finished part. Another source of difficulty can be carbon deficiency, which results in a brittle condition known as eta phase. These conditions must be avoided because neither can be tolerated in the finished parts.

Porosity can also be caused by improper techniques employed in processing the powders and by the introduction of impurities during the production of the cemented carbides. Most porosity caused by impurities or reasons other than improper carbon balance can be eliminated by the employment of better manufacturing techniques. If porosity or voids are not at the surfaces, hot isostatic pressing (HIP) after sintering will close the voids. This process also improves the average transverse rupture strength as well as the surface integrity after grinding. Tools made from carbides pressed isostatically are particularly desirable for applications in which pits cannot be tolerated on the tool surfaces, such as high-stress applications and those in which thin materials are formed.

The HIP technique, being used extensively as a standard production operation, especially for forming dies, is generally done with an inert gas (such as argon) in a pressure chamber at 5000 psi to 20,000 psi (34.5 MPa to

138 MPa), with temperatures ranging from 2200°F to 2550°F (1200°C to 1400°C).

Cemented Carbide Properties. The cemented carbide materials used most extensively for forming operations are the so-called straight tungsten carbides with cobalt binder, a family of two-phase WC-Co compositions. Occasionally, tantalum carbide is added for lubricity, for increased hot strength, or for inhibited grain growth, but other additives are normally avoided for die materials.

Tungsten carbide grains in these materials range from less than 1 micron (0.00004 inch) to 10 microns (0.0004 inch), and cobalt contents vary from 3 percent to 25 percent.

Desirable properties of the straight tungsten carbides used for forming-die applications include high hardness at room and elevated temperatures, high abrasion or wear resistance, high modulus of elasticity, high compressive strength (much higher than its tensile strength), and low rate of thermal expansion. Some of the major properties of tungsten carbides compared to tool steels are presented in Table 11-1.

The hardness of a straight tungsten carbide depends primarily upon the percentage of cobalt binder that it contains. In general, the more binder, the lower the hardness. Fine-grain tungsten carbides have a higher hardness than coarse-grain materials. There is a linear relationship in the increase in hardness from R_A 84 (R_C 65) for a material with 25 percent cobalt to R_A 94.3 (R_C 83.5) for one with less than 1.5 percent cobalt. Fine-grain compositions with the same cobalt content have a higher hardness of between one and two points on the Rockwell A scale than materials with coarse grains. The coarse-grain carbides are generally stronger.

At elevated temperatures, tungsten carbides retain their hardness much better than steels and most other tool materials. Above 1000°F (538°C), however, the cobalt binder can melt out (oxidize). The success of hot-forming applications with tungsten carbide tooling depends upon the workpiece material and temperature, the contact time, and the type and application method of the coolant/lubricant.

The wear or abrasion resistance of straight tungsten carbides also depends primarily upon the percentages of cobalt binder they contain. In general, the lower the binder content, the higher the wear resistance, and the finer the grain size, the better the wear resistance (see Figure 11-1). Maxi-

Table 11-1
Comparison of Approximate Properties of Straight Tungsten Carbides and Tool Steels

Material Property	Tungsten Carbides	Tool Steels
Hardness, R_C	65-83.5	66
R_A	84-94.3	
Abrasion resistance	to 825	14
Tensile strength, ksi (MPa)	to 200 (1379)	to 290 (2000)
Compressive strength, ksi (MPa)	to 900 (6205)	to 290 (2000)
Modulus of elasticity, ksi x 10^3 (GPa)	to 94 (0.65)	30 (0.21)
Specific gravity	to 15.0	7.7-8.7
Impact resistance	poor to good	good
Hot hardness at 1200° F (649° C), R_C	45-77.9	33-38
R_A	73-90.8	
Corrosion resistance	good	fair

mum wear or abrasion resistance is obtained with materials having fine grains and a cobalt content in the range of 3 percent to 6 percent.

Cobalt content and grain size, however, also affect the material strength. The higher the cobalt content, the higher the strength; and for some compositions, the coarser the grain size, the higher the strength. For most forming operations, high strengths are needed and materials with low cobalt contents can only be used for fine-wire drawing and low-impact applications. Also, coarse grains are not generally used for materials with a cobalt content less than 6 percent, but are often employed for carbides with higher cobalt contents, particularly in the 20 percent to 25 percent range.

The addition of tantalum carbide improves the abrasion resistance of these materials slightly because it inhibits grain growth and maintains controlled grain sizes. In general, the small quantities of tantalum carbide (as little as 0.2 percent) sometimes added to materials are added to control grain size, and larger quantities (usually a maximum of 5 percent) are added to increase lubricity or to increase the resistance of the carbide to deformation in hot forming operations.

The abrasion or wear resistance of tungsten carbides is measured with dry sand, wet sand, or metal-to-metal tests. In the metal-to-metal test, a carbide ring is run against a test block. The area of the groove worn in the block is used to determine the volume of material removed by abrasion.

Transverse Rupture Strength. Since standard tensile tests give erratic results with tungsten carbides because of the notch sensitivity of the materials, the carbide industry has chosen the transverse rupture test as a standard for determining relative strengths. The test measures the strength of the material by determining the maximum stress at the extreme fiber when using a three-point loading on a standard test specimen. The tensile strengths of tungsten carbides have been determined to be 45 percent to 50 percent of their transverse rupture strengths, and these values can be used in stress analysis calculations.

Cobalt content and grain size have measurable effects on transverse rupture strengths. As the cobalt content increases, the strengths increase. With respect to fracture toughness and resistance to chipping, however,

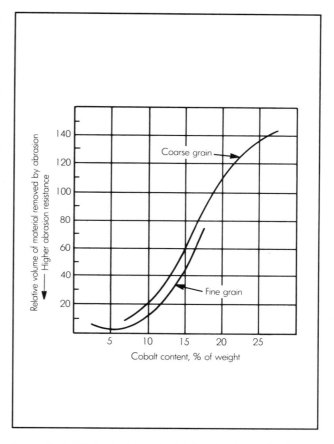

Figure 11-1. *Effect of cobalt content and grain size on the abrasion resistance of straight tungsten carbides.*

coarse-grain materials are far superior to fine-grain materials with the same cobalt content. As a result, generally most materials with the same cobalt content above 9 percent to 10 percent are made with coarse-grain tungsten carbide when high strength is needed.

Certain compositions exist that belie this general rule, but the majority of cemented tungsten carbides produced for forming-die operations are made from mixed, medium, or coarse grains. The use of coarse grains increases as the cobalt content rises from about 12 percent to the maximum of 25 percent.

Elastic Modulus. The elastic modulus of tungsten carbide is two to three times that of steel. If the same load were applied to a tungsten carbide punch and a steel punch (having the same geometry), the tungsten carbide punch would deflect only a third to half as much as the steel punch.

Corrosion Resistance. Since carbides themselves are virtually inert, the corrosion resistance of tungsten carbide is largely determined by the corrosion resistance of the binder metal. In a corrosive environment, such as that created by some grinding and metalforming lubricants or coolants, cobalt binder can be leached from the tungsten carbide. What remains is a skeletal surface structure of tungsten carbide particles. Since there is little cobalt binder left to hold the particles together, the particles can abrade away more easily. This can lead to edge chipping, accelerated wear, galling, and even breakage in some cases. Lubricant/coolant suppliers should be asked if their products are compatible with tungsten carbide. Lubricants containing active sulfur usually attack the cobalt binder. Normally, corrosion of metalforming punches and dies, such as those used for extrusion tooling, is not a serious problem. However, it is a problem with punches and dies used in the manufacturing of drawn and ironed beverage cans and some other stampings.

Design Considerations. Special design techniques are required for the successful application of tungsten carbide materials. Sharp edges, notches, or abrupt changes in cross section are stress risers and should be avoided. Dies should be designed to keep the carbides in compression because the compressive strengths of these materials are much higher than their tensile strengths. Draw radii or approach angles, punch and die clearances, and reliefs are similar to those for steel dies. Bearing lengths for carbide draw dies, however, may be lower (see Table 11-2), and back relief is required for best performance.

Table 11-2
Typical Dimensions for Carbide Draw Dies

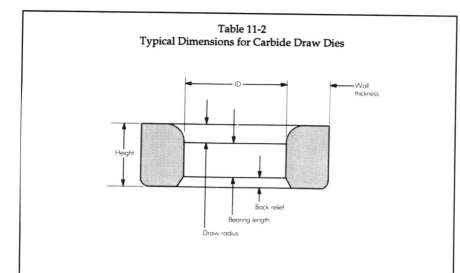

Inside Diameter, in. (mm)	Wall Thickness, in. (mm)	Bearing Length, in. (mm)	Back Relief Length, in. (mm)
to 1/2 (12.7)	5/16 (7.9)	1/8 (3.2)	1/16 (1.6)
1/2 to 1 (12.7 to 25.4)	3/8 (9.5)	3/16 (4.8)	3/32 (2.4)
1 to 1 1/2 (25.4 to 38.1)	1/2 (12.7)	1/4 (6.4)	1/8 (3.2)
1 1/2 to 2 1/2 (38.1 to 63.5)	9/16 (14.3)	5/16 (7.9)	1/8 (3.2)
2 1/2 to 5 (63.5 to 127)	5/8 (15.9)	3/8 (9.5)	5/32 (4.0)
5-10 (127 to 254)	3/4 (19.1)	7/16 (11.1)	3/16 (4.8)
10-15 (254 to 381)	7/8 (22.2)	1/2 (12.7)	1/4 (6.4)

When a carbide die insert is subjected to high-impact loads and internal bursting pressures, it must be adequately supported externally by pressing or shrinking the carbide ring into a hardened steel case. Suitable steels for die cases include SAE 4140, 4340, and 6145, as well as AISI Type H13 tool steel, hardened to R_C 38-48. The outside diameter of the steel case should be two to three times the outside diameter of the carbide ring when high internal pressures are involved. While carbide can be shrunk into steel successfully, steel cannot generally be shrunk into carbide. With the thermal expansion of steel being about three times that of carbide, the steel can break the carbide with only a moderate increase in temperature.

Calculations for shrink allowances should be performed for any new or unusual designs, such as dies for forming parts to complex geometry, for drawing or ironing thin-walled cylinders, for operations at elevated temperatures, and for applications that exert high internal pressures on the dies.

A die for blanking discs (see Figure 11-2) incorporates a punch which is a carbide ring shrunk into a hardened steel case. When the carbide ring does not require a case or holder, it can be bolted or clamped directly in place (see Figure 11-3).

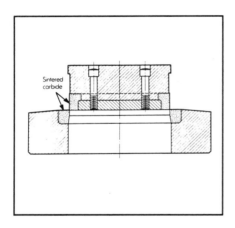

Figure 11-2. *Punch and die with tungsten carbide inserts for blanking discs.*

For blanking dies, carbide inserts may incorporate soft nickel-iron or steel plugs (usually brazed in place), which are drilled and tapped for hold-down screws and dowels (see Figure 11-4). A high-nickel (35 percent to 40 percent), low-expansion alloy is recommended for plugs or tapping inserts. Another method of retention sometimes employed is to tap holes directly in the carbide by electrical discharge machining (EDM). This method is only employed when space does not allow the use of plugs or tapping inserts, because EDM often results in chipping of the carbide unless extreme care is exercised.

Draw dies have a solid carbide punch and a carbide die ring shrunk into its case. In designs with sleeved carbide draw punches, retention is generally accomplished with an inner steel shank mechanically holding the carbide in place. The carbide dies are shrunk into steel cases.

Dies using carbide inserts or segments are similar in design to sectional steel dies. Such dies are generally used for applications in which irregular shapes are to be blanked or punched, or in which maintaining close relationships between holes in the die segments is necessary.

Carbide Progressive Dies

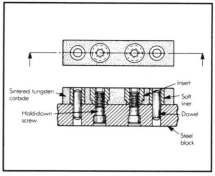

Figure 11-4. *Carbide blanking die having soft steel plugs for screws and dowels.*

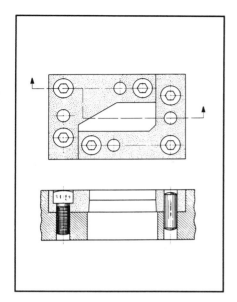

Figure 11-3. *Carbide die held with screws and dowels.*

Finishing of Carbides. Unsintered, compacted carbide parts can be formed to the required shape by conventional machining. These parts are dewaxed and generally presintered at a low temperature, 350°F to 850°F (662°C to 1562°C). However, because precise tolerances cannot be maintained during sintering, shrinkage encountered is variable, and the as-sintered surface finish is generally between 50 μinch and 100 μinch (1.27 μm and 2.54 μm), subsequent finishing is normally required for metalforming punches and dies. Finishing can be done by diamond-wheel grinding or carefully applied EDM.

Polishing Die Components. In most instances, especially for drawing and extrusion dies as well as many forming, shaping, and PM compacting dies, a polished finish is needed on the carbide components. Polishing is generally done with diamond paste or diamond polishing compounds. For flat surfaces, diamond lapping operations are most suitable. For contoured surfaces, special tools and fine-diamond polishing compounds are used. Surfaces of the die components should be free of nicks, grinding grooves, and cracks. Surface finishes of 1 μinch to 2 μinch (0.025 μm to 0.051 μm) are sometimes specified.

12

Materials Selection

A wide variety of materials is used for dies and molds, and many dies and molds contain several materials, most of the more common of which are discussed in this chapter. No single material is best for all forming applications because of the extensive range of conditions and requirements encountered in various operations. Selecting the proper material for a specific application can improve workpiece quality, increase productivity, and reduce costs.

Factors affecting the selection of a proper die or mold material for a specific application include:

- The operations to be performed, including their severity, forces applied, temperatures encountered, and lubricants used;
- The workpiece material, including its hardness, thickness, and condition, as well as the size of the workpiece;
- The production rate and quantity, accuracy, and finish requirements;
- The press or machine to be used, including its type and condition;
- The design of the die or mold;
- The accuracy and rigidity of the setup;
- The cost per part produced, based upon the material, manufacturing, heat treatment, and maintenance costs, as well as the life of the die or mold;
- The current availability of the die or mold material;
- The properties of the material, including resistance to wear, heat, and deformation, and the ease with which it can be machined, heat treated, and ground.

Different applications require specific characteristics and properties for the material to be used. For many materials, the chemical analyses and heat treatments can be adjusted to change the properties. The relative evaluation of properties for various materials, however, is of necessity qualitative, and the proper choice for a specific application cannot always be made with assurance. Consultation with the material supplier, die or mold producer, and heat treater is recommended, to advise them of the specifics of the application.

In many instances, the choice is not limited to a single material that can be used for an application. It is desirable, however, to select the one material that provides the most economical overall performance, based on the factors just discussed. A continuing evaluation of the materials used, employing accumulated performance data, is important.

Carbon and Low-alloy Steels

Wrought plain-carbon and low-alloy steels, in the form of plates, rounds, and shapes, are often used in the fabrication of auxiliary die components, as well as some die parts. Applications for dies are mainly those in which strength and weldability, rather than wear resistance, are the primary requirements.

Some of the more common plain-carbon and low-alloy steels used for die components are listed in Table 12-1. Advantages of using these materials include economy, availability, easy machinability in their annealed condition, and their capability of being heat treated to provide a high surface hardness and fairly tough core. The alloy steels can be heat treated to higher strength levels with optimum impact properties. These grades are often carried in stock in the heat-treated condition for the convenience of users, and resulfurized grades are available for improved machinability in the hardened condition.

A limitation to the use of plain-carbon and low-alloy steels is their poor hot hardness. Their hardness decreases with increasing temperatures encountered in forming operations, and care is required when grinding components made from these materials. Care is also required in heat treatment to maintain dimensional stability and to prevent cracking. The plain-carbon steels generally require straightening after heat treatment, but alloy steels, requiring a less severe quench to achieve hardness, suffer less distortion.

For severe forming operations, steel die components are sometimes hard chrome plated to prevent galling; however, the plate may spall off, especially if the die components have small radii. Nitriding of alloy steels that contain chromium and molybdenum generally minimizes or prevents

Materials Selection

Table 12-1
Wrought Plain-Carbon and Low-Alloy Steels Commonly Used for Die Components

AISI/SAE No.	UNS No.	Composition, %*					
		C	Mn	Ni	Cr	Mo	Si
1010	G10100	0.08/0.13	0.30/0.60				0.10/0.20
1012	G10120	0.10/0.15	0.30/0.60				0.10/0.20
1015	G10150	0.13/0.18	0.30/0.60				0.10/0.20
1017	G10170	0.15/0.20	0.30/0.60				0.10/0.20
1020	G10200	0.18/0.23	0.30/0.60				0.10/0.20
1040	G10400	0.37/0.44	0.60/0.90				0.15/0.30
1060	G10600	0.55/0.65	0.60/0.90				0.15/0.30
1080	G10800	0.75/0.88	0.60/0.90				0.15/0.30
4140	G41400	0.38/0.43	0.75/1.00		0.80/1.10	0.15/0.25	0.15/0.30
4150	G41500	0.48/0.53	0.75/1.00		0.80/1.10	0.15/0.25	0.15/0.30
4340	G43400	0.38/0.43	0.60/0.80	1.65/2.00	0.70/0.90	0.20/0.30	0.15/0.30
4615	G46150	0.13/0.18	0.45/0.65	1.65/2.00		0.20/0.30	0.15/0.30
6150**	G61500	0.48/0.53	0.70/0.90		0.80/1.10		0.15/0.30
8620	G86200	0.18/0.23	0.70/0.90	0.40/0.70	0.40/0.60	0.15/0.25	0.15/0.30
8640	G86400	0.38/0.43	0.75/1.00	0.40/0.70	0.40/0.60	0.15/0.25	0.15/0.30

* Plain carbon steels: Maximum P 0.040%, maximum S 0.050%.
** Contains 0.15% minimum V.

galling, but the nitrided surfaces can spall off in some severe applications, especially with die components having small radii and/or complex contours.

Forming dies subjected to high stresses are sometimes made from forged and heat-treated carbon and alloy steels. One large manufacturing concern has replaced a water-hardening tool steel with AISI/SAE 1060 steel, hardened to R_C 58-60, for blanking and trimming steel to 0.090 inch (2.29 mm) thick. This provides the advantages of lower cost, increased toughness, and the capability of performing more regrinds because of the deeper hardened case attained.

Hot-rolled Steels. Hot-rolled low-carbon steels are relatively inexpensive and are used extensively for die components for which machining and/or welding are required. These materials can be purchased in standard-size bars and plates from stock. Large plates may be cut to required sizes and contours with torches and templates, thus substantially reducing machining costs during die construction. These materials can be case hardened to provide limited hardness and toughness for short-run dies.

Hot-rolled steels hold their shape well when machined and welded because surface stresses caused by hot rolling at the steel mill are minimal. A limitation of hot-rolled steels is their poor wear resistance. When used for die components, these materials gall and cause scoring of the workpiece surfaces in applications subjecting the dies to wear.

Because of their low cost, good machinability and weldability, and minimum distortion during machining or welding, hot-rolled steels are often preferred for many die components. Applications include welded die bases and holders, support members, blankholder plates, die-shoe plates, mounting plates for trim steels, and parallels for use under or over dies. Other uses include fabricated strippers and stripper stops, punch fastening plates, fabricated slide drivers, guide blocks for pads, strippers and stripper stops, fabricated ejectors, nitrogen and oil reservoir manifolds, and shape gages for locating blanks or stampings in dies.

Cold-rolled Steels. Cold-rolled low-carbon steels have smoother surface finishes, closer dimensional tolerances, and higher strengths than hot-rolled low-carbon steels because of their cold-roll processing at the steel mill. They are generally used for die components for which hardening is not required, but wear surfaces are often cyanided.

The major advantage of cold-rolled steels is that they can often be used without machining, thus reducing costs. Such cost reductions frequently exceed the initial high price of the materials — about 50 percent more than hot-rolled steels. Cold-rolled steels, however, have internal stresses which are relieved by machining or welding. This can cause sufficient bowing, warping, or twisting to require additional costly operations, which often negates the use of cold-rolled steels for die components needing machining or welding.

Applications of cold-rolled steels for die components include keepers, stock guides, knockout bars and rods, and keys that withstand the thrust of trim and flange steels. Other uses for these materials include pad stop plates, air-line header blocks, and studs for spring locations and springs to return slides.

Cast Irons and Steels

Wrought irons are used only occasionally for dies, one example being bases for welded composite tool-steel cutting sections for trim dies. Castings of iron or steel, however, are used extensively for large dies to form, draw, or trim sheet metal. While such dies are sometimes made in one piece, they are often of composite construction. Composite dies have inserts made from carbon, alloy, or tool steel or other materials, or liners placed at sections most subject to wear or breakage.

In recent years, the advent of styrofoam patterns has decreased casting costs considerably. Patterns are now easy to cut and assemble and they are light in weight, thus reducing handling costs. They are incinerated by the cast molten metal, thus eliminating the need for pattern storage. These cost reductions have decreased the need for fabricated, hot-rolled steel weldments. Castings are normally less expensive, require less machining, and are structurally strong. Most cast irons and all cast steels can be hardened conventionally or by induction or flame hardening.

Composite dies and castings with inserts increase flexibility. The inserts can generally be changed to accommodate alterations in workpiece design or to produce different parts. Dies with inserts, however, are more costly; and because the inserts wear less than the softer casting, the uneven joint lines between inserts and castings can cause marking of the workpieces, thus necessitating reworking of the dies.

A possible limitation to the use of castings is the time required between starting to make the patterns and receiving castings from the foundry. Production scheduling may not allow time for this delay, and weldments are often substituted for castings.

Cast Irons for Dies. Irons are comparatively low in cost and are easily cast and machined. For uniform properties and improved machinability, they should be free of excessively large flake graphite, large primary carbides, and excessive phosphates. Another advantage of irons is their ability to resist galling. These materials, however, have relatively poor weldability. Irons used for casting dies include unalloyed, alloyed, and ductile irons.

Advantages of unalloyed cast irons include their low cost (the least expensive of all irons), their ready availability, and their very good machinability. A major limitation is that their structural strengths are only fair. As a result, they are generally used on simple applications. The weldability of unalloyed cast irons is very poor, making any necessary repairs difficult. Most castings are also fairly porous, which can create problems.

Castings of unalloyed irons are used for many die applications in which no actual operations are performed on the irons themselves. These applications include upper and lower die shoes, upper die holding pads, slides, slide adapters, and slide drivers (with tool steel inserts).

Alloyed cast irons are used extensively for heavy-duty dies to form, flange, or restrike sheet metal on the irons themselves. Castings are also used to hold tool steel inserts for trimming, forming, or combined piercing and forming operations. Wear resistance of these materials is excellent, and metals can be moved over their surfaces with a minimum of scratching or galling. Machinability is good and the castings can be repaired by welding with proper care.

Again, a major limitation of alloyed cast irons is their lack of structural strength. The brittle nature of these materials requires careful consideration when designing die sections that must have high strength. For some applications, stronger materials, such as ductile irons, cast steels, or Meehanite, must be used. High-strength, wear-resistant iron castings produced by the Meehanite licensed process have only limited use for dies.

Applications of alloyed cast irons include form punches, upper and lower blankholders for double-action draw dies, upper and lower stretch forming dies, upper pads for use where metal movement exists or forming is done, and master and holding surfaces of redraw dies. Other applications include the master surfaces of flanging dies and collapsible slides for forming dies.

Ductile irons, sometimes called nodular irons, have most of the desirable properties of unalloyed and alloyed cast irons, with added features of higher structural strength and toughness levels approaching those of steel because of their spheroidal free graphite. These materials are available in two grades: unalloyed and alloyed.

Unalloyed ductile irons are used where added strength — to 80,000 psi (552 MPa) — is required. Applications include die shoes having thin sections, slides, and slide adapters.

Alloyed ductile irons are used where even higher strength — to 90,000 psi (621 MPa) — and more wear resistance are needed. Applications include punches, thin-section die pads, cams for dies, blankholder rings, and lower die posts for collapsible cam dies. Alloyed ductile iron dies have replaced more expensive cast iron dies with steel inserts.

A limitation of the use of ductile iron castings is that they must be stress relieved, which increases their cost. Castings of these materials also cost about 30 percent more than unalloyed iron castings, and weldability is poor. Steel castings should be considered for high strength requirements if repair welds are anticipated.

Steel Castings for Dies. Steel castings used for dies include medium-carbon, high-carbon, and alloy steels.

Steel castings having a *medium carbon* content are used for die components that require higher structural strengths and toughness than can be obtained with cast irons. They have the lowest cost and best machinability of the various steel castings used for dies, but their cost is about twice that of unalloyed cast irons. Flame hardening properties of medium-carbon steels are poor, but these materials can be welded readily, similar to hot-rolled steels. Their uses include general purpose applications, such as die shoes, pads, keepers, and other components in which wear is at a minimum and no need exists for flame hardening.

Limitations of these materials are that they cannot be used when forming is done on the castings themselves, and the castings must be annealed or normalized and tempered to attain the required hardness. Availability is limited to foundries in which steel is poured, and in these foundries, styrofoam cannot be used as a disappearing pattern; the styrofoam pattern must be removed from the sand before the molten steel is poured.

Castings made from *high-carbon* steels are used for punches, inserts, and other die components in which savings in material and machining costs are realized in comparison to the use of tool steels. They are usually flame hardened in localized areas, but are sometimes hardened by annealing or normalizing and tempering. Such castings are not recommended for applications in which there is a tendency toward galling, seizing, or metal pickup, and they should not be used for delicate dies which might break or distort during heat treatment.

Castings made from *alloy* steels have good machinability and wear resistance, very good toughness, high strength, and excellent flame hardening properties. These characteristics, together with good weldability for repairs, make them versatile die materials. Applications for alloy steel castings include punches, die inserts, collapsible slides, and 45-degree clinching dies requiring the toughness of steels which can be hardened in critical areas.

Such castings are also applied for low and medium production requirements in forming and cutting operations, often with savings in material and machining costs. Another application is the flanging of thin, long surfaces where the joint surfaces of adjacent tool steel inserts can cause marks on the workpieces. By making long inserts of alloy steel castings, fewer joint surfaces exist and possible damage to the workpieces is minimized.

High-alloy steel castings, such as the high-carbon, high-chromium type, are also used for die components. Excellent wear resistance is the primary advantage of these materials, resulting in little or no maintenance under high-production conditions.

Uses for high-alloy steel castings include inserts for blanking, trimming, forming, and drawing dies in high-production applications in which galling or wear are problems. A possible limitation is high initial cost of the materials. The need for heat treatment before and after machining and the need for rework after hardening to remove scale and any distortion add to the cost.

Stainless and Maraging Steels

Martensitic stainless steels and maraging steels are being used for some dies and molds, especially those of intricate design and requiring long life. While these materials have a high initial cost, the cost per part formed or molded in long production runs is often lower than with dies or molds made from other materials.

Stainless Steels. Stainless steels of martensitic, hardenable metallurgical structure are used for dies and molds. These materials have a ferritic structure in the annealed condition; but when they are cooled quickly from above the critical temperature range, which is about 1600°F (870° C), they develop a martensitic structure.

The martensitic stainless steels that are most commonly used for dies and molds, in order of increasing chromium content, strength, and abrasion resistance, are American Iron and Steel Institute (AISI) Types 410 (UNS S41000), 420 (UNS S42000), and 440C (UNS S44004). Type 410 stainless steel, which contains 11.50 percent to 13.50 percent chromium, can be hardened to about $R_C 41$, with a tensile strength of 195 ksi (1344 MPa). Type 440C, which contains 16.00 percent to 18.00 percent chromium and a maximum of 0.75 percent molybdenum, can be hardened to about $R_C 57$, with a tensile strength of 285 ksi (1965 MPa).

Maraging Steels. Maraging steels, generally containing 18 percent nickel, are used for aluminum die-casting dies and core pins, intricate plastic molds, hot-forging and extrusion dies, punches, and blanking and cold-forming dies. Applications to plastic molds are most common for compression molds requiring high pressures. A major advantage of maraging steels, especially for intricate dies and molds with close tolerance requirements, is the simple precipitation-hardening (aging) treatment. Steels supplied in the solution-annealed condition are relatively soft ($R_C 30$-35) and readily machinable. Depending upon the specific type of steel, hardnesses to $R_C 60$ can be produced after machining.

Full hardening of maraging steels is attained by means of a simple aging treatment, generally about three hours at 900°F (482°C) and requiring no protective atmosphere. Since quenching is not required, cracking or distortion from thermal stresses is eliminated. Shrinkage during heat treatment is uniform and predictable. All these factors reduce the cost of manufacturing dies and molds. The steels can be nitrided to increase both surface hardness and wear resistance.

The original maraging steels contain 7 percent to 12 percent cobalt as their strengthening agent. More recently, maraging steels containing no cobalt have been introduced, with titanium as the primary strengthening agent. Mechanical properties and processing of both types of maraging steels are essentially the same.

Tool Steels

Tool steels are special grades of carbon, alloy, or high-speed steels capable of being hardened and tempered, and are the most widely used materials for dies and molds. They are usually melted in electric furnaces and produced under high-quality, tool-steel practice to meet special requirements. Tool steels are produced in the form of hot- and cold-finished bars, special shapes, forgings, hollow bar, hot extrusions, wire, drill rod, plate, sheets, strip, tool bits, powdered metal products, and castings. They are made in small quantities compared to the high-volume production of carbon and alloy steels.

Tool steels are used for a wide variety of applications, including those in many nontooling areas, in which strength, toughness, resistance to wear, and other properties are selected for optimum performance.

Classification of Tool Steels. A method of identification and type classification of tool steels has been developed by the AISI to follow the most commonly used and generally accepted terminology. The present commonly used tool steels have been grouped into eight major classifications, with the tool steels under each classification assigned a prefix letter, as indicated in Table 12-2.

Alloying Elements. The type of alloying elements added to tool steels and the amount added affect the properties of the various tool steels. Some elements are added to enhance specific properties for certain applications.

The most important alloying element affecting the properties of tool steels is *carbon*. Carbon enables a tool steel to harden through austenitic transformation. This transformation occurs by heating the steel above its critical temperature, followed by martensite formation upon cooling with sufficient speed through the martensite temperature range to about 150°F (60°C). In general, increased carbon content provides higher hardnesses

Table 12-2
Tool Steel Groups and Prefix Letters

Tool Steel Headings	Identifying Prefix
Standard high-speed tool steels:	
Molybdenum types (except M50-M59)	M
Tungsten types	T
Intermediate high-speed tool steels:	
Molybdenum types	M50-M59
Hot-work tool steels:	
Chromium types	H1-H19
Tungsten types	H20-H39
Molybdenum types	H40-H59
Cold-work tool steels:	
High-carbon, high-chromium types	D
Medium-alloy, air-hardening types	A
Oil-hardening types	O
Shock-resisting tool steels	S
Mold steels	P
Special-purpose tool steels, low-alloy types	L
Water-hardening tool steels	W

after heat treatment and improved wear resistance in service, accompanied by some sacrifice in toughness.

The addition of *manganese* increases the hardenability of tool steels. Even small amounts have significant effects on depth of hardening in carbon tool steels. The addition of 2.00 percent manganese in Type A6 tool steel enables this cold-work die material to be air hardened strongly at a relatively low austenitizing temperature.

Silicon improves the toughness of low-alloy tool steels of the shock resisting group. When added to hot-work tool steels, silicon raises the critical points and reduces scaling tendencies. Silicon also increases hardenability and resistance to tempering. This element is added to the graphitic free-machining steels to promote the formation of free carbon.

Both *tungsten* and *molybdenum* are crucial alloying elements for hot-work and high-speed steels. This is because they provide hot hardness (the ability to maintain hardness at elevated temperatures), increased resistance to tempering, and wear-resistant carbides that are harder than chromium carbides. Molybdenum has about double the potency of tungsten in its effect on hot hardness. Relatively small amounts of molybdenum are frequently added to low-alloy tool steels for improved hardenability.

Chromium, a moderately strong carbide former, contributes to wear resistance in the cold-work die steels. This is especially true for the high-carbon, high-chromium types, the D group. Chromium also promotes resistance to tempering and hot hardness in hot-work and high-speed tool steels. Additions of chromium also improve the hardenability of tool steels.

Vanadium, a very strong carbide former, is added to hot-work and high-speed tool steels for increased wear resistance. This alloying element also improves hot hardness and tempering resistance, particularly in high-speed steels, and promotes grain refinement.

Cobalt is added to improve hot hardness and resistance to tempering in both high-speed and hot-work tool steels. This alloying element remains entirely in solid solution in the steels and does not form carbides.

Like cobalt, *nickel* goes into solid solution and does not form carbides. Improved toughness and lower critical points normally result from the addition of nickel. In general, high nickel contents (above about 2 percent) are not desirable in tool steels because of the element's strong tendency to stabilize the austenite, thus increasing difficulty in annealing.

Production Variables. There are many variables in the production of tool steels that influence the properties of the materials. These variables include melting practice, hot reduction, annealing, straightening, and the use of powder metallurgy (PM) processes.

Optimum performance from forming dies and molds depends upon superior cleanliness, good chemistry control, freedom from harmful gases (particularly oxygen and hydrogen), and minimum porosity and segregation in the tooling material used. This is particularly true when unblemished, highly polished surfaces are required for applications such as forming rolls and plastic molds, or when high operational stresses are involved, particularly in the transverse direction of larger die or mold sections.

The advent of argon-oxygen decarburization (AOD) and vacuum-arc degassing (VAD) melting has resulted in improved cleanliness and reduced gas content, as well as closer composition control, in the production of tool steels. Special melting techniques, such as consumable-electrode, vacuum-arc remelting (VAR), and electroslag remelting (ESR) are used where further improvement in cleanliness and quality is required. These methods also ensure greater internal soundness (minimum porosity and center segregation).

The VAR and ESR practices are particularly applicable to the production of larger die and mold sections, in which the retention of internal quality becomes more difficult with conventional air melting practice. Internal quality refinement attained in larger bars or plates results in improved toughness in the transverse direction. This is highly beneficial when critically stressed working surfaces must be close to the center areas of the tooling.

Hot working in the production of tool steels is usually accomplished by pressing, hammering, or rolling, or by a combination of these. Normally, either pressing or hammering is used for the initial ingot breakdown of the higher alloyed tool steels, such as the high-speed and high-carbon, high-chromium types. These processes more effectively break up segregated structures through their kneading action.

Sufficient hot reduction (from original ingot to finished shape) is essential to provide normal grain size, minimize internal porosity, and effectively refine carbide size and distribution. This is especially critical in higher alloyed tool steels. Internal quality is necessary to maintain adequate strength, toughness, and working surface integrity in dies and molds. In general, optimum internal quality and grain refinement in tool steels are accomplished with as few reheatings as practical during hot reduction to finished size.

Proper *annealing* of tool steels is essential to provide optimum machinability and/or formability during production of the dies or molds. Normally, low annealed hardnesses and uniformly spheroidized microstructures are desirable for best machinability, formability, and size stability during heat treatment (discussed later in this chapter).

Care must be exercised in the final *straightening* of tool steel bars and plates to avoid excessive cold-working stresses that could lead to abnormal distortion during subsequent heat treatment. Stress relieving, within the temperature range of 1000°F to 1300°F (538°C to 704°C), should be employed on any bar or plate suspected of having been overly cold worked during straightening.

Some tool steels are produced by *powder metallurgy* (PM) processes. The use of PM processing is generally applied to highly alloyed grades, for which the benefits imparted by the process may be needed the most. The AISI Type A11 is an air-hardening, cold-work tool steel produced using PM techniques. This grade would be almost impossible to manufacture using normal methods.

Tool steel products made by PM are characterized by minimum segregation and fine-grain, uniform microstructures usually containing small carbide particles. Major advantages of these products include better size stability during heat treatment and improved grindability, which can reduce the costs of producing dies and molds.

Selecting Tool Steels. Selecting the proper tool steel for a specific application requires careful consideration of many factors. One approach is to use a material that has proved successful in the past for a certain die or mold and operation. It is essential, however, to have an understanding of the reasoning behind using different tool steels for various applications, thus permitting some judgment in the selection process.

Many different properties are desirable for tool steels used to make dies and molds for various operations. These properties can be separated into the following two groups:

1. *Primary or surface properties.* These are inherent performance characteristics which pertain directly to the properties the steel possesses to perform the required operation.
2. *Secondary or fabricating properties.* These are material characteristics which affect the manufacture of the die or mold.

Primary properties include wear resistance, toughness, and heat resistance. An additional primary property required for hot-work tool steel applications is resistance to heat checking or thermal fatigue cracking. Heat checking is characterized by a network of fine cracks that appear on the working surfaces of tooling as a result of stresses associated with alternate rapid heating and cooling during service.

Although hardness is a property developed in heat treatment, rather than an inherent property, it is still very important. Without adequate hardness, the steel would not be able to withstand the loads imposed on the die or mold. There is a direct relationship between resistance to deformation and hardness of the steel, as illustrated in Figure 12-1. Variations in the steel grade and heat-treated condition, however, cause variations in this relationship so design values should not be scaled from this graph.

Secondary properties include the many characteristics that influence the ability to make a die or mold, as well as the cost of manufacturing. Some of these are machinability, grindability, polishability, hardenability, and distortion and safety in heat treatment. Availability and cost of the tool steels are also important considerations.

In examining these properties, it becomes apparent that some compromising is required. As the alloy content of a tool steel is changed to provide greater wear and heat resistance, the toughness usually decreases. If the alloy content is adjusted to provide increased toughness, the wear and heat resistance may decrease. In addition, the secondary properties change, possibly resulting in higher alloy steels that may be more difficult to machine and grind. The challenge of selecting tool steels is to get an optimum blend of both primary and secondary properties, matching these properties to the requirements of the job.

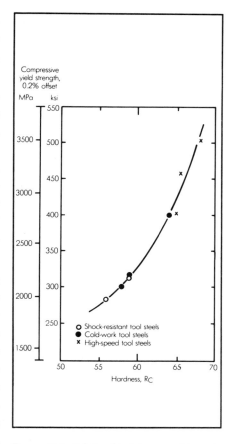

Figure 12-1. *Relationship between resistance to deformation (compressive strength) and hardness of tool steels.*

If a punch is being manufactured to have an intended life of only 100 holes, it would be wasteful to make it from a tool steel having properties to produce 10,000 holes. The added alloy in such a steel would cause more difficulty than necessary to manufacture the punch.

Heat Treatment of Tool Steels. Development of the optimum properties in any tool steel is dependent on adherence to correct heat-treating procedures. Proper heat-treating methods take into account metallurgical considerations as well as practical factors affecting such aspects as distortion and tool finish.

The important metallurgical factors in the heat-treating procedure include preheating, heating to the proper hardening temperature (also termed austenitizing temperature), adequate holding time, proper quenching, and tempering.

Preheating prior to hardening is recommended for most tool steels. The primary purpose of preheating is to reduce thermal shock when the tool enters the high heat employed in hardening and to reduce the soaking time required at the high-heat temperature. Reduced thermal shock minimizes cracking propensity and distortion effects. Lowered soaking times at high heat, resulting from preheating, reduce decarburization and scaling in air.

Important as well is the choice of the proper *hardening temperature*, which is dependent primarily on the service requirements of the die or mold. The most desirable balance of strength and toughness is a prerequisite for optimum tool life. Hardening temperatures at the high ends of the recommended ranges favor development of maximum hardness with reduced toughness. Hardening temperatures at the lower ends of the ranges favor better toughness with slightly lowered attainable hardness.

The extremities of the recommended hardening ranges are fixed by two metallurgical considerations. The minimum hardening temperature must exceed the critical temperature of the individual tool steel type in order to transform the internal structure. This transformation brings about the condition resulting in hardening during subsequent cooling. The maximum recommended hardening temperature is established as the temperature above which adverse grain growth and incipient melting occur.

Choice of the specific hardening temperature within the recommended range is best established on the basis of actual tool life results. When no prior tool life data is available, a hardening temperature is recommended at the midpoint of the range. Proper soaking times at the hardening temperature are fixed by the tool steel type, the size of the die or mold, and the heating medium. Tool steels hardened below 2000°F (1093°C) require 20 minutes to 30 minutes soaking time per inch (0.8 min/mm to 1.2 min/mm) of thickness, whereas those tool steels hardened above 2000°F require but 2 minutes to 5 minutes per inch (0.1 min/mm to 0.2 min/mm) soak time.

Tempering is a necessary part of the heat-treating process. Tempering is performed to relieve the locked-in stresses resulting from the volumetric changes of hardening. Without tempering, the die or mold would be brittle and would fail prematurely in service. Tempering relieves the stresses of hardening and imparts added toughness and shock resistance to the tool. Tempering is accomplished by heating to the desired temperature for an appropriate length of time (generally two to three hours) and then air cooling to room temperature. Tool steel compositions such as high-speed and hot-work grades require two or even three separate tempers because of the greater amount of retained austenite in the microstructure.

In all dies requiring heat treatment, design considerations for the die or mold are important in reducing the risk of cracking. Die designers should:

- Use generous radii to eliminate sharp corners,
- Avoid abrupt section changes,
- Avoid deep stamp marks, and
- Remove decarburization resulting from processing at the steel mill.

Sharp corners, drastic section changes, and deep stamp marks are the most common design factors contributing to cracking in heat treatment. These factors have the common characteristic of acting as stress raisers in the heat-treated tool. The intensified stresses can exceed the tensile strength of the tool, causing it to rupture.

Tool steel mill bar in the hot-rolled and annealed condition contains a layer of decarburization which must be removed before heat treatment. The presence of decarburization on the finished tool surface results in a soft surface having excessively high tensile stresses that promote cracking. Information pertaining to the required amount of surface removal from a hot-rolled and annealed mill bar can be obtained from the producer.

An extensive range of equipment to heat-treat die metals is available to toolmakers. Most heat-treating furnaces used for dies or molds are either salt-bath, vacuum, controlled-atmosphere, or muffle type.

Salt-bath furnaces offer the advantage of protecting the tool from the harmful effects of the atmosphere during heat treating. Dies or molds immersed in molten salt do not develop an oxide skin or harmful decarburization. In addition, heating is accomplished quickly and uniformly from all sides due to the intimate contact of the tool and heating medium. The molten salt mixture is contained in a brick-lined receptacle. Heating is usually accomplished by two submerged electrodes, or the furnace may be gas fired. Electrical resistance of the salt generates heat as voltage is applied to the electrodes.

Use of *vacuum furnaces* for heat treating tool steels has gained wide acceptance. Reasons accounting for the success of vacuum procedures include versatility, efficient utilization of energy, improved tool quality, and improved environmental factors. In the process of vacuum heat treating, the die or mold is heated in a vacuum and the tool surface is completely protected from the harmful effects of the atmosphere.

A variety of vacuum furnace equipment is available. The simplest type consists of a single chamber used for both heating and cooling. The single-chamber furnace may be sufficient for most air-hardening tool steels, but it is inadequate for those tools requiring rapid cooling. A three-chamber vacuum furnace represents maximum flexibility in respect to heating and cooling cycles. A three-chamber furnace is constructed with movable doors between each compartment and doors at both ends of the furnace. The arrangement facilitates a steady flow of work from one compartment to another, thus promoting efficiency.

In the operation of a vacuum furnace, vacuum levels of 0.050 mm of mercury (0.067 Pa) pressure or lower are normally used. Dies or molds heat treated at vacuum levels below this level are clean, bright, and free of decarburization. Vacuum furnaces are designed to include a gas or oil quench. In gas quenching, the furnace is back-filled with nitrogen. The gas quenching method has been successfully used for air hardening dies or molds of nominal size. Oil quenching is required for tools for which rapid cooling to 1100°F (593°C) is necessary to prevent grain boundary precipitation.

Controlled atmospheres are often used in the heat treating of tool steels in sealed furnaces. Protective gases used for heat treating dies or molds are in the form of a generated atmosphere, dissociated ammonia, or bottled inert gas. Generated atmospheres are made by the partial combustion of a hydrocarbon gas such as methane. The resultant gas is primarily nitrogen with controlled amounts of carbon monoxide for carbon potential control. Carbon dioxide and water vapor must be removed from generated gas because of their high decarburizing potential.

An atmosphere of 75 percent hydrogen and 25 percent nitrogen is produced when ammonia is passed over a heated catalyst. The gas mixture is used for bright annealing many types of steels and has been successfully used for the bright hardening of tool steels.

The long-used method of heat treating in *muffle furnaces* is still employed when some minor amount of decarburization or scaling can be tolerated in the heat-treated dies or molds. A retort is sometimes used to protect the tool from direct exposure to the heating medium, whether gas or electric. In this method, sufficient grinding stock must be left on the die or mold to allow for decarburization removal after heat treatment.

Surface Treatments for Dies and Molds. Various surface treatments of tool steels have been used to improve the performance of dies and molds, principally with respect to wear resistance. The most popular surface treatments used in industry for tool steels have been nitriding, oxidizing, and chromium plating. Carburizing is also a popular surface-hardening method, used primarily for steels with low to medium carbon contents. Because of their high carbon content, most tool steels are not normally carburized.

The *nitriding* process imparts a hard surface to the die or mold by the penetration of nitrogen atoms into the material and the formation of hard nitrides. Nitriding can prevent or minimize galling, but the nitrided surface can spall from small radii.

A nitrided case has a hardness of R_C 70-74. Depth of the nitride case varies considerably, depending upon the nitriding process and the time/temperature parameters employed. Case depth must be adjusted to avoid brittleness. Nitriding is accomplished either in a molten salt bath, in gaseous ammonia, or by the glow discharge method (ion implantation).

Molten salt baths composed of cyanide salts impart nitrogen to the steel at temperatures in the range of 950°F to 1200°F (510°C to 649°C). Case depths produced by the molten salt method are typically in the range of 0.0001 inch to 0.001 inch (0.003 mm to 0.03 mm).

In the *ammonia gas process*, nascent reactive nitrogen is released when the gas comes in contact with hot steel. The nitrogen diffuses inward, developing a hard case. The process is conducted typically at 980°F to 1000°F (527°C to 538°C) for times of 10 to 80 hours. The gaseous method accomplishes deeper nitrogen penetration than the molten salt method can and is therefore especially useful where heavy case depths are required.

In *glow discharge nitriding*, the workpiece is made into the negative electrode in a vacuum furnace. A glow discharge is formed around the workpiece at a potential of 300V in a vacuum of 0.001 mm to 0.010 mm of mercury (0.0013 Pa to 0.013 Pa) pressure. Nitrogen is introduced in controlled amounts, and nitrogen ions are implanted in the workpiece by virtue of their kinetic energy. The principal advantage of the glow discharge nitriding process is the control of nitrogen concentration and case depth.

Some dies and molds are produced with a black oxide film which extends tool life. The oxide film reduces direct metal-to-metal contact between tool and workpiece and retains cutting lubricant, thus promoting improved tool life. The oxide film can be developed by a steam treatment or by immersion in a molten oxidizing salt.

In using the *steam oxidizing* method, a sealed retort furnace is employed which incorporates the addition of steam in controlled amounts. The steam treatment is normally performed simultaneously with the tempering operation, typically in the 1000°F to 1050°F (538°C to 566°C) temperature range.

The technique of *oxidizing* dies or molds in a molten *salt bath* employs a mixture of sodium nitrite and sodium hydroxide. Tools develop a tenacious oxide film when immersed in the oxidizing salt mixture at 300°F (149°C) for approximately 15 minutes. Most oxide films applied to tools are typically 0.0001 inch to 0.0002 inch (0.003 mm to 0.005 mm) deep.

Chromium plating is employed to advantage for some dies or molds on which friction is a critical factor in tool life. A chromium-plated surface has a reduced coefficient of friction as compared to a machined steel surface. It is also useful in minimizing or preventing galling for some severe applications, but the plating may spall from small radii. This method can also be used to build up worn areas of a die or mold. A typical chrome plate thickness used for antifriction purposes is 0.001 inch to 0.005 inch (0.03 mm to 0.13 mm) and has a hardness of R_C 65-75.

In chromium plating, the workpiece is made into the cathode in an electrolytic cell and chromium metal is plated on the surface from an anodic source. Hydrogen is also released at the cathode, which makes the die or mold sensitive to hydrogen embrittlement. A post stress-relief treatment must be performed on chromium-plated tools to ensure the removal of hydrogen. Exposure to temperatures of 400°F to 500°F (204°C to 260°C) is employed for periods of three to four hours to accomplish hydrogen removal.

Among many available coating method options are those based on *vapor deposition*. Extensive testing of tools coated with titanium nitride shows promising results for metalforming tools such as punches, dies, and rolls. Increased tool life is attributed to the lubricous hard qualities of titanium nitride which resists galling and metal pickup.

In the process termed *physical vapor deposition* (PVD), the finished die or mold is placed in a chamber where it is bombarded with titanium ions in the presence of nitrogen, producing a thin layer of titanium nitride. The PVD process is conducted at temperatures below the tempering temperature.

The *chemical vapor deposition* (CVD) process requires a high temperature and therefore necessitates subsequent vacuum reheat treating after coating. In CVD, the die or mold is heated to approximately 1900°F (1038°C) in the presence of titanium tetrachloride gas and methane or nitrogen. Two separate layers of first titanium carbide and then titanium nitride are deposited on the tool.

A deposited layer thickness of 0.0002 inch to 0.0003 inch (0.005 mm to 0.008 mm) is typical for the vapor deposition processes. Both the PVD and CVD processes can be adapted to a variety of coating materials, such as hafnium nitride, aluminum oxide, tungsten carbide, and nickel borides.

Steel-bonded Carbides

Steel-bonded carbides belong to the family of cemented carbides and are produced by the powder metallurgy (PM) process. These materials differ, however, from conventional cemented carbides in that they have variable physical properties (particularly hardness) obtained by heat treatment of their matrices.

Grades Available. These sintered ferrous alloys are made in several grades. Different grades contain from 20 percent to 70 percent (by volume) titanium carbide, tungsten carbide, titanium-tungsten double carbides, or other refractory carbides as their hard phase. The balance of the content, the heat-treatable matrix or binder, is a carbon or alloy steel containing at least 60 percent iron. By controlling the composition, the material grade can be tailored to specific property requirements.

The grade used most commonly for dies and molds contains neither tungsten nor cobalt. Main constituents of this grade are about 45 percent titanium carbide and approximately 55 percent alloy tool steel. The tool steel contains 3 percent molybdenum, 10 percent chromium, and 0.80 percent carbon. This grade has a compressive strength in excess of 450,000 psi (3100 MPa) and a transverse rupture strength of about 200,000 psi (1380 MPa).

Advantages. Advantages of steel-bonded carbides for dies and molds include machinability in the annealed condition, hardenability, good wear resistance, minimum friction, and the ability to withstand heavy compressive loads at high temperatures. Annealing of these materials to hardness levels of R_C 43-46 permits machining (discussed later in this section) with conventional steel-cutting tools.

The wear resistance of steel-bonded carbides is much better than that of most tool steels and approaches that of some cemented tungsten carbides with cobalt binder. This is the result of the titanium carbide particles, about 0.0002 inch to 0.0003 inch (0.005 mm to 0.008 mm) diameter and having a Vickers hardness of 3300, embedded in the hardened matrix. These materials are less brittle than cemented tungsten carbides, thus reducing the possibility of chipping.

For many applications, dies and molds made from steel-bonded carbides are providing higher quality parts with less scrap, increased productivity as a result of less downtime, and reduced costs. In comparison to production runs in which dies made from high-carbon, high-chromium tool steels were used, production runs between regrinds of 10 times as long and increases in die life of 50 or more times have been reported. While steel-bonded carbides cost about 20 percent more than the more expensive tool steels, they are less costly than sintered tungsten carbides.

Applications. Steel-bonded carbides are being used primarily for single-station and progressive dies employed in severe stamping operations, such as forming, drawing, notching, and blanking, including the production of laminations. Rigid presses with good parallelism between moving and stationary members are essential for the use of these die materials. Excessive deflection reduces the productivity and life of the dies. The applications of steel-bonded carbides for plastic molds include gate and mold inserts, and nozzles.

Die Design and Construction. Sectional construction is generally preferable for dies made from steel-bonded carbides because it permits easier manufacture of complex sections and facilitates the replacement of worn components. For high-production dies, it is most economical to provide die sections with tapped holes in their bottom surfaces. This allows the sections to be clamped to die shoes from below and permits the maximum number of regrinds.

Draw-die inserts or rings require suitable steel retainer rings that are precompressed to absorb and counteract the stresses exerted radially on the bore of the die. This can be accomplished by providing an interference shrink fit or by means of a taper fit (see Figure 12-2), drawing the rings together onto a support plate with bolts. If shrink fitting is performed, care is required to minimize or prevent the absorption of heat from the expanded ring, which could reduce the hardness of the steel-bonded carbide.

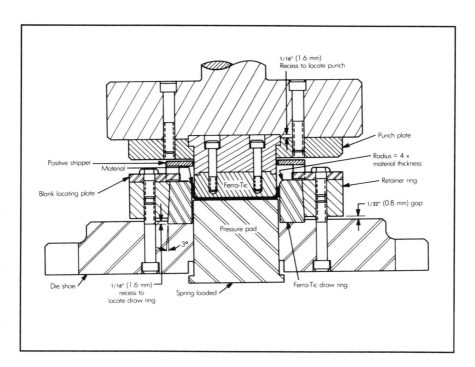

Figure 12-2. *Draw die with punch and taper-fit draw ring made from steel-bonded carbide.*

When die sets are used, they should be of four-post design for increased rigidity. Hardened, precision-ground bushings and pins are preferable for accurate parallelism. Strippers and knockouts must also be sturdy and well guided, and pressure pads should be spring loaded instead of the positive type. Safety devices to detect misfeeds, buckling of the stock, and other malfunctions are essential for die protection.

Optimum rigidity is also important for punches made from steel-bonded carbides. Punches of straight design with nested bases are preferable to L- or T-shaped designs because less material is required and they are easier to grind. Brazing or welding is not generally recommended, and punches are usually attached to the punch plates with screws.

Clearances between the punches and dies are generally slightly larger than when tool steels are used. A clearance of 7 percent to 8 percent of the stock thickness per cutting side is recommended for blanking or piercing carbon or alloy steels, as well as nonferrous materials. For stainless steel or prehardened spring steel, the clearance is generally 9 percent to 10 percent of stock thickness per cutting side.

When sharp outside corners are required on stampings, the die sections should be split at the corners. Corners on the punches, however, must be rounded with radii of 0.003 inch to 0.004 inch (0.08 mm to 0.10 mm) to avoid chipping or rapid wear.

Machining. With only a few variations, the machining of steel-bonded carbides is essentially the same as for other materials. In fact, it is possible to remove about 50 percent more stock per minute than is removed in machining high-carbon, high-chromium tool steels, such as AISI Type D2.

When steel-bonded carbides are being machined, low cutting speeds are essential to preserve the cutting edges of the tools and to avoid overheating, which can cause premature hardening of the metals. A maximum cutting speed of 30 sfm (9.1 m/min) is recommended for turning, milling, and sawing. For drilling, the cutting speed can vary from 125 sfm to 250 sfm (38 m/min to 76 m/min). Tapping should be done by hand or with a slow machine.

Relatively heavy cuts are also important to avoid glazing and undesirable work hardening. Depths of cut should never be less than 0.003 inch (0.08 mm) but can be as heavy as 0.25 inch (6.4 mm) or more on rigid machines. Feed rates for turning vary from 0.003ipr to 0.012 ipr (0.08 mm/rev to 0.30 mm/rev) and for milling, 0.003 inch to 0.010 inch (0.08 mm to 0.25 mm) per tooth. Down (climb) milling should be used whenever possible in preference to up (conventional) milling.

Tungsten carbide tools, classification numbers C-1 and C-2, are used extensively for turning operations. The tools or inserts generally have a 0- to 5-degree negative rake, 5-degree side rake, and nose radius of 0.030 inch (0.76 mm). High-speed steel tools are common for milling, drilling, and tapping. Carbon steel blades having 8 or 10 teeth per inch are satisfactory for sawing. Good results have been obtained in cutting threads with five-flute, high-speed steel taps having a negative rake angle, shallow flutes, and narrow lands.

Cutting fluids should not be used in machining steel-bonded carbides. This is because the fluids have a tendency to combine with the carbide grains in the chips and form an undesirable lapping compound that can quickly destroy the cutting edges of tools.

Heat Treatment. The heat treatment of steel-bonded carbides is conventional. However, only the steel matrix is hardened, with the titanium carbide particles being unaffected. The cycle consists of heating to 1975°F (1079°C), preferably in a vacuum furnace, quenching with nitrogen or another inert gas, tempering by heating to 950°F (510°C) for one hour, air cooling to room temperature, retempering at 950°F, and again air cooling. This treatment produces a hardness of R_C 66-70.

The need for decarburization-free heat treatment is essential. If a vacuum furnace is not available, the metals can be heated in an atmosphere-controlled furnace and quenched in oil. If an atmosphere-controlled furnace is not available, the die components can be wrapped in airtight bags of stainless steel foil for heating. Quenching should be done with the components still in the bags.

Rough-machined, annealed components are often stress relieved prior to finish machining and hardening. This is accomplished by heating the components to 1200°F (649°C) for one hour and cooling in air. By finish machining the material in the annealed and stress-relieved condition, the need for grinding after hardening becomes minimal.

Like tool steels, steel-bonded carbides undergo a slight growth during hardening in the conversion of the matrix from austenite to martensite.

Aluminum Bronzes

One common application of aluminum bronzes is deep-drawing dies, especially when forming the tougher grades of stainless steels. In addition to the improved quality of parts produced, smaller blanks can be used because draw beads are not necessary. Higher pressure-pad pressures, however, are necessary. In one application for deep drawing stainless-steel cooking utensils, die life was doubled, maintenance was reduced 45 percent, and workpiece finishing costs were lowered 25 percent. Aluminum bronzes are not recommended for drawing copper, brass, bronze, or unpickled steels.

Draw rings, pressure pads, and noses of steel punches are sometimes made of aluminum bronzes, especially when they are to be used in forming tough stainless steels. Galling and scratching of the workpiece surfaces are eliminated because there is no metal pickup, and long life results due to the low friction between the dissimilar metals.

Forming rolls for the production of stainless steel tubing and other shapes are often made from aluminum bronzes. In addition to improving workpiece finish and wear resistance, such rolls resist adhesion of welding spatter. Straightening dies for roll-formed aluminum house siding are also made from aluminum bronzes.

Other applications for aluminum bronzes include bending dies for press brakes, wing dies for vertical tube benders, dies for rotary swaging of stainless steel and aluminum alloy tubing, and gang arbors (snakes) for bending tubing.

Die Design. The design of dies made from aluminum bronzes varies with the operation to be performed, workpiece material (thickness and condition), type of press or machine used, ram and hold-down pressures, surface finish desired, lubricant employed, and quantity of workpieces required.

When used for drawing, the die must be solidly supported and well seated in a cavity of a steel or iron member. A shrink-type interference fit of 0.001 inch/inch (0.03 mm/mm) of diameter is suggested. The usual procedure is to shrink the aluminum bronze die by packing it in dry ice, rather than to expand the backup member.

The radii on draw dies and punches must be generous. As mentioned, for draw dies, the radii should be about four times the thickness of the metal to be drawn; for punches, the radii should be about eight times the metal thickness. Clearance between the punch and die should be proportional to the metal thickness plus an allowance (generally 7 percent to 20 percent of the metal thickness) to minimize wall friction.

Composite or segmental dies consist of rings or narrow strips of aluminum bronze placed in retainers at wear areas on cast-iron or steel dies. For long production runs, full rings or blocks of aluminum bronze are sometimes used. Larger radii are required and higher pressures must be exerted on the pressure pads when composite dies are employed. Solid dies permit smaller radii, lower pressures on the pressure pads, and the use of smaller blanks. With solid dies, draws can be made closer to the edges of the blanks, thus reducing material losses when trimming.

Machining and Finishing Dies. Aluminum bronzes have the inherent ability to develop very smooth surface finishes when machined properly. Turning, boring, facing, and similar operations are generally performed with conventional carbide cutting tools. For roughing operations, cutting speeds to 350 sfm (107 m/min) are employed, with a feed rate of 0.005 ipr to 0.020 ipr (0.13 mm/rev to 0.51 mm/rev). For finishing, cutting speeds to 800 sfm (244 m/min) and feed rates of 0.002 ipr to 0.005 ipr (0.05 mm/rev to 0.13 mm/rev) are recommended. A soluble-oil cutting fluid is generally used.

Drilling and tapping of aluminum bronzes are difficult and should be avoided if other methods of assembling the die components can be used. If drilling is required, carbide-tipped, straight-flute drills should be employed, with a cutting speed of 70 sfm to 150 sfm (21.3 m to 45.7 m/min) and a feed rate of 0.002 ipr to 0.007 ipr (0.05 mm/rev to 0.18 mm/rev). Holes should be chamfered before tapping is performed to prevent edge breakout. Fair results have been obtained by using taps having a 0-degree rake angle, a 10- to 15-degree chamfer for a length of two to three threads, and a spiral point extending beyond the first full thread.

Grinding of aluminum bronzes is generally done with vitrified-bond silicon-carbide or aluminum-oxide wheels, with wheel speeds of 5000 sfm to 6000 sfm (25.4 m/s to 30.5 m/s) and work speeds of 25 rpm to 150 rpm. Polishing can be done with silicon-carbide abrasive cloth having a grain size of 240 or 320. Crocus cloth is sometimes used for final finishing.

Die Operation. For optimum performance, the first 35 or so workpieces produced with a new die made of aluminum bronze should be formed using a strained, thick slurry of water and unslaked lime as a buffing compound. Alternatively, this run-in can be done using a silicon-Teflon dry lubricant. After the run-in period, conventional sulfur-free lubricants can be used, with a dry lubricant added if desired. For drawing stainless steels, the lubricant should have a higher film strength than that used for drawing carbon steels. It is essential that smooth blanks, without burrs or turned edges, be used with aluminum bronze dies.

Beryllium Coppers

Cast alloys of beryllium, cobalt, and copper have characteristics comparable to those of proprietary aluminum bronzes. These alloys are sometimes used for molds to form plastics, plunger tips for die-casting dies, and other components. Ample exhaust ventilation is essential in making such components to minimize concentrations of beryllium in the air, which can cause a health hazard.

Zinc-based Alloys

Zinc alloys are used extensively for punches, dies, and molds to form, draw, blank, and trim steel and aluminum alloys, plastics, and other materials. Applications are predominant in the automotive and aircraft industries for producing prototypes and limited quantities of large parts. Frequently, one member of a two-piece die set is made of zinc alloy and the other (usually the punch) is made of a softer material such as antimonial lead, especially for drop hammer operations on soft sheet metals. Zinc alloy construction is generally required for both die members when steel sheets are being formed, when sharp definition is needed, when production runs are long, or when binder rings are necessary.

Tools made from these alloys are often capable of forming 10,000 or more parts before they have to be replaced or repaired. When abraded and worn areas are repaired by welding, 25,000 or more parts can be produced. There may be some creep of the material under extreme-pressure conditions, and close tolerances cannot be maintained in blanking operations.

Advantages. Zinc alloys provide a low cost and fast method of making punches, dies, and molds having a dense, smooth working surface. These materials are easy to melt, cast, machine, grind, polish, weld, remelt, and recast. Casting provides sharp definition of contours because of the fluidity of the alloys, and accuracy of the castings minimizes the need for costly finishing.

Other important advantages of these alloys include no scratching of the workpiece material, good abrasion resistance, inherent self-lubricating properties, and high impact and compressive strengths. Their low melting temperatures reduce energy costs, and the tools can be remelted and recast a number of times without loss of mechanical properties. Care must be taken, however, to ensure that contaminants, such as iron or lead, are minimized and that excessive casting temperatures, which might cause dealloying and immoderate grain growth, are avoided.

Alloys Available. Various proprietary zinc alloys, some called Kirksite alloys, are available. Many contain about 3.5 percent to 4.5 percent aluminum, 2.5 percent to 3.5 percent copper, and 0.02 percent to 0.10 percent magnesium. Some contain additional magnesium or copper, or minor amounts of nickel or titanium. As with zinc die-casting alloys, practically pure zinc (a minimum of 99.99 percent) is used in these alloys to keep the impurities of lead, tin, iron, and cadmium low.

Mechanical properties vary with the specific alloy. Typical property values include tensile strengths of 30 ksi to 40 ksi (207 MPa to 276 MPa), compressive strengths of about 65 ksi (448 MPa), and Brinell hardnesses to approximately 105.

Casting Practice. Zinc alloys are generally sand cast in plaster (sometimes wood) patterns or in preformed plaster, but rarely in steel molds. For superior results, shrink patterns are made for all die members (allowing for the thickness of the workpiece material), which are cast in sand. Mounting surfaces are machined, critical die surfaces may be checked against model surfaces, and hand grinding is performed for final accuracy and clearances. In some cases, only one die member is cast and it is used as a mold for casting the opposite die member, usually of lead. A separating layer of insulating material must be used in such cases. Urethane, cast in place, is also used for some applications.

Machining the Castings. With accurate patterns or molds and correct casting techniques, little machining or finishing of zinc alloy castings is required. Machining of the castings, however, presents no problems. It should be done at high cutting speeds with light cuts, using polished tools having generous rake and clearance angles. Drills used should have large spiral flutes, thin webs, an included point angle of 100 degrees, and a lip clearance angle of at least 10 degrees. When made from high-speed steels, the drills should operate at a cutting speed of 300 sfm (91 m/min) or more.

For turning and boring the castings, high speed steel or cast alloy tools are generally used. A top rake angle of 10 degrees, an end clearance angle of 15 degrees, and a side clearance angle of 50 degrees are recommended for the tools. If carbide tools are used, the end clearance should be under, not over, 8 degrees. A cutting speed of 300 sfm (91 m/min) is suggested for roughing operations, and 600 sfm (183 m/min) for finishing.

Cutters with staggered, coarse teeth, 10-degree clearance angles, and 10-degree rake angles are used for milling at high speeds. Coarse-toothed blades (6 teeth per inch) are suggested for sawing, with a blade speed of 150 fpm (45.7 m/min) or more. Feed pressure exerted during sawing should be sufficient to removed about 9 inches^2/min (58 cm^2/min) of stock.

Zinc alloy castings can be plated with any common metal coating, including hard chromium.

Antimonial Lead

Punches made of antimonial lead are sometimes used with dies made of zinc alloy, especially for drop hammer operations on soft metals. Lead-antimony alloys are available with various percentages of antimony to suit specific requirements. The balance of the contents of these alloys is lead, but from 0.25 percent to 0.75 percent tin is often added to improve the casting properties. The alloys generally best suited for forming operations contain 6 percent to 7 percent antimony, which provides the best combination of mechanical properties, including adequate ductility, hardness, and tensile strength.

Antimonial-lead punches are made by casting into the cavity of zinc alloy dies. The antimonial lead is sufficiently ductile to accurately assume the dimensions of the zinc alloy die under impact. These alloys, like zinc alloys, can be remelted and recast.

Bismuth Alloys

The alloys of bismuth, often called low-melting-point alloys, are used chiefly as matrix material for securing punch and die parts in small die sets, and as-cast punches and dies for short-run forming and drawing operations.

13

Die Engineering — Planning and Design

Factors in Pressworking Stamping Materials

Early in the product design process, the decision must be reached whether the contemplated component, as-designed or with permissible redesign, will be stamped or produced by some other process. The decision calls for the concerted attention of the design, materials, methods, tooling, manufacturing, and any other functions with responsibility for producing the product. Certain practical guiding criteria follow.

Product design factors:

- *Shapes* are limited to those which may be produced by the cutting, bending, forming, compressing, compacting, or drawing of materials.
- With the exception of high-velocity explosive forming operations, which are limited only by the types and sizes of available materials, maximum *sizes* are limited chiefly by press capacity. There are few limitations upon minimum size; sections as thin as 0.003 inch (0.08 mm) are possible, with parts so small that 10,000 may be held in one hand.
- *Tolerances* are very good. For small stampings made with progressive or compound dies, plus or minus 0.002 inch (0.05 mm) is common, and closer limits are possible on small and thin parts. Precision fineblanked parts typically are stamped to much closer tolerances.
- The *weight* factor is highly advantageous. Parts formed from sheet metal and stampable plastics have favorable weight-to-strength ratios.
- Surface *smoothness* is excellent, since surface condition usually is not affected by the forming operation. Prepainted steels are in common use as stamping materials. The surface finish of sheet steels can be optimized by special texturizing of the finish mill rolls for maximum paint luster on the finished stamping.
- A wide *choice of materials* is available, including any in sheet form and not so brittle as to break.

- *Design changes* are usually costly, if required after the original tooling has been completed. Careful planning for ease of manufacturability early in the design stage is essential.

Production factors:

- Except for temporary or low-production tooling, the *lead time* required for tooling is lengthy compared with some other production methods. Die design, tryout, and development may take months.
- *Output* is very high; more than 600 large automotive body stampings per hour are commonly produced on automated transfer presses. Small stampings produced on high-speed equipment often have production rates well in excess of 10,000 pieces per hour.

Economic factors:

- *Stamping-materials costs* are comparatively low. A favorable cost factor is the minimum scrap loss achieved through careful selection of stock and skilled strip layout.
- *Tool and die costs* are high, often higher than tooling for comparable parts that are to be die-cast. Costs are most favorable where large production is planned.
- *Direct labor costs* depend upon the part size and shape and the extent of automation. These costs are characteristically very low.
- *Presses*, except for the small manual punch presses, are typically more costly than standard machining equipment such as lathes and grinders, and require a higher machine-hour rate.
- *Finishing costs* are low. Often no finishing is required other than normal painting or plating.
- *Inventory costs* can be quite low. Quick die change equipment and techniques permit a different part to be produced with a changeover time of under 10 minutes.

Basic Procedure for Pressworking Process Planning

When a decision has been reached to produce a part by stamping, the process planner may have three major areas of responsibility.

1. Planning the sequence of operations, and specifying the metalworking equipment and gaging necessary to produce high-quality parts economically at the specified production rate.
2. Coordination of the allied processes such as heat treatment, metal finishing, and plating.
3. Integrating the required material handling and operator movement paths.

The second and third responsibilities are often performed by experts in plant facilities and layout.

The following steps illustrate a typical procedure for the planning of a pressed-metal manufacturing process.

1. ANALYZE THE PART PRINT. To aid this step it may be desirable to have enlarged layouts, additional views, experimental samples, models, and limit layouts. It will be helpful to chart the assigned responsibilities for carrying out the specifications.

    a. *What Is Wanted?* The product designer must establish explicit detailed specifications for size, shape, material type, material condition, and allied processes. The process planner must be left in no doubt as to all the specifications and how they relate to each other.

    b. *List Manufacturing Operations and Allied Processes.* A typical listing would be:

    1. Pierce hole, 0.501 inch (12.72 mm), + 0.002 (0.05 mm), - 0.000.
    2. Flange 90 degrees ± 2 degrees.
    3. Buff external surfaces.
    4. Blank.

    At this point, there is no need to list the operations in their proper sequence, or to combine the operations, but only to make a preliminary survey of basic operational requirements. Each listed item should be checked off on the part print drawing.

    c. *Determine Manufacturing Feasibility.* Consider the possible die operations that could produce the part with the specified surface relationships. Computer-aided formability analysis should be employed at this point. Close cooperation with the product designer is required at this point. A hole close to a flange, an unreasonably small radius, a draw requiring annealing, an area of a drawn part having an excessive percentage of elongation, a blank that cannot be economically nested—these and other conditions can frequently be improved by the product designer without affecting the functional requirements.

    d. *Make Recommendations to Product Engineering.* Upon completion of part print analysis, recommendations should be written to the product designer. All accepted recommendations call for necessary engineering changes in the part print.

2. DETERMINE THE MOST ECONOMIC PROCESSING. For the stamping of a single part, there are usually several alternative production methods. The method selected should be the one which will result in

the lowest overall cost of producing a finished part of the required quality. The cost includes material, tooling, direct and indirect labor, and overhead burden.

Determination of the most economical processing can be accomplished by comparing two or more feasible processes for producing the given pressed-metal part. The comparison of unit costs for each such process, for equal production quantities, will give a breakeven point which is a guide to selecting the most economical tooling. Productive labor costs and burden rates are estimated from past performance and by the use of standard time data.

A graphic presentation of the breakeven point is useful where the spread between processes is small. Where the spread is small, but increased production requirements in the future are likely, it may be preferable to use the higher-cost process.

Unless new pressworking equipment can be amortized over a rather short period, or has future value as standard equipment, it may prove more economical to use existing available equipment even though production costs would be higher.

Likewise, simple dies may be favored over the high-production dies which seem indicated by the anticipated requirements, due to a lack of the special skills required to design, construct, and maintain high-production dies. Also, the simpler single-operation dies may permit interchangeability of tooling for different parts which have several common shape and/or size specifications.

3. PLAN THE SEQUENCE OF OPERATIONS. Operations planning done only on the basis of past experience can prove costly if seemingly minor details are overlooked.

   a. *Determine Critical Specifications.* Dimensional specifications which have comparatively close tolerances, or the limitations placed upon the specifications for allied processes, are known as "critical" specifications. Study of the comparative effects of specifications upon surface relationships, with the aid of a limit layout, will reveal the critical specifications from a manufacturing standpoint.

   b. *Select Critical Areas.* Most critical specifications pertain to the measurement of surface relations within specified close tolerances. Critical areas are those areas or surfaces from which the measurements for all specifications can be taken to determine the geometry of the part. Limit layouts serve also to determine the critical areas. In ideal planning, critical areas should be established first, provided that they are "qualified" as surfaces of registry in a location system, for subsequent operations and allied processes.

c. *Determine Critical Manufacturing Operations.* An operation is designated "critical" when it is required to establish a critical area from which subsequent operations or allied processes can be gaged or located. The required degree of control over such stock variations as width, thickness, camber, and mechanical properties, should be specified within obtainable commercial limits. Stock variability results in product variation, especially as regards the amount of springback. These are basic factors in defining a critical manufacturing operation. The ideal critical manufacturing operation would establish the critical areas in a single operation from the sheet, strip, or coiled stock.

d. *Accomplish Critical Manufacturing Operations.* This is the major responsibility of the die designer, working to the process plan. However, the process planner must know the basic types of dies and their general applications. Close cooperation with the die designer and production facility is desirable to get needed input regarding the ability to accomplish critical operations.

e. *Determine Secondary Manufacturing Operations.* These operations are intermediate between the critical manufacturing operations and the finished part. Limitations in these operations are imposed by the workpiece specification, and by the amount of metal flow and/or movement. Additional secondary operations, such as restrike, may sometimes be necessary to coordinate with an allied process or to reestablish a critical area for subsequent operations.

f. *Accomplish Secondary Manufacturing Operations.* This follows the same procedural pattern as for critical manufacturing operations, and involves the same responsibilities for the planner.

g. *Determine Allied Processes.* These processes are determined during part print analysis, except as they arise through emergency or necessity. Thus, an annealing operation may become necessary when secondary manufacturing operations have been determined. In some cases, the process planner may avoid annealing by recommending a change in material specifications. All possible elimination of annealing, plating, cleaning, and other allied processes will appreciably reduce total manufacturing costs.

h. *Accomplish Allied Processes.* These are often the function of specialists, but the pressed-metal process planner must cooperate by delivering a workpiece in suitable condition for the allied process. The specialist should advise the processor of the effect the allied process will have on the workpiece.

4. SPECIFY THE NECESSARY GAGING. Gaging here includes (1) the gaging of material as received, and (2) gaging of the workpiece in process. Gaging of material will include width, thickness, camber for specified tolerances, and mechanical properties. The planning of in-process gaging for use during manufacture follows the same procedure as used to select critical areas, or areas from which measurements can be taken to define the geometry of the workpiece. The workpiece should not be allowed to continue through the sequence of operations if it is defective from a previous operation.
5. SPECIFY THE NECESSARY PRESS EQUIPMENT. A press should be specified according to the size and actuation requirements of the die, the type of pressworking operation to be performed, properties of the workpiece material, required tonnage, and the required production accuracy.
6. PREPARE ROUTING OF PROCESS. The operational machine routings vary in form throughout industry but must meet two common essential requirements: (1) description of the operation must be accurate and complete; (2) the nomenclature used should be according to accepted practice. A pictorial sketch of the part, shown on the operational machine routing sheet, aids considerably.

3-2-1 System for Locating and Tests for Qualified Areas

Qualified areas are those areas which fulfill the requirements of arithmetical, mechanical, and geometrical tests to serve as surfaces of registry.

1. ARITHMETICAL TESTS. The selected surface of registry must not cause a limit stack. If surfaces of registry cannot be selected, they must be qualified, i.e., produced to tolerances closer than those required and specified by the product engineer.
2. MECHANICAL TEST. The size, shape, and finish of the selected surfaces of registry must permit a seat of registry design so each will withstand the operating forces exerted, and also the necessary holding forces.
3. GEOMETRICAL TEST. This test pertains to the distribution of the surfaces of registry so that the workpiece will be positionally stable. If surfaces of registry are not thus qualified, the process planner must consider suitable redesign with the product engineer. In the 3-2-1 locating system (Figure 13-1), six is the minimum number of points required to fix a square or rectangular shape in space: three points establish a plane, two points define a straight line, and one point for a point in space. A small pyramid symbol is used to designate a locating point. In Figure 13-2, this symbol is used to illustrate a locating system for a

rectangular solid. Variations of the illustrated system can be used to fix the location of a cylinder, cone, disk, or other geometric shape. The surface of the device used by the die designer to establish the locating point specified by the process planner is known as a "seat of registry." The corresponding area on the workpiece is known as a "surface of registry." Process planning symbols can be used to avoid lengthy writing in the preliminary stages of planning to utilize critical areas for critical and secondary manufacturing operations. Figure 3-2 illustrates the use of such symbols.

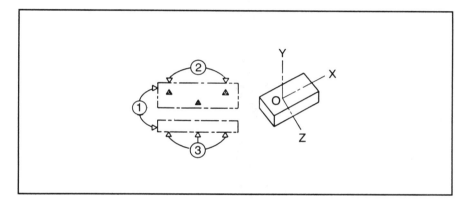

Figure 13-1. *3-2-1 locating system.*

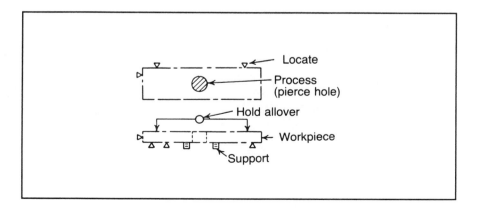

Figure 13-2. *Application of process symbols.*

Application of Process Planning to Specific Pressed Parts

Sound general principles are always of value. An example of how good process planning is applied to a pressed steel part that is being successfully produced illustrates the method.

Cross-shaft Bracket. Figure 13-3 is the part print of a cross-shaft bracket, an approved component of an assembly with a forecast production of approximately 100,000 per year. In some plants the process planner might be expected at the outset to furnish a preliminary routing so that unit costs may be estimated. Such routing might indicate: (1) "blank and pierce," (2) "first flange," (3) "finish flange," and (4) "inspect," together with other normally required information and specifications. Although such preliminary routing may satisfy estimating needs, it does not always ensure quality

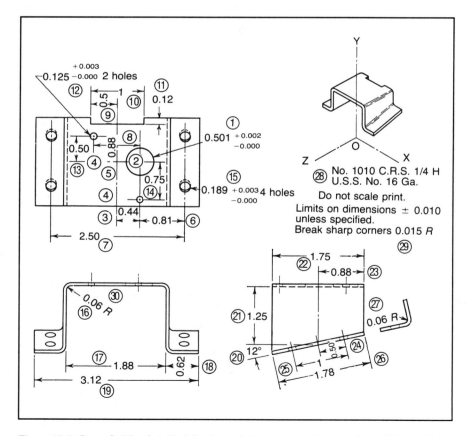

Figure 13-3. *Cross-shaft bracket. Circled values refer to entries on tabular analysis of Figure 13-4.*

parts, economically produced. For such goals, it is essential to apply an accepted planning procedure.

Analyze Part Print. Each specification, dimensional or noted, must be studied to understand exactly what is specified by the product engineer. One method used by the experienced planner is to check off each specification on the print, once it has been interpreted to his or her satisfaction. A tabular analysis such as that in Figure 13-4 may prove more effective, because it is a graphic record of information on all specifications shown or implied on the part print.

Terms used as column heads, "Material," "Die," and "Processing" in Figure 13-4 are listed in the following annotation.

"Material" refers to the material of the part to be pressworked, and it should be considered when a surface relationship depends upon a variable of the material. For example, variation in thickness would affect a forming operation.

"Die" refers to the tool which is anticipated for obtaining a required surface relationship. For example, the size of a pierced hole and the obtainable limits depend upon the skill of the diemaker in producing the punch to proper size.

"Processing" refers to the work of the process planner as it affects surface relationships. For example, if two holes were pierced simultaneously, the accuracy of surface relationship would depend upon the die; but if one hole were pierced and the second hole is located from the first, then the accuracy of surface relationship would depend, in part, on the skill of the process planner.

This technique in process planning classifies all specifications concerned with surface relationships and reveals whether the relationships depend upon the material, the die, the processing, or some combinations of these.

The process planner is primarily concerned with the "Die" and "Processing" columns but, to plan an acceptable sequence of operations, the items in the "Remark" column must be clarified in consultation with the product engineer. The product engineer must also be consulted on any changes in materials specifications which might avoid manufacturing difficulties without affecting the part's functioning requirements. In extreme cases, a study of the materials considerations might show the wisdom or even the necessity of changing to some process other than pressed metalworking. In short,

TABULAR ANALYSIS
PART NAME—*Cross shaft bracket*

No.	Specifications	Between Surfaces — Depend On			Remarks	Operational Requirements
		Material[1]	Die[2]	Processing[3]		
1	0.501 $^{+0.002}_{-0.000}$	✓	✓		Squareness implied 90°	★
2	℄ Part to ℄ 0.501 hole			✓		
3	0.44 to 0.501 & 0.125 holes			✓	Possible limit stack with spec No. 6	
4	℄ to ℄			✓	Implied coincident	
5	℄ ⊥ ℄			✓	Implied 90°	
6	0.81	✓	✓	✓		
7	2.50	✓	✓	✓		
8	0.88		✓	✓		
9	0.50		✓	✓		
10	1	✓	✓		Squareness implied 90°	★
11	0.12	✓	✓	✓		
12	0.125 $^{+0.003}_{-0.000}$ (2) holes	✓	✓			★
13	0.50 to 0.125 hole		✓	✓		
14	0.75 to 0.125 hole		✓	✓		
15	0.189 $^{+0.003}_{-0.001}$ (4) holes	✓	✓		Squareness implied 90°	★
16	0.06 Radius (2)	✓	✓		Implied 90° bend	
17	1.88	✓	✓			★
18	0.62	✓	✓		Possible limit stack with spec. No. 17	

Figure 13-4. *Analysis of cross-shaft bracket specifications.*

discussions between the process planner and the product engineers should firmly resolve the difference between "What Is Specified" and "What Is Wanted."

Die Engineering — Planning and Design

No.	Specifications	Between Surfaces — Depend On			Remarks	Operational Requirements
		Material[1]	Die[2]	Processing[3]		
19	3.12	✓	✓		Implied symmetrical	◀
20	12°		✓	✓	No tolerance on L	
21	1.12		✓	✓	No tolerance on flatness or parallelism	
22	1.75	✓	✓	✓		◀ — ★
23	₵ to 0.88			✓		
24	₵ to 0.50			✓		
25	1		✓	✓		
26	1.78	✓	✓	✓		◀
27	0.06 Radius (2)	✓	✓		Implied 90° bend	★
28	1010 C.R.S. ¼ H U.S.S. No. 16 Ga.	✓			Tolerance on thickness	
29	Break sharp corners 0.015 R	✓		✓	Question need	Not needed

Figure 13-4. (*Continued.*)

Operational Requirements. The basic pressed-metalworking operations such as cutting, forming, or drawing, which are used to obtain the surface relationships, must be resolved after all engineering decisions have been made on specifications which were noted in the "Remarks" column.

Operational requirements for the cross-shaft bracket, without regard to final determined sequence, are indicated by stars in the tabular analysis. The basic operations are:

1. Cut: (1 notch) . . . 0.12 inch x 1 inch (3.0 mm x 25.4 mm)
2. Cut: (1 blank) . . . 3.12 inches x 1.75 inches x 1.78 inches (79.2 mm x 44.4 mm x 45.2 mm)
3. Cut: (pierce 1 hole) . . . 0.501 inch +0.001 inch, -0.000 inch (12.72 mm +0.02 mm, -0.00 mm)
4. Cut: (pierce 2 holes) . . . 0.125 inch +0.003 inch,-0.000 inch (3.18 mm +0.08 mm, -0.00 mm)
5. Cut: (pierce 4 holes) . . . 0.189 inch +0.003 inch, -0.001 inch (4.80 mm +0.08 mm,-0.02 mm)
6. Form: (2 bends). . . 0.06 inch (1.5 mm) radius
7. Form: (2 bends). . . 0.06 inch (1.5 mm) radius

A further analysis must now be made of each listed operation in light of "Feasibility for Manufacturing" and "Economics of Tooling," previously discussed. The operational requirements (*excluding 1 and 2*) are here examined briefly, in the order listed.

3. (Pierce one hole, 0.501 inch +0.002 inch, -0.00 inch [12.73 mm =0.05 mm -0.0 mm]): No problems are apparent in producing this hole. Tolerance is close, but not impossible with properly maintained tools. The natural break from piercing will provide adequate surface in the hole to meet functional requirements.
4. (Pierce two holes, 0.125 inch +0.003 inch, -0.000 inch [3.18 mm +0.08 mm, -0.0 mm]): No problems are apparent in producing the holes. The natural break from piercing will provide adequate surface in the hole to meet functional requirements.
5. (Pierce four holes, 0.189 inch +0.003 inch, -0.001 inch [4.80 mm + 0.08 mm, -0.02 mm]): No problems are apparent in producing the holes. The product engineer objected to the suggestion of keeping all hole axes in the same plane, which might have simplified tooling and processing.
6. and 7. (Form four bends, 0.06-inch [1.5 mm] radius): These forming operations are similar in some respects and, combined, would constitute the complete forming of the part. Therefore, they evidently merit being analyzed together.

The forming operations consist of working a flat blank into a shape which will meet part print specifications (Figure 13-3), plus any decisions reached between the process planner and the product engineer.

Controllability of the blank, the metal flow and movement, and quality of finish are factors in the "Feasibility for Manufacturing" and the "Economics of Tooling" for the bracket-forming operations.

Figure 13-5 shows three possible methods of forming the flat blank to obtain the final shape of the cross-shaft bracket.

In the single-operation method shown at *A*, the developed blank would be placed in the die having a suitable locating system. In a single stroke, the metal would be worked over almost the entire surface in forming the blank to shape. In such a method, the surface on the die radius would be subject to excessive wear and, under production conditions, the part might have a distorted surface. Also, it would be difficult to control springback and the part symmetry because of variables of the material, even though the pad would hold the blank securely against the punch face. The 12-degree flange

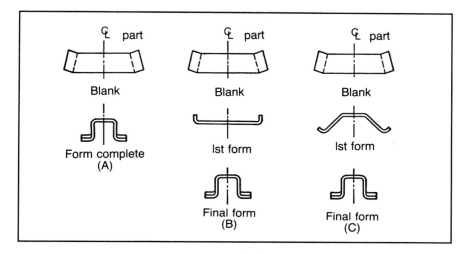

Figure 13-5. *Alternative methods of forming the cross-shaft bracket.*

on the part (Figure 13-3) would cause a localized flow of material, upon initial closing of the die, which would also distort the part.

In the two-die method shown at B of Figure 13-5, the developed blank would be flanged on the ends in one die; then the sides would be formed by a single-pad flanging die. Methods A and B both provide fair control of the part, but springback control is difficult in both methods because the flanges are parallel.

In the method shown at C, using the same blank, a single solid-type die could both form the end flanges and establish the break lines for the sides. In the final forming stage, the sides are flanged in a single-pad die. Blank control would be good, and springback in the end flanges could be compensated for by overbending. Since both the break lines are established in the same operation, with minimum metal flow and movement in the die, better dimensional control should be secured in the part.

It therefore appears to the process planner that the method shown in C of Figure 13-5 would be the most satisfactory and would permit combining the blank and pierce operations if hole location tolerances are sufficiently large.

1 and 2 (blank and notch): The general dimensions of the outline of the part were listed in the tabular analysis (Figure 13-4). In studying a part print, the process planner will determine if an initial blanking operation must be

developed to prepare a flat piece of material for the subsequent operations (i.e., a blank). The blank for the cross-shaft bracket is shown in Figure 13-6. When the blank has been developed, the process planner must immediately consider utilization of the material, grain direction as related to blank nesting, and any other pertinent factors.

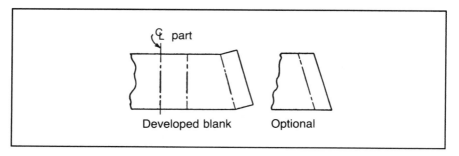

Figure 13-6. *Developed blank for the cross-shaft bracket.*

The shape of the ends of the blank requires a skeleton of material between the blanks. However, if the product engineer can be induced to decide that the "optional" blank end, shown in Figure 13-6, will not affect the part's functional or structural requirements, then both the material and the die costs can be lowered.

Grain direction will not be a problem in this plan, because the specified quarter-hard SAE 1010 cold-rolled steel will form satisfactorily either across or parallel to the grain direction.

The 0.125-inch x 1-inch (3.18-mm x 25.4-mm) notch is, in effect, a portion of the outline of the blank. There being no metal flow or movement in the notch area, the decision will be to include the notch with the blanking operation.

Planning the Operations Sequence. The major decisions on the basic operations for the cross-shaft bracket have now been made. The next step is to determine the critical specifications, and so establish the critical areas which can be used as surfaces of registry after they have been accomplished by the proper critical manufacturing operations.

Since close tolerances are an indicator of critical specifications, the process planner will first consider the 0.501-inch +0.002-inch, -0.000-inch

(12.72-mm +0.05-mm, -0.00-mm) hole and the two 0.125-inch +0.003-inch, -0.000-inch (3.18-mm +0.08-mm, -0.00-mm) holes. The four 0.189-inch +0.003-inch, -0.001-inch (4.80-mm +0.08-mm, -0.02-mm) holes might also be considered on the basis of close tolerance, except for the facts that there is metal flow between the planes of the two groups of holes and that the planes of the 0.501-inch +0.002-inch,-0.000-inch (12.72-mm +0.05-mm, -0.00-mm) hole were established as common to all flanging operations in the form of the part. Hence, the surfaces of the 0.501-inch and 0.125-inch (12.72-mm and 3.18-mm) holes are decided to be the critical areas from which may be selected the surface of registry which will constitute the locating system required.

Since a minimum of six points or surfaces of registry are required to locate the workpiece, the process planner will consider the inside surface of the top side of the cross-shaft bracket (provides three points), the 0.501-inch (12.72-mm) hole (provides two points), and the 0.125-inch (3.18-mm) hole (provides one point).

This system should be checked to determine whether the selected surfaces of registry are qualified (1) arithmetically, (2) mechanically, and (3) geometrically:

1. *Arithmetically*, the selected areas are qualified because of their close tolerances. The holes have a maximum 0.004-inch (0.10-mm) tolerance, as compared with location dimensions having plus or minus 0.010-inch or 0.020-inch (0.25-mm or 0.51-mm) tolerance.

2. *Mechanically*, the surface of the 0.501-inch (12.72-mm) hole in the YOZ and XOY planes (Figure 13-3) qualifies satisfactorily, since a 0.50-inch (12.7-mm) diameter pin is known to be structurally adequate for the two seats of registry, one in each plane. The surface of the hole will be only 30 percent of the thickness of the stock, because of the metal action in piercing the hole, but this will not disqualify the surface, since only one point in each plane is needed.

   The 0.125-inch (3.18-mm) hole qualifies for the other surface of registry in the YOZ plane, except that so small a hole does not permit a structurally adequate locating pin for a seat of registry. Therefore, it is necessary to select some other surface of registry in the YOZ plane, such as the edge of the blank. This edge is qualified because it is to be produced in the same die as will produce the holes.

   Although the 0.125-inch (3.18-mm) hole should theoretically be used, it was necessary to move the symbols of the surfaces of registry to the edges. Since a workpiece of this type would be nested, the equivalent

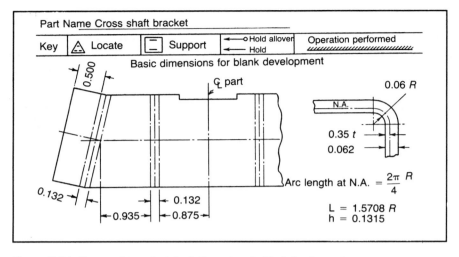

Figure 13-7A. *Process picture sheet; basic dimensions for blank development.*

of a nest is indicated in the operation diagrams (Figures 13-7C, D, and E) by a dashed "Locate" symbol.

3. *Geometrically,* the 0.501-inch (12.72-mm) hole would be qualified to provide the necessary surfaces of registry to locate the workpiece in the YOZ and XOY planes, since the inside metal surface in the ZOX plane provides the required three points.

The 0.125-inch (3.18-mm) hole is geometrically qualified in relation to the 0.501-inch (12.72-mm) hole, because it provides an adequate distance between surfaces of registry for locating purposes. However, it was previously disqualified mechanically because of small size.

Critical manufacturing operations are those required to obtain critical areas for secondary operations. The holes could be pierced first, and then used to locate for a blanking operation. However, it is obviously practical to blank and pierce in the same operation.

Since the same locating system can be used both for the forming operations and for piercing the 0.189-inch (4.80-mm) holes, this information is now passed along to the die designer in the form of process picture sheets (Figures 13-7A to E, inclusive), which supplement the usual machine and tool routing sheet.

Process picture sheets show the workpiece for each operation. Only those views are shown which are necessary to specify the surfaces of registry and any dimensions needed other than those given on the part

Die Engineering — Planning and Design

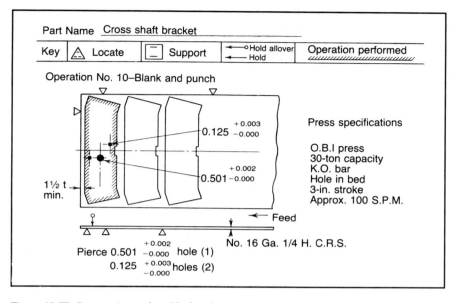

Figure 13-7B. *Process picture sheet; blank and pierce.*

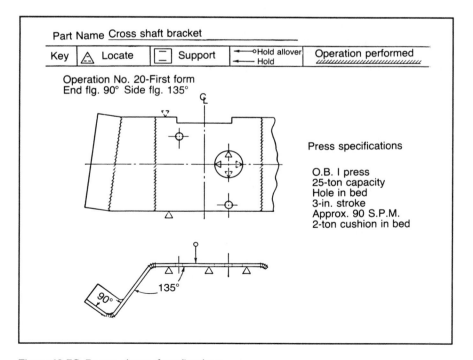

Figure 13-7C. *Process picture sheet; first form.*

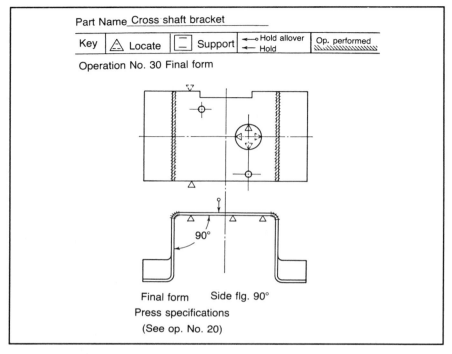

Figure 13-7D. *Process picture sheet; final form.*

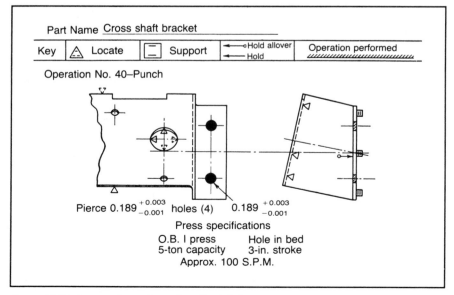

Figure 13-7E. *Process picture sheet; pierce flange holes.*

print. On each sheet showing an operation, the required press specifications are given.

The machining and tool routing sheet will indicate the gages which must be either designed or selected from commercial standard sizes.

Allied processes, such as heat treating or tumbling, were considered and determined unnecessary in producing the cross-shaft bracket.

Die Dimensions

Die-block dimensions are governed by the strength necessary to resist the cutting forces and will depend on the type and thickness of the material being cut. On very thin materials, 0.5-inch (12.7-mm) thickness should be sufficient, but except for temporary tools, finished thickness is seldom less than 0.875 inch (22.22 mm), which allows for blind screw holes and resharpening and also builds up the tool to a common shut height for quick die change considerations.

Design of Large Die Sets. Studies of stress distributions in dies weighing more than 100 tons (890 kN) have been made to determine their optimum thicknesses and width.

Nearly all the vertical stress components are contained in the stress boundary diagrammatically shown in Figure 13-8A. The stress studies were made for both edge-to-edge and circular loading. For assumed uniformly applied loads at the top of a die, the vertical component P_Y at the bottom is a maximum at the center ($x = 0$) and a minimum where $x = w/2$, as shown in Figure 13-8B.

Optimum dimensions of a large die can be determined with the use of Figure 13-9. Examples involving edge-to-edge and circular loading are as follows:

EXAMPLE 1: (circular loading) Given the circular loading on an area of 40 inches (1 m) diameter, $c = 20$ inches (0.5 m), load $P_o = 50{,}000$ psi (345 MPa), find the size of the die, so that the maximum stress transmitted to the press bed shall not exceed 30,000 psi (206 MPa).

Solution:

1. Compute ratio $P_Y/P_o = 30{,}000/50{,}000 = 0.600$ and locate on ordinate in Figure 13-9.

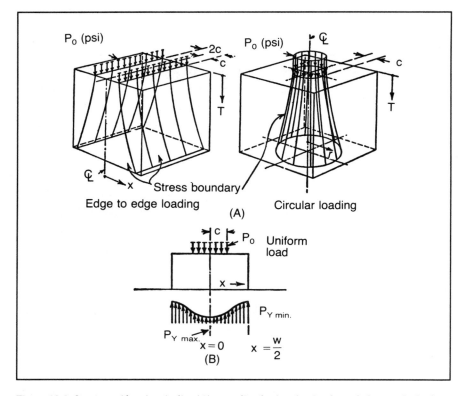

Figure 13-8. *Stress considerations in dies: (A) stress distributions for circular and edge-to-edge loading; (B) maximum and minimum vertical stress components.*

2. Follow along horizontal line from point $P_Y P_O = 0.600$ to intersect with curve P_Y (max).

3. Extend vertical line down to intersect with curve P_Y (min). The point of intersection corresponds to a value of $r/c = 1.51$ or $r = 1.51c$, $r = 151 \times 20 = 30.2$ inches (767 mm), and

   $d = 2r = 60.4$ inches (1.53 m)

   This is the required diameter or width of the die.

4. To determine the thickness of the die set, extend the vertical line farther to intersect abscissa T/c. This point of intersection corresponds to $T/c = 1.78$ or

   $T = 1.78c = 1.78 \times 20 = 35.6$ inches (904 mm)

   This is the required thickness of the die.

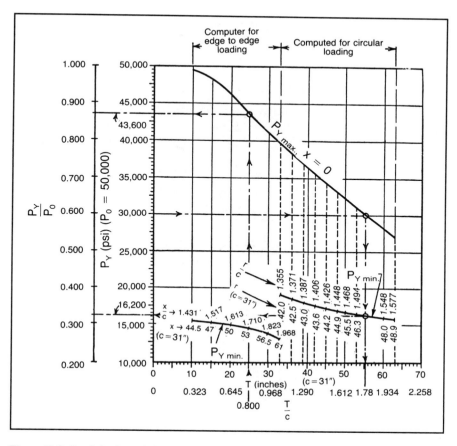

Figure 13-9. *Graph for determining optimum die proportions.*

5. From intersection with curve $P_Y(\min)$ read horizontally the ordinate value

   $P_Y/P_O = 0.324$

   or $P_Y = P_Y (\min) = 0.324 \times P_O = 16,200$ psi (112 MPa). This is equal to the minimum stress at the edge of the die being transmitted to the press bed.

EXAMPLE 2: (edge-to-edge loading) Given the loading on rectangular area of $2c \times L$ as shown in Figure 13-8, intensity of load $P_O = 50,000$ psi (345 MPa). Assume the thickness of die is limited to a value of $T/c = 0.800$ and $c = 31$ inches (787 mm). Required to find the width of the die, w, and the maximum stress P_Y transmitted to the platen below the die for these conditions.

Solution:

1. Ratio $T/c = 0.800$. Locate this point on abscissa of Figure 13-9.
2. Extend vertical line from this point upward to intersect curve P_Y (min). This point corresponds to a value of $x/c = 1.700$. Hence $x = 1.700c$ or $w = 2xc = 3.4c = 105.4$ inches (2.68 m). This is the required width of the die.
3. Extend vertical line $T/c = 0.800$ farther to intersect curve P_Y (max).
4. Read horizontally on ordinate P_Y/P_O = value equal to 0.872, or

$P_Y = P_Y$ (max) $= 0.872 \times P_O = 43{,}600$ psi (301 MPa)

This is the maximum stress transmitted to the platen for the specified loading condition. The maximum stress is at the center of the die at the plane of contact with the press platen.

14

Design Practice

Strip Development for Progressive Dies

The individual operations performed in a progressive die are usually simple, but when they are combined in several stations, the most practical and economical strip design for optimum operation of the die requires careful analysis. The sequence of operations on a strip and the details of each operation must be carefully developed to ensure that the design will produce good parts without production or maintenance problems. The following strip development sequence is applicable to both manual and computer-aided design.

Although a method for determining proper sequence of operations that would apply universally cannot be formulated, a tentative sequence should be established with the following items considered as the final sequence of operations is developed:

STEP 1. Analyze the part.
a. What is the material and thickness?
b. What hole dimensions — size and location — are critical?
c. What surfaces are critical?
d. What forms are required?
e. Where can carriers be attached?
f. Is direction of material grain important?

STEP 2. Analyze the tooling required.
a. What production is required per month, per year, total?
b. What presses are available? It is important to know bolster area, shut height, feed height, strokes per minute, air cushion, etc.
c. What safety conditions must be met?

STEP 3. Make dummy drawings. A part dummy drawing shows the finished part and all positions in which the part will be formed to produce the final form. Provide for proper metal movement for drawing or forming operations.

a. Tips for dummy drawings:
   1. Show the complete part.
   2. Show all positions necessary to form the part in plan and elevation views.
   3. Show over-bend positions if they are critical.
   4. Show all necessary views to achieve clarity.
   5. Show work lines and setup lines.
   6. Provide vertical and horizontal center lines for measuring when assembling the strip layout.
   7. Show the strip carriers, if known, before strip layout is assembled.
   8. Trace the dummy from part print, if part print is dimensionally accurate (less time and fewer errors).
   9. Use design aids:
       a. Wax for sample parts and carriers.
       b. Rubber skins.
       c. Plastic skins.
       d. Models.
   10. Check accuracy.

STEP 4. Make a strip layout. A strip layout for a progressive die is a series of part dummy drawings marked up to indicate the die operation to be performed in each station of the die. To construct a strip layout:

a. Determine the proper progression for the part.
b. Tape on a drawing board a series of prints of the plan view of the part in die position, using the horizontal center line for alignment.
c. Apply clear tape over prints to hold prints together after removal from the drawing board.
d. Mark all die operations on each station that will be performed in the die. (Use red color pencil for cutting operations and green or blue for forming operations.)
e. Mark operations directly on prints or use an overlay sheet for the preliminary layout, but mark directly on the prints for final layout to prevent something being missed during die design.

STEP 5. Review proposed process with another person to check for errors. Use a checklist when analyzing the preliminary strip layout.
a. What stations can be eliminated by combining with another station?
b. Are good die steel conditions maintained?
c. Does movement of part between stations require a stretch web?
d. Are idle stations provided to permit "breathing" of strip, if stretch web is not feasible?
e. Provide for pitch notch(es), if possible, to maintain proper progression.
f. Avoid sight stop for first hit, if possible.
g. If possible, pierce in first station and pilot in second station to establish pitch control.
h. Provide adequate pilots for all subsequent stations.
i. Is there room for stock lifters to permit free flow of strip during feed?
j. Are close tolerance holes pierced after forming to eliminate development of hole location?
k. Use an overlay sheet to run a simulated strip through the die to check each operation and to spot any loose pieces of scrap which might be left on the die.

STEP 6. Draw plan views by tracing proper part positions for each station from the strip layout.

STEP 7. Draw the plan of the punch over the plan of the die to permit tracing as much as possible and to reduce scaling errors.

STEP 8. Make views and notes to communicate properly to the die maker. (Remember: to assume is to blunder.)

STEP 9. Problem areas to watch:
a. Part lifters.
b. Part gages.
c. Part control — pilots, etc.
d. Pad travel.
e. Scrap ejection.
f. Part ejection.
g. Poor die steel conditions.
h. Will the die fit the press?

i. Will the die fit production requirements?

STEP 10. Receive preliminary design approval.

STEP 11. Finish design layout.

STEP 12. Detail as required.

Computer-aided Design and Machining

The principal advantage of computer-aided progressive die design and machining is the ability to build precision tooling in less time and at a lower cost. Integrating the part and die design process with the generation of cutter path information also greatly reduces the chance for errors.

Traditional Design Procedure. To fully understand the ramifications of being able to cut tool steel directly from an electronic drawing data base, the major faults of traditional methods need to be examined. The following sequence lists the major tasks that occur in moving a progressive die project from conception to machine:

1. An engineered part drawing is given to the die designer to begin work on the development of a new die.
2. The die designer then begins the detailed task of laying out the design of the progressive die that will in the end produce the finished part that is defined by the part drawing.
3. A detailer takes the finished die design and produces the detailed drawings of each of the die parts; e.g., punches, punch holders, die steels, etc.
4. The finished details are then passed on to an NC programmer who determines the tool path coordinates for each line, circle, and arc that defines the die part geometry.
5. The NC programmer then writes out the complete scripted program including all tool-path coordinates and any "G" or "M" codes that will be required by the machine to perform such tasks as offset cutting compensation, feed-rate settings, coolant on/off switching, etc.
6. The written program must then be typed directly into the machine controller, or into an input device that will transfer the cutting instructions to the controller or onto a medium such as paper tape that will be read by the machine controller.
7. The machine is now ready to cut the die part that will be required to manufacture the designed part from step one.

Error Sources. Close examination of each step of the die design procedure as outlined above clearly shows that there is much potential for error to be introduced into the machine instructions. During the course of steps one through seven, the original design data of the part goes through five transformations.

The first transformation occurs when the die designer transfers the dimensional data from the new part drawing to the drawing board when developing the design of the new die. It is in this transformation that most geometric errors occur. Complex shapes can make calculations of offset arcs and angular lines very difficult when designing clearances between punches and die steels. Such a miscalculation can lead to scrap parts and unworkable dies.

Transformation of the data happens again when the detailer takes information from the die design drawing and begins drafting the die part details. Even if the die design had been created flawlessly, there is still a good chance of an error occurring in the drawing of the die part, as the detailer reads the design drawing and transfers the drawing data for the second time.

The third transformation of the drawing data begins as the NC programmer starts to calculate all of the endpoint coordinates of the lines and arcs of the geometry on the die part drawings. In most NC applications accuracy can be carried out to three or four decimal places. Complex shapes, again, may make for some very difficult calculations of precise endpoint or tangency point locations. Miscalculation of a single point could mean the difference between a workable and a nonworkable die.

Transformation number four occurs when the NC programmer transfers the coordinate information from the drawing to a written program.

Error Generation. The most frequent occurrence of errors happens in step four. To understand why errors occur easily, consider the amount of data that must be entered into a CNC machine tool program just to machine 10 progressive die sections.

On each detail drawing, there are 100 points that define the tool path for the section, and each of these points must be written into the NC program. Each point consists of an X coordinate and a Y coordinate, and precision is to four decimal places. An example of a point on the drawing might be:

$$(X, Y) = (1.3574, -3.3468)$$

and the written line in the NC program might look like this:

G01 X13754 Y-33468

Notice that two numbers in the NC program have been transposed. The transposing of numbers can occur in the misreading of a point coordinate from the drawing or the miswriting of a point coordinate in the program, and the possibility of error increases with the greater number of points that have to be transferred. The number of point coordinates that must be transferred is as follows:

(10 drawings) x (100 points) x (2 coordinates) = 2000

and the total number of digits would be:

(2000 point coordinates) x (5 digits) = 10,000

To machine 10 die sections without error, 10,000 number digits must be read and written without transposing a single one.

In addition to the potential of transposing the point coordinate digits, there is still another potential for error in transformation number four. Besides the point coordinate information, the NC machine will need some other instructions to properly cut the part. These instructions are generally handled in the "G" and "M" codes written into the NC program along with the point coordinates. The codes tell the machine such things as which side of the line to offset to compensate for cutter kerf, or which direction to travel on an arc, i.e., clockwise or counterclockwise. The misplacement or misuse of one of these commands will produce erroneous cutting instructions.

For example: "G02" tells the cutter to travel along an arc in a clockwise direction and "G03" in a counterclockwise direction. If the NC programmer inadvertently used a "G02" where a "G03" should be used, the cutter would travel off the desired tool path.

The fifth and final data transformation is the keypunching of the NC program into the machine controller or any other input device that serves as the communicator of the cutting data to the NC machine. The possibility of error here is obvious as all the lines of the written NC program, including all point coordinates, machine codes, and cutting comments, have to be keyed in. Once again, errors may result from transposing digits either in reading or writing the point coordinates or by inadvertently striking the wrong key inputting.

In any one of these transformations, there is a chance for error. Some errors may be detectable before machining, others may not become evident until the part has already been cut wrong. In either case errors expend both energy and time, and may result in the production of unsalvageable die details. Reduction of the number of data transformations will greatly reduce the possibility of error.

Integrating CAD with CNC Machining. One proven solution to the reduction in data transformations is the integration of computer-aided design and CNC programming. The concept-to-machine procedure can be simplified through the use of computer-based aides. For example:

1. A CAD engineered part drawing is given to the die designer to begin work on the development of a new die.
2. The die designer electronically copies the geometry of the part drawing onto his or her die design drawing and adds any additional geometry to complete the die drawing.
3. The detailer electronically dissects the die design drawing into the necessary drawings of each of the die parts.
4. The finished CAD detail drawings of the die parts are then tool-pathed by an NC programmer.
5. An NC post-processor program on the CAD system automatically transforms the tool-path information into the NC program.
6. The NC program is then electronically uploaded to the machine controller.
7. The machine is now ready to cut the die part that will be required to manufacture the designed part from step one.

The reason one can be confident that the program in the machine is correct is that the program itself is generated from the data that originated in NC form in step one. Through all of the steps, the exact same CAD data used in the part design is used in the die design.

When the die designer specifies die clearances for punches, it is not necessary to calculate offsets for complex shapes with arcs and lines at odd angles. The only instruction that the computer requires is the offset clearance dimension.

The detailer no longer has to transfer the dimensional data of the die design to create die section drawings, but simply pulls the geometry and dimensional data off the die design drawing electronically. The detail drawing is also a byproduct of the same geometry that defines the part drawing.

Calculation of the endpoint coordinates is virtually nonexistent for the NC programmer, because the endpoints are automatically defined to the CAD system, by the presence of geometry. The NC programmer needs only to define to the CAD system the order of the coordinates he or she wants programmed and does this simply by tracing a tool path on the CAD drawing itself by selecting the geometry in the order of desired cutting. "G" and "M" codes can be added to the tool path at this time as well, and some commands will be automatically determined by the computer.

Since there is no manual writing of the program, there is very little potential for the errors that were possible with the manual programming methods. A program which can reside as part of the CAD system post-processes the NC tool path information, and generates the NC program automatically and electronically.

Finally, keypunching the data into the controller or any other input device is eliminated. The electronic NC program can be directly uploaded to the NC machine controller, without the keying in of a single point coordinate or machine code.

Cost Effectiveness. The following are actual times recorded at one company before and after the CAD and CNC integration for wire EDM machining:

	Task Description	Time
	Drawing of the die design	16 hr
	Detailing of 10 parts	20 "
BEFORE	NC point calculations	10 "
	NC programming	10 "
	NC program keypunching	10 "
		66 hr

	Task Description	Time
	Drawing of the die design	8 hr
	Detailing of 10 parts	5 "
AFTER	NC tool path defining	5 "
	NC postprocessing	1 "
	NC program uploading	1 "
		20 hr

No allowance was made for the savings resulting from machining error reduction. If records of the cost of machining errors are available, this data can be used to further justify the cost of integrating CAD with a CNC machining system. Shortening the cycle time for getting a progressive die from the design concept into production also enhances flexible manufacturing capability.

Good Design Practices

Designing for Strong Die Blocks. Figure 14-1 illustrates the use of a three-station die to avoid weak die blocks. At *A*, the pierced hole is near the edge of the part where it is cut off, thereby weakening the die block at this point. If an idle station is added so that the piercing operation is moved ahead one station, the die block is stronger and there is less chance of breaking. At *B*, the pierced holes are centered on the strip but close together. In this case the holes should be pierced in two stations to avoid thin sections in the die block between the holes. The addition of an idle station also permits a larger, more robust punch retainer to be used.

Layout for a Flanged Channel. The layout of the strip for a flanged channel shown in Figure 14-2 illustrates several of the points to be considered in preliminary layouts of progressive dies. A hole is pierced in station 1 which can be used in any succeeding stations as a pilot hole. Notching the strip with two punches in the first station and trimming the rounded edges of the part in the second station avoided the use of a delicate J-shaped punch. The first forming operation is done in station 5, with two idle stations on each side to allow the strip to drop below the level of the other stations to

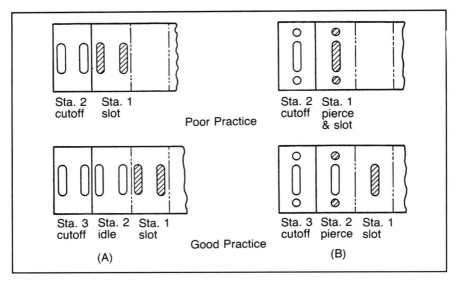

Figure 14-1. *Use of an additional station to avoid weak die blocks: (A) pierced hole close to the edge of part; (B) pierced holes close together.*

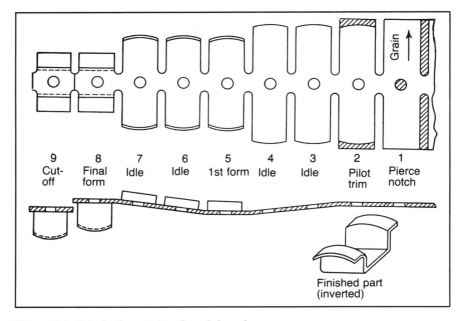

Figure 14-2. *Strip development for a flanged channel.*

form the flanges upward. Final forming is done at station 8 and a shear cutoff is achieved at station 9. The carrier strip is through the center of the strip development.

The need to have the grain across the width of the stock strip means that the die cannot be fed from a coil and that strips no longer than available coil width must be fed by hand. This layout is not recommended when reproduction requirements are high. An improvement in steel quality may permit the use of coil stock with the grain rotated 90 degrees.

Forming U-channels. The strip development for a channel-shaped part, and a section through the die, is shown in Figure 14-3. Slots are cut in station 1 of sufficient length to allow the flanges of two adjoining parts to be formed at one time in station 3. The strip is lanced and one flange partially formed at one time in station 2. The part is cut from the two side carrier strips in station 4. The use of a lance rather than a notching punch conserves material.

Part Orientation Can Affect Die Cost. The layouts of Figure 14-4, using wide and narrow stock widths, illustrate the use of oblong commercial punches as compared to special heeled punches. Commercially available piercing punches should always be considered where possible to avoid the cost of special punches. Figure 14-4A shows a part made from a wide stock strip using two special heeled punches and the same part made from a

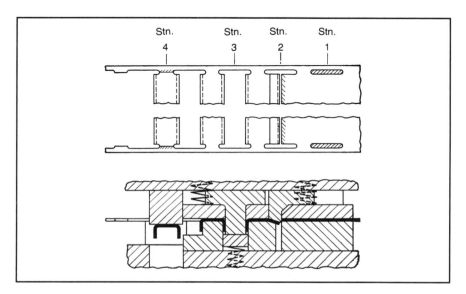

Figure 14-3. *Channel forming progressive die.*

Progressive Dies

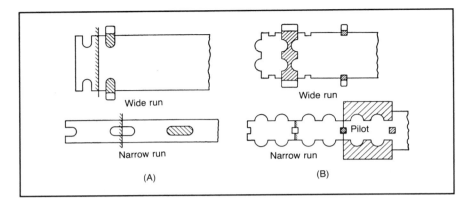

Figure 14-4. *Strip layouts for using commercial punches: (A) oval punch; (B) square punch.*

narrow strip using a purchased oval punch. In both cases a shear-type cutoff punch is used. Figure 14-4B shows a wide strip using two heeled rectangular notching punches and a slug-type trim and cutoff punch. The narrow strip development for the same part uses a square piercing punch, heeled trimming punches, and a shear-type cutoff punch. Costs of various stock widths and the greater number of pieces that can be produced between coil changes may outweigh the savings of using commercially available punches.

Multi-operation Layout. A layout of the sequence of operations for producing an actuator bracket is shown in Figure 14-5. In the first station the center hole is pierced. The hole is used to pilot the strip for each successive operation. In the second station, a circular stretch flange is formed around this hole to part print dimensions, and the strip is notched to a partial outline of the left leg. In the third station similar notching for the right leg and cutting of three slots in the left leg occurs. In station 4, similar slots are cut out and the central portion, as well as the completed outline of the left leg. The right leg outline is completely cut out and the central portion is trimmed to length at station 5. The bulges in the sides are formed up and tabs on the center section are formed down in the sixth station. In station 7 the part is severed from the strip, the carrier tab on the left side is cut off, and the legs are formed downward.

The die for this part is made of high-carbon, high-chrome steel with hardened backup plates for the punches and die blocks. An air-operated ejector removes the part from the form block. The die operates at a speed of 75 strokes per minute producing about 150,000 parts per sharpening.

Design Practice

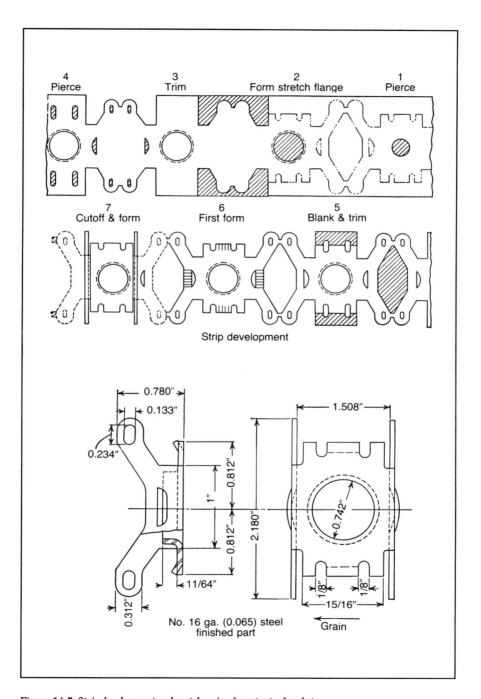

Figure 14-5. *Strip development and part drawing for actuator bracket.*

289

Progressive Dies

Punch Back into Carrier Strip. The nature of the forming or the size or shape of a part sometimes requires that the blank be completely severed from the strip before forming. The blank is then pushed back into the strip so that it can be advanced properly to the succeeding stations. A strip development of this type is shown in Figure 14-6. This is sometimes referred to as a cut-and-carry operation.

The strip is pierced and notched in the first two stations and the blank is severed from the strip in station 3; then, by the action of a spring-loaded pressure pad, it is pushed back into the strip. In station 4, the blank is spanked to flatten and secure it into the strip. The first forming is done at station 5, and finish forming and removal from the strip are achieved at station 7. Station 6 is idle to add strength to the die. In addition to the piloting hole, a notch is cut along the edge in the first station. The purpose of the notch is to engage a locator which operates a limit switch. The switch is connected to the electrical circuit controlling the press clutch to prevent press operation if the strip is improperly positioned.

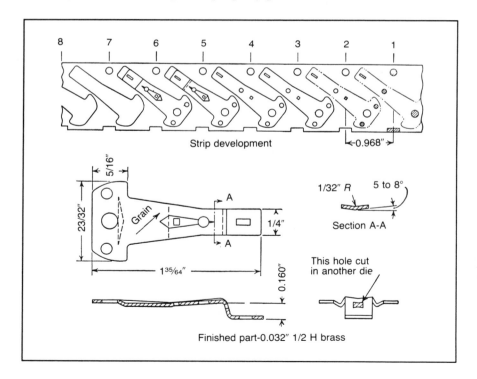

Figure 14-6. *Strip development with a push-back blank.*

The strip development and part drawing for a camera diaphragm plate are shown in Figure 14-7. The stock is 0.020-inch- (0.51-mm-) thick by 1.875-inch- (47.62-mm-) wide 6061-T6 aluminum. The strip development allows the periphery of the part to be trimmed in several steps to simplify punch shapes and also to prevent concentration of stresses which would tend to

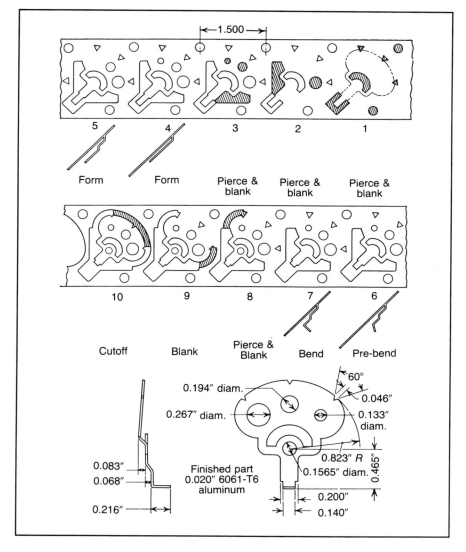

Figure 14-7. *Strip development for a die to pierce, form, and trim a diaphragm plate.*

warp the workpiece. This arrangement facilitates fabrication of the die sections. The dies for the triangular-shaped cutouts are round inserts made in two pieces for ease and accuracy of grinding.

Avoiding Pitch Growth. During normal die operations, die components are deflected and compressed as defined by the laws of physics. For example, a 1.0-inch (25.4-mm) steel cube subjected to a force of 14.5 tons (129 kN) will compress 0.001 inch (0.02 mm), or one part per thousand. This is based on a modulus of elasticity of 29,000 ksi (200 MPa). The sides of the cube will expand 0.0003 inch (0.008 mm) based upon a Poisson's ratio of 0.3.

This size change of steel die sections when subjected to pressure can cause a troublesome growth of pitch. Figure 14-8 shows how pitch growth can be lessened by providing slots for expansion between the stations.

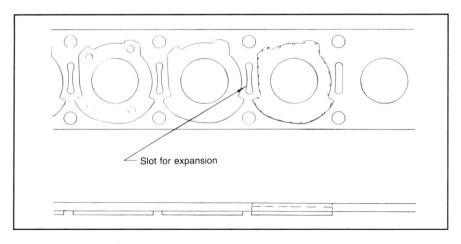

Figure 14-8. *Providing a slot for expansion to avoid pitch growth when cutting heavy stock.*

Good General Practices. The following are good progressive die design practices.
1. Pierce piloting holes and pitch notches in the first station. Other holes may also be pierced that will not be affected by subsequent noncutting operations.
2. Use idle stations to avoid crowding punches and die blocks together. Idle stations also permit the use of larger die blocks and punch retainers. An added advantage is that future engineering changes can be incorporated at low cost.

3. Use solid spring-loaded stock guide rails rather than spool lifters where possible.
4. Plan the forming or drawing operations in either an upward or a downward direction, whichever will assure the best die design and strip movement.
5. The shape of the finished part may dictate that the cutoff operation should precede the last noncutting operation.
6. Locate cutting and forming areas to provide uniform loading of the press slide. If this is not practical, and the press is large enough to permit the die to be offset, determine the required offset and have instructions to the diesetter placed on the die.
7. Design the strip so that the scrap and part can be ejected without interference. The best way to eject the part is to cut it off and drop it through the lower die shoe.

Pilots

Since pilot breakage can result in the production of inaccurate parts and jamming or breaking of die elements, pilots should be made of good tool steel, heat-treated for maximum toughness and to hardness of Rockwell C57 to 60.

Retaining Pilots. Pilots are usually retained by conventional heads as well as standard ball lock retainers. Several other methods are shown in Figure 14-9. A threaded shank, shown at view A, is recommended for high-speed dies; thread length X and counterbore Y must be sufficient to allow for punch sharpening. For holes 0.75 inch (19 mm) in diameter or larger, the pilot may be held by a socket-head screw, shown at B; dimensions X and Y again must provide for sharpening. A typical press-fit type is shown at C. Press-fit pilots, which may drop out of the punch holder, are not recommended for high-speed dies but are often used in low-speed dies. Pilots of less than 0.25 inch (6.4 mm) diameter may be headed and secured by a socket setscrew, as shown at D.

Indirect Pilots. Designs of pilots which enter holes in the scrap skeleton are shown in Figure 14-10. A headed design, at A, is satisfactory for piloting in holes of 0.2 inch (5 mm) diameter or less. A quill is used to support the pilot shown at B.

Spring-loaded pilots are desirable for stock exceeding 0.062 inch (1.57 mm) in thickness. A bushed shouldered design is shown in Figure 14-10C.

Progressive Dies

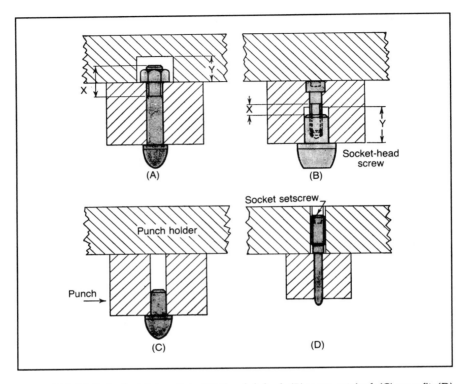

Figure 14-9. *Methods of retaining pilots: (A) threaded shank; (B) screw- retained; (C) press-fit; (D) socket setscrew.*

A slender pilot of drill rod shown at D is locked in a bushed quill, which is countersunk to fit the peened head of the pilot.

Tapered slug-clearance holes through the die and lower shoe should be provided, since indirect pilots generally pierce the strip during a misfeed.

Combined Piloting and Stock Lifting. Figure 14-11A illustrates how a pilot guide bushing can be used to support the stock at the feed level to aid pilot entry. Provision should be made for slug clearance as shown at B in the event of a misfeed.

Electrical Die Protection. Figure 14-12 illustrates the use of an electrical limit switch to stop the press in the event that the spring-loaded pilot (D1) and limit switch (D2) detect a misfeed. For this system to be successful, the press must stop very quickly, which limits speeds at which this method is applicable. The pilot should be as long as possible to provide extra stopping time.

Design Practice

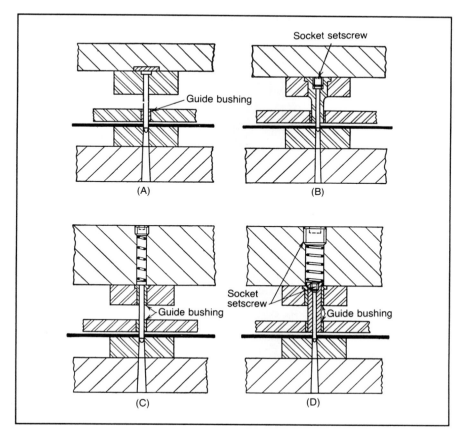

Figure 14-10. *Indirect pilots: (A) headed; (B) quilled; (C) spring-backed; (D) spring-loaded quilled.*

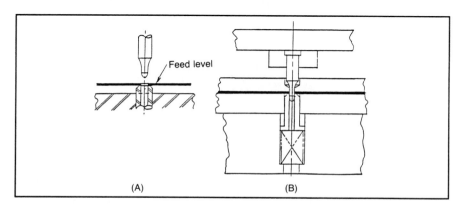

Figure 14-11. *Using a spring-loaded pilot bushing to support stock: (A) stock supported at feed level; (B) view showing spring and slug clearance.*

295

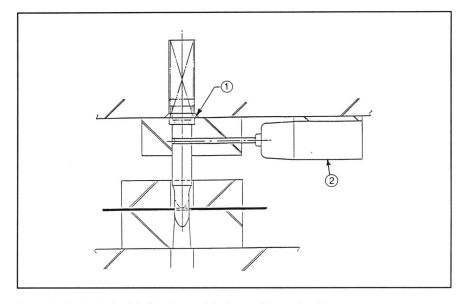

Figure 14-12. *Spring loaded pilot trips a switch when stock is out of position.*

Miscellaneous Piloting Methods. The pilot need not always be affixed to the upper die shoe. Figure 14-13 illustrates a pilot (*D*1) attached to the lower die shoe. A spring ejector (*D*5) assists in lifting the stock off the pilot on the ascent of the ram. Pilots may also be attached to pressure pads, provided the pads are precisely guided. A pilot (*D*2), attached to a pad (*D*3) which is guided (*D*4) by pins and bushings is much stronger than a longer pilot attached to the upper die shoe.

Stops

Solid blocks are often used with final blanking and cutoff operations to position the end of the stock or workpiece. Solid stops may be incorporated in the final cutoff station of a progressive die to gage one end of a completed part which is attached to the advancing strip. The part is then severed to the desired dimensions by a cutoff punch. The finished part drops through the die.

Another design incorporates a cutoff punch having a heel which bears against the vertical surface of a stop, gaging the part and confining it (attached to the strip) to the station. The punch prevents the part from cocking while being sheared.

Design Practice

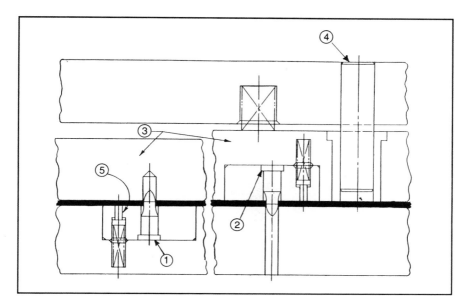

Figure 14-13. *Pilots mounted to the lower die shoe and guided pad.*

Pin Stops. One design incorporates a small shouldered pin, pressed in the die block, to engage an edge of blanked-out portions of manually fed strip stock. Since the operator must force the stock over the shoulder to secure a desired feed length, this stop is not suitable for high production or high speed dies.

Another design that may be used with a double-action blank-and-draw die, or a die in which very little or no scrap is left between blanked-out areas in strip stock, also has a small pin pressed in the die block, but the pin has a sharp edge which faces the incoming stock strip. The edge cuts through and thrusts aside the thin fins of stock left between successive blanked-out openings as the strip is fed into the die. The pin functions as a stop when it engages the trailing edge of a blanked out area before, and only before, the next succeeding area is cut out.

Latch Stops. A latch pivoting on a pin, as shown in Figure 14-14, is held down by a spring. The latch is lifted by the scrap bridge and falls into the blanked area as the stock is fed manually into the die. It is then necessary to pull the stock back until the latch bears against the scrap bridge. This design is suitable for low production only.

Progressive Dies

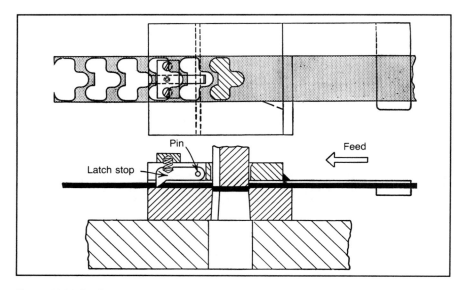

Figure 14-14. *Latch stop.*

Starting Stops. A starting stop, used to position stock as it is initially fed to a die, is shown in Figure 14-15A. Mounted on the stripper plate, it incorporates a latch which is pushed inward by the operator until its shoulder (D1) contacts the stripper plate. The latch is held in to engage the edge of the incoming stock; the first die operation is completed, and the latch is released. The stop will not be used again until a new strip is fed to the die.

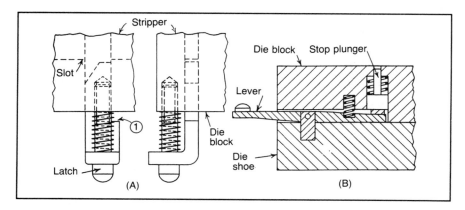

Figure 14-15. *Starting stops.*

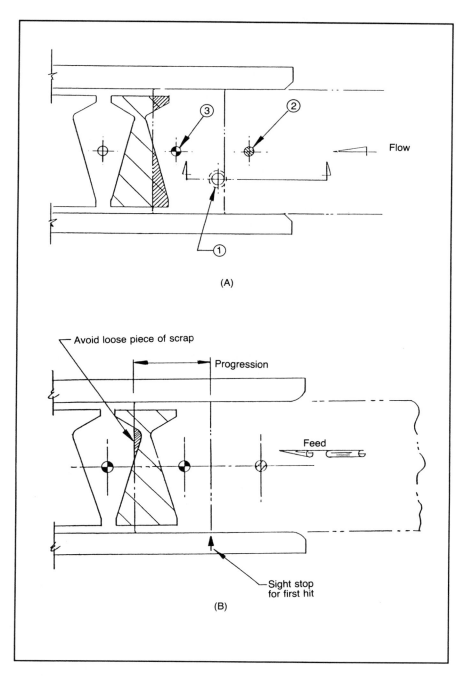

Figure 14-16. *The location of the first hit is important: (A) the first hit does not produce loose pieces of scrap; (B) the production of a loose piece of scrap.*

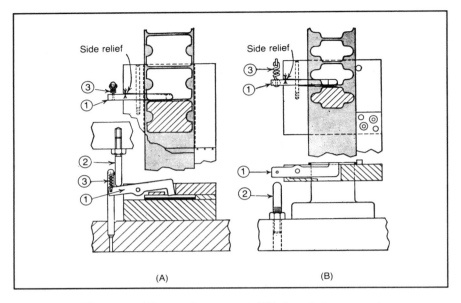

Figure 14-17. *Trigger stops: (A) top stock engagement; (B) bottom stock engagement.*

The starting stop shown at view B, mounted between the die shoe and die block, upwardly actuates a stop plunger to position incoming stock. Compression springs return the manually operated lever after the first die operation is completed.

It is important to plan the starting sequence of a strip or coil to avoid damage. In some cases when a partial cut is made, excessive punch deflection caused by an unbalanced load results in a sheared die section. Figure 14-16A illustrates an acceptable starting sequence. The stock is first positioned against a stop pin (D1) and the press cycled to pierce the pilot hole (D2). The second hit does not produce loose pieces of scrap. The production of a loose piece of scrap is shown at (B).

Trigger Stops. Trigger stops incorporating pivoted latches (D1, Figure 14-17A and B) at the ram's descent are moved out of the blanked-out stock area by actuating pins (D2). On the ascent of the ram, springs D3 control the lateral movement of the latch (equal to the side relief) which rides on the surface of the advancing stock and drops into the blanked area to rest against the cut edge of cutout area.

Stops for Double Runs. The stops shown in Figure 14-18 were designed for double runs in the same direction. The stock is turned over for the second

Design Practice

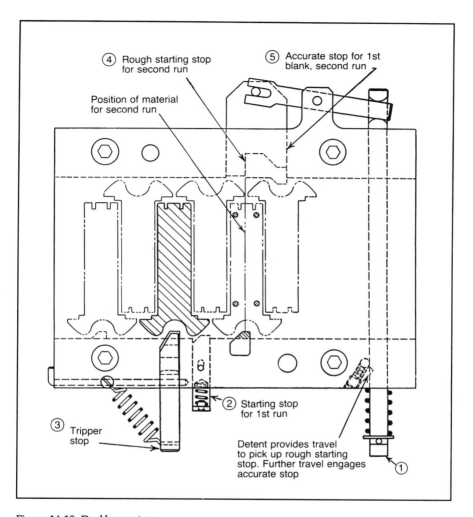

Figure 14-18. *Double run stops.*

run. A conventional automatic trigger stop (D3) functions continuously after the starting stop (D2) for the first run is pushed in to engage the notch in the strip. The rough starting stop (D4) for the second run and accurate stop (D5) are actuated by handle D1 for the cutting of the first blank of the second run; the automatic trigger stop then functions for the remainder of the run.

Another stop design for double runs in the same direction is shown in Figure 14-19, in which flat spring-returned pin stops (D1) are actuated by a

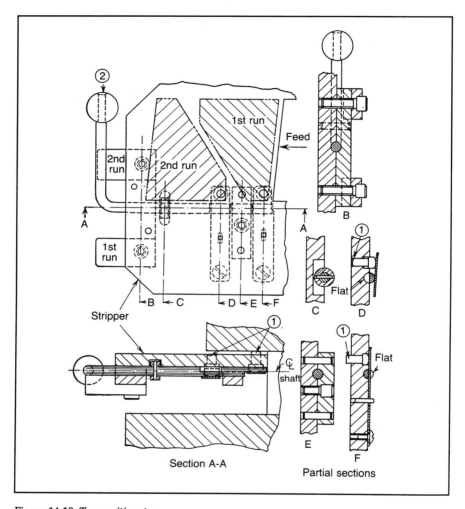

Figure 14-19. *Two-position stop.*

rod terminating in handle D2, which is swung to either of the two positions for corresponding stock runs.

Trim Stops. Trimming or notching stops bear against edges previously cut out of strip edges. These stops are also known as pitch stops. A trimming punch cuts the strip to the exact width desired and to a length equal to the feed distance or pitch. The punch length is slightly greater than the feed advance so that no scrap can remain attached to the strip to impede proper stock travel.

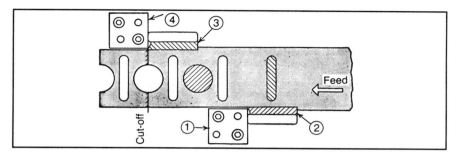

Figure 14-20. *Trim stop.*

Double trimming stops are shown in Figure 14-20; stop D1 is a starting stop as well as a running stop. It bears against the trimmed edge cut by the first notching or trimming punch (D2).

The second punch (D3) trims the stock to width and provides a cut stock edge for the second stop (D4) to ensure proper stock positioning for severing the workpiece from the scrap skeleton.

Preferred Pitch Notch Design. In cases where a notching punch is combined with the punch used to cut a pitch notch, there is a chance that the scrap may lift with the punch and jam the incoming strip. Figure 14-21A illustrates the preferred method of accomplishing the combined operations in a single station. A poor method is illustrated at B.

If the condition illustrated at B cannot be avoided, the portion of the punch that cuts the pitch notch should be sharpened as shown at C to curl the slug and break the oil film suction.

Providing for Strip Removal. A solid pitch stop block will not permit the progression strip to be withdrawn from the end of the die where the parts exit. In some cases, the strip cannot be withdrawn from the other end either because the partly formed parts jam in the stock guides. A swing-away stop (Figure 14-22) permits the strip to be withdrawn from the part discharge end of the die.

Pitch Stop Limit Switch. The pitch stop shown in Figure 14-23 pivots a small amount to permit actuation of a limit switch when the feed advance is completed. It is important to provide a long beveled lead on the edge of the stop to avoid shearing the pitch notching punch when the press is cycled in the event of a short feed or no feed.

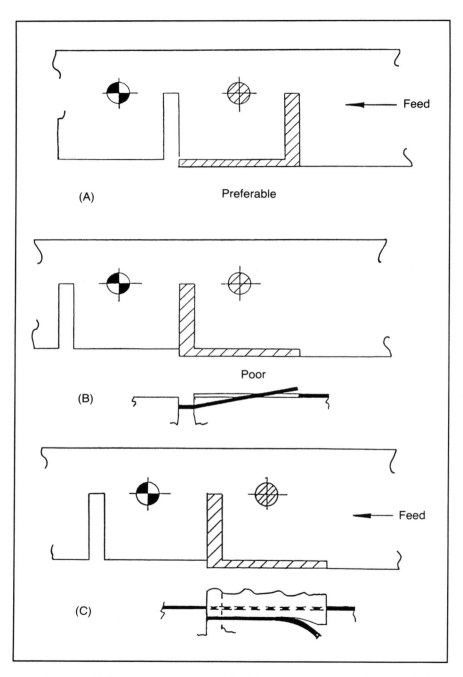

Figure 14-21. *Avoid slug tipping when cutting pitch notch: (A) preferred method; (B) poor method; (C) curved surface on punch used to break oil suction.*

Design Practice

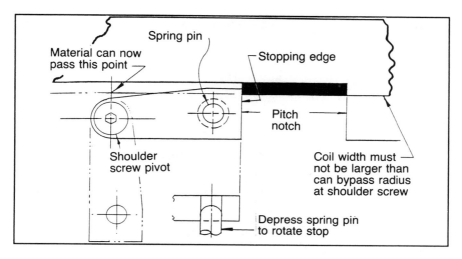

Figure 14-22. *Swing-away stop to facilitate strip removal.*

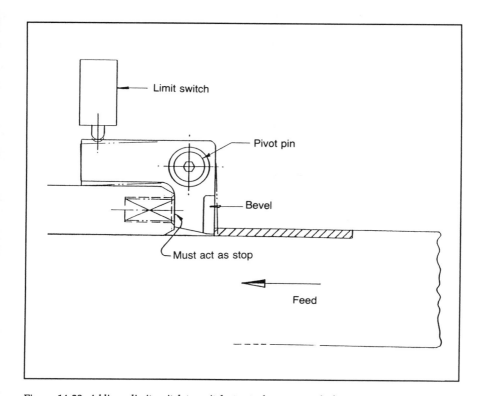

Figure 14-23. *Adding a limit switch to a pitch stop to detect proper feed.*

Progressive Dies

Detecting Both Under- and Overfeed. The stop shown in Figure 14-24 has several advantages when compared to a conventional pitch stop. The small semicircular cutout on the edge of the strip saves stock, and both under- and overfeeds are detected. The stop acts as a detent to retain the stock in position. Another advantage to this type of protection system is that a misfeed is detected much sooner than is possible with a pilot-actuated limit switch.

The usual practice with a die protection system of this type is to stop the press during that portion of the stroke between completion of forward feed until just after feeder pilot release. This will serve to detect both proper forward feed and stock slippage upon release by the feeder.

Since a partial cut is made by the notching punch, some method to heel up the punch is required.

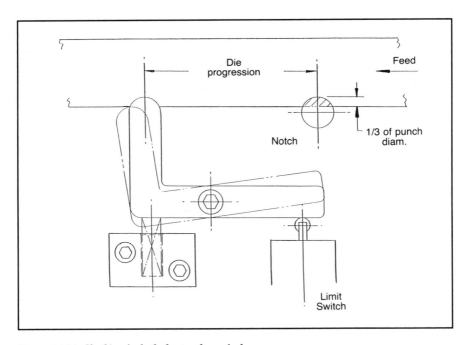

Figure 14-24. *Checking for both short and over feed.*

15

EDM and Progressive Dies

Introduction

Of the many techniques in the field of metalworking — chemical milling, ultrasonic machining, etc. — none bears the unique relationship to progressive dies that electrical discharge machining (EDM) does. Progressive dies are being built to tolerances and at speeds unknown just a few short years ago. In progressive die design, cutting stages which are absolutely necessary in conventionally built progressive dies can be omitted in an EDM-built die. This is because the electrical discharge process permits die design which is impossible from the standpoint of mechanical manufacture.

The Concept of EDM

EDM is similar to short circuiting a spark plug with a screwdriver. Those who have done this will recall the deep pitmark burned into the screwdriver. If it is used often enough, it is conceivable that its entire tip will soon be eroded away. This is electroerosion of steel by electricity and it represents EDM in its simplest form.

That a spark of electricity, in striking a piece of steel, will vaporize the area of contact is the basis of electrical discharge machining.

Machines which utilize this principle send a high-frequency shower of sparks into a piece of conductive material. These sparks can erode as much as 20 cubic inches (312 cubic cm) of metal per hour depending on the machine used. This amount of metal is small compared to the machining ability of a milling machine or lathe, but it is more than adequate when one considers the purpose of EDM. This technique is used for cutting complex contours and configurations in tungsten carbide, the superalloys, and hardened steel and does it without any of the workpiece distortion normally associated with high temperatures.

EDM Cycle

Figure 15-1 is a schematic diagram of the basic electric circuit of EDM. The electrode — source of the sparks — is marked with a minus sign. It can be positioned manually but in operation it is actuated by a servo mechanism. The workpiece — the part to be machined — is represented by a positive sign.

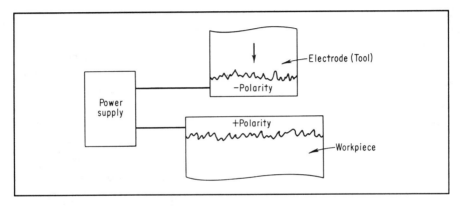

Figure 15-1. *Schematic diagram of an electrical discharge circuit.*

When current is applied and the electrode fed toward the workpiece, a point is soon reached where the intervening air ionizes. (Ionization is caused by the electrical pressure at the electrode.) When this happens, an electronic avalanche — a spark — crosses the ionized air gap.

This spark, almost infinitesimal in size, has an enormously high temperature. Upon striking the workpiece, it heats the area of contact to a point well above the temperature of vaporization. The spot thus contacted swirls away as a cloud of metallic vapor and a machining cycle is complete.

In practice, a hydrocarbon oil rather than air is used as a spark carrier. This oil serves a dual purpose in that it keeps the workpiece cool and provides a flushing action in the work area. Like air, the oil must ionize, and the ease with which it ionizes is called its dielectric strength. Various types of oil, each with a different dielectric strength, are used in EDM. In some cases distilled water is used. However, in tool and die applications, a standard oil much like kerosene is used on all applications.

Chip Formation

Although the dielectric coolant used in EDM keeps the workpiece at room temperature, it has absolutely no cooling effect on the spark or upon the point of spark contact. The area of contact is vaporized and swirls up through the coolant as a vaporous cloud. However, the natural cohesive forces in the metallic cloud draw it into a tight little ball, at which point the cooling action of the dielectric begins to take place. The outer portion of the ball quickly solidifies, although its center is still in an expanded gaseous-molten state. As the center of the ball cools, it adheres to the already solidified outer wall, creating a hollow sphere. This is the chip produced by EDM.

In this machining cycle a spark vaporizes an area of metal which is converted into a hollow sphere and washed away. As described, the process may sound slow, cumbersome, and involved. However, consider that the process:

1. Can be rigidly controlled dimensionally;
2. Can be repeated from 10,000 to 1,000,000 times in one second in a conventional electrical discharge machine.

Overcut

It is erroneous to infer that an electrical discharge machine throws out a stream of sparks which is directed like a stream of water from a hose.

First of all a spark will not jump very far through either air or a dielectric fluid. Generally 0.0035 inch (0.089 mm) is the maximum distance that a spark can traverse in either medium. As an electrode is fed to a workpiece, the distance the spark jumps is dependent on the applied voltage. (Voltage in electrical circuitry is pressure.) For illustration, assume that an M-500 machine operating at full capacity (75 volts) is being used. In this case, gap or side clearance will be 0.0035 inch (0.089 mm), which means that when the electrode approaches within 0.0035 inch (0.089 mm) of the workpiece, the first spark will jump. Thereafter sparks continue to jump between the two closest points, always at a distance of 0.0035 inch (0.089 mm). If a 0.500-inch (12.7-mm) dowel pin is used as an electrode (Figure 15-2), it will soon create a perfectly round hole in the workpiece. If the operation continues, it will soon erode a hole completely through the workpiece. The size of the hole will be 0.507 inch (12.88 mm).

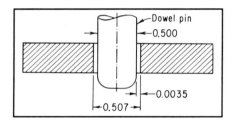

Figure 15-2. *Dowel pin used to machine a hole in a plate. This drawing illustrates the phenomenon of overcut.*

Sparks do not emanate from the end of the electrode only. They travel the shortest distance possible. Thus, as the EDM operation starts, a spark jumps from the end of the dowel pin only, because the point closest to the workpiece is at the end of the dowel pin. As the pin sinks into the workpiece, the sparks start coming from the sides. Because the machine is preset to a 0.0035-inch (0.089-mm) gap, the hole will be 0.0035 inch larger — on each side — than the electrode. This is defined as "overcut" and in die work it is equated to die clearance.

This is at the heart of the theory of EDM. Some overcut is essential. Because of overcut, the hole is always larger than the electrode. Further, the electrode never contacts the workpiece. When the EDM operation is finished, there is a uniform dimensional clearance between the work and the electrode.

It should be noted that although a dowel pin is referred to as an electrode, EDM equipment is not used to machine dowel holes. Whether or not EDM is utilized in building a die, such operations as doweling are performed conventionally.

Overcut is a controlled variable in EDM. Its dimensional value can be established at any point between 0.0005 inch and 0.0035 inch (0.127 mm and 0.089 mm) simply by turning a dial.

The Punch as an Electrode

The preceding example in which a dowel pin was utilized as an electrode presented the basic principle of EDM. It is an accurate theoretical picture, but it omits one variable found in actual practice — the wear on the electrode. As an electrode performs a machining operation, it also erodes, although generally at a slower rate than the workpiece. Because a finished punch can be used as an electrode in creating a die section contour, this wear on the punch must be considered.

Compensating for this wear is simple: the punch is made longer than necessary and cut off to size after the EDM operation has been completed.

Thus, if die design indicates that a punch 0.75 inch (19.1 mm) long is required, and preliminary estimates indicate electrode erosion of 1.0 inch (25.4 mm), the diemaker makes his punch 1.75 inches (44.5 mm) long.

The efficiency of this type of operation can be easily seen. Unlike the dowel pin, most punches have complex contours which take a great deal of machining time. In addition to machining time in the die section, maintenance of die clearance is difficult. This is especially true in dies with complex contours and small clearance. By use of the finished punch as an electrode, tedious filing or spotting operations can be dispensed with. Die clearance is obtained by controlling the overcut — a dialable machine function.

Wear Ratios

Measurement of electrode erosion, which is to say the amount the punch will wear in machining a die opening, is accomplished by a study of wear ratio. As such, there are three types of wear ratio the designer must consider before designing a die. First in importance is volumetric wear ratio.

Volumetric wear is defined as the ratio of the metal machined to the volume of electrode eroded. It is the wear ratio always referred to when "wear ratio" is spoken of and is dependent upon:

1. The type of power supply unit used,
2. The characteristics of both the electrode and the work,
3. The amount of current used, and
4. The type of coolant.

The type of power supply is important. For example, if a tube-type power supply is used when a brass electrode is machining steel, the wear ratio will be 3:1. However, the same setup powered by a resistor-condenser type of power supply unit will result in a 1:1 wear ratio.

Material characteristics (in the workpiece) which influence wear ratio are (1) melting point and (2) conductivity. Because the machining operation involves the melting and vaporization of metal, it is obvious that the easier the metal melts, the more readily it will machine.

Low electrical conductivity in the workpiece also increases machinability, whereas high conductivity in the electrode is desirable as a means of preventing wear.

Two other types of wear ratio important in spark machining are corner wear ratio and end wear ratio. These ratios are measured linearly and can be defined as follows:

Corner wear ratio is the ratio of the thickness of the metal pierced to the corner length of electrode required to pierce through.

End wear ratio is the ratio of the depth of workpiece cavity (or depth of through hole) to the end length of electrode consumed in producing that cavity or hole.

Surface Finish

Another consideration important to diemakers is surface finish, and here again electrical discharge machining produces exceptional results. Top-quality finish is mandatory in cavity work, but in progressive die work EDM is used primarily for machining cutting areas. In these cases surface finish does not assume the importance it would have in machining forging dies, for example. However, two machining cuts are sometimes taken even in the blanking areas. For this reason it is important to consider surface finish as a machine function.

Surface finish is a function of cyclic frequency, because the volume of metal machined is a function of amperage.

The amount of metal machined by EDM is dependent on amperage alone; voltage is important only as it relates to overcut or machine stability. Assume that in one hour's time at A amperage a volume of metal equal to V is machined. If in the next hour $2A$ or twice as much amperage is used, then $2V$ or twice the volume of metal will be machined. Should amperage be tripled, then volume of metal removed will also be tripled. This is always true regardless of voltage or frequency. Metal removed is directly proportional to amperage.

Now consider the effect of the frequency of the current. If a machine makes 10 discharges per second, 10 pitmarks will result. If the frequency is doubled to 20 discharges per second — with no change in amperage — 20 pitmarks will result (Figure 15-3). However, the size of each pitmark at 20 discharges will be equal to one-half the size of each pitmark at 10 discharges per second. This is because the volume of metal removed per second does not change as long as the amperage does not change. Clearly, if the frequency were increased to 100 cycles per second (cps), the same amount

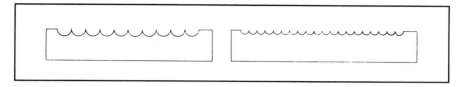

Figure 15-3. *Effects of frequency on surface finish. The higher the frequency, the smoother the finish.*

of metal would be removed, but from 100 pitmarks. Each pitmark being 10 times smaller than in the first example, surface finish would be 10 times smoother.

Therefore three fundamental rules emerge:

1. Overcut and surface finish are both related to machining current and machining frequency. As current increases for the same frequency, overcut increases and surface finish becomes worse. As frequency is increased for the same machining current, overcut decreases and surface finish becomes better.
2. Metal removal rate or volume of metal machined for a particular type of power supply is a function only of machining current or amperage. As amperage increases, the rate of metal removal increases in direct proportion.
3. The electrode wear is a function of the electrode material used, the workpiece material, the coolant, and the particular power supply used.

Advantage of EDM in the Diemaking Trade

It is important for die designers to learn where and when the electrical discharge process is applicable. This can best be done by constantly studying the process itself. Because the process continues to be refined, new applications and techniques are constantly emerging. In many cases EDM techniques which are new today will be obsolete tomorrow. For this reason, a specific EDM machine should be regarded as an evolutionary tool which will be superseded by more advanced units in the future.

It is noteworthy, however, that although this technique is evolutionary, the economies it now effects in diemaking make it a factor of contemporary importance. It can be said without exaggeration that no blanking die can be built as inexpensively with conventional techniques as it can with EDM.

In addition to the economy inherent in EDM, the increased accuracy of the technique simply cannot be matched by traditional diemaking practice. Clearance is uniform and because of this uniformity, production runs are greatly extended.

EDM provides other advantages as well:
- Dies need not be sectionalized;
- Die sections are heat-treated before the EDM operation;
- Complicated strippers can be easily machined;
- Die design is, in many cases, simplified.

Disadvantages of EDM are few, but chief among them are tolerance limitations and inferior surface finish. These criticisms are made principally by lamination die builders who specialize in dies with duplicate sections. Because of the almost unbelievably tight tolerances maintained by some of these manufacturers, EDM is as yet impractical in their applications. Normally EDM is not accurate beyond 0.0002 inch (0.005 mm).

As for surface finish, EDM finishes produced by the highest possible frequencies are still not fine enough for many lamination die applications. Again, because this technique is still in a state of development, surface finishes can be expected to improve. The practical upper limit of frequency is now around 130,000 cycles per second (cps). In research laboratories, EDM has been performed at frequencies as high as 2 to 3 million cps. However, until laboratory frequencies of this order become shop realities, lamination diemakers will need to use lapping operations on spark-machined surfaces.

A final disadvantage of the technique is the slight angle produced in a cutting edge (Figure 15-4). For practical purposes this angle is negligible, but some in the trade find it objectionable merely because it exists. Curiously, a closing angle is regarded by others as a desirable feature. It holds slugs and it compensates for die wear caused by slug passage. However, many who object to the angle find EDM a practical and necessary tool. Their solution is to turn the die sections over and machine

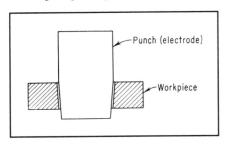

Figure 15-4. *Negative angle caused by electrical discharge machining. For practical purposes this angle is negligible.*

from the underside. This, of course, reverses the angle but it does not use EDM to its full advantage.

Die Plates

A basic difference between EDM and conventional die design is that heavy die sections are replaced with comparatively thinner die plates. The die plates are backed up by heavy backing plates which may or may not be hardened, depending on the production run of the die. Many designers specify "Ohio Knife" for backing plates because of its toughness and general ability to withstand the pounding inherent in a blanking operation. An example of this design is illustrated in Figure 15-5, which is the cross section of the part in Figure 15-6. The solid (unsectionalized) plate is screwed and doweled to the backing plate as shown, and the backing plate is screwed and doweled to the shoe. The advantage of this design comes out of a description of the die itself.

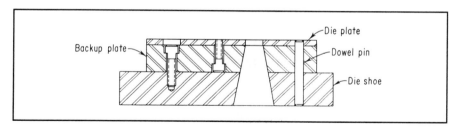

Figure 15-5. *Cross section of the cutting stage in an EDM progressive die.*

This is the blanking stage of a high-production automotive progressive die. It works on a 24-hour-per-day schedule, five and sometimes six days a week, cutting metal 0.162 inch (4.11 mm) thick. Under this workload, die sections are exhausted in several weeks' time, which means that replacement of cutting sections is a never-ending job. Before adaptation of the die to an EDM design, four die sections were used to create the blank outline (Figure 15-6). They are, of course, interchangeable. Replacement of these sections was expensive and time-consuming because of their size and the complexity of their contour. However, with an EDM design, sectionalizing is unnecessary. The contour is machined in the die plate which is then taken direct from the EDM unit to the die.

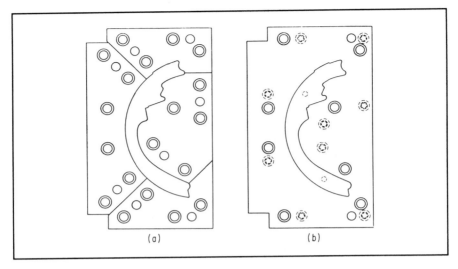

Figure 15-6. *Conventional (a) and EDM (b) methods of design. Principal advantage of EDM in this example is the elimination of thick, sectionalized die sections.*

Diemaking

One important consideration in making these replacement die plates is that the punch cannot be used as an electrode. This is because clearance (on one side) between punch and die for 0.162-inch (4.11-mm) stock must be at least 0.008 inch (0.20 mm). But, as noted, the maximum overcut possible on precision machines is 0.0035 inch (0.89 mm). Therefore, if the punch were used as an electrode, clearance would be 0.0045 inch (0.114 mm) too small.

To successfully spark machine the opening in the die plate, an electrode must be made. The manufacturer in this case uses a brass electrode machined to within 0.0035 inch (0.89 mm) of the die opening size. This electrode is then mounted and doweled on a holder at known dimensions from the edge of the holder (Figure 15-7).

The die plate is made by first machining the plate to overall size plus grind stock, then jig boring screw and dowel holes and two construction holes in what will be the opening. The dowel holes should be reamed from 0.005 to 0.010 inch (0.13 to 0.25 mm) undersize to provide grind stock for jig grinding. It is not absolutely necessary to jig bore the screw holes, however. If their locations are center drilled on the jig borer, they can be finished more efficiently on a drill press.

After all holes have been made, a template is used to scribe off the cutting contour. This is accomplished as follows. The template has two jig-bored holes which correspond to the two construction holes in the die plate. The plate can be colored with layout fluid or, preferably, blue vitriol, after which the template is doweled in place. The cutting contour can then be scribed from the template.

When the contour has been scribed, it can be sawed in the conventional manner with approximately 0.031 inch (0.79 mm) left on the line. The die plate is then ready for heat treatment. After hardening, the plate is ground top and bottom. Next the edges are ground, after which the dowel holes are jig ground. At this point the plate is complete except for the cutting contour, which is finished by spark machining.

EDM Setup

The setup used to spark machine the die plate is made by setting the ground edges of the plate in relation to the edges of the electrode holder, as shown in Figure 15-7. The electrode is doweled to the holder at a known distance from the edges of the holder. To complete the setup, it is necessary only to determine the reference dimensions and to set the workpiece accordingly. The electrode will then be directly over the sawed opening and EDM can begin. The electrode is lowered manually, the servo feed is cut in, and machining of the contour is completed automatically. The setup is typical of the technique used to produce duplicate sections. It is also practical for producing "one-shot" die plates which will last for the useful life of the die.

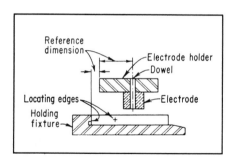

Figure 15-7. *Method of mounting the die plate in relation to the electrode. All such setups are made from matched edges.*

A Second Type of Setup

Another setup technique involves first mounting the finished punch in the upper die. The hardened but unfinished die plate is then mounted in position in the lower die.

The entire die is then mounted in the electrical discharge machine and the die plate is spark-machined. This technique provides an efficient method of building a die because the die plate is not merely machined; it is machined in place with perfect clearance between it and the punch.

This process can also be used when a large number of piercing punches are mounted in the upper die. It is of particular value when die plates are made of tungsten carbide and is, in fact, the only method whereby a multiplicity of holes can be produced in a solid carbide plate.

In such setups coolant must be supplied through the backup plate. This is accomplished by drilling a hole through the plate and forcing coolant in under pressure (Figure 15-8).

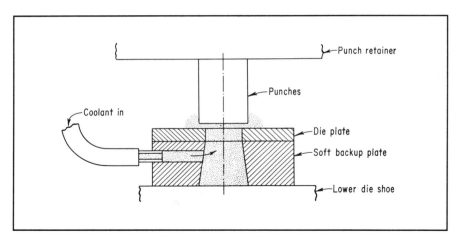

Figure 15-8. *An adequate supply of coolant is necessary. Because of small gap (overcut) it must be supplied under pressure.*

Another consideration in this setup is proper insulation of the upper die from the lower die. A review of the setup will undoubtedly raise the question of why electricity does not travel through the guide pins and short-circuit the EDM operation. In fact, a short circuit *will* take place unless one of three steps is taken.

The first — and best — solution is the use of die sets with ball sleeves working between pins and bushings. Once upper and lower dies have been securely mounted in the EDM machine, the sleeves can be removed and all circuitry broken except at the point of spark machining.

Another way to prevent short circuiting is through the use of a die set with removable guide-pin bushings. Many companies now specify such die sets for conventionally built dies as well as for those designed for EDM. In these sets, guide-pin bushings can be inserted and withdrawn manually and held in place with toe clamps. To use this type for EDM, the set need only be mounted in the machine and the bushings removed to sever effectively the electrical connection. It is also possible to purchase insulated sets made especially for EDM.

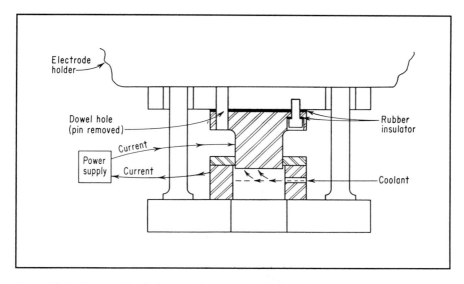

Figure 15-9. *Diagram of insulating procedure required when standard uninsulated die sets are used.*

If a standard die set is used — an unusual practice in EDM diemaking — it is necessary to insulate each punch retainer and punch section from the die shoe. This can be done with sheet rubber, removal of dowel pins, and undercut cap screws seated on rubber washers (Figure 15-9).

The electrical lead is then placed on the punch and spark machining can begin. However, this is a clumsy setup and should be avoided in preference to the first two methods specified. Also, the tightness of pin and bushing in a conventional die set inhibits the "hunting" action of the servo feed.

Hunting is not a desirable aspect of EDM. It is most apt to occur when coolant in the work area is contaminated. It consists of an up-and-down motion of the electrode as the servo strives to maintain correct spark gap. However undesirable, if conditions dictate that the servo must hunt, it should be free to do so without any drag from the pins in the bushings.

Fewer Progressive Stations

Once acquainted with the diemaking possibilities inherent in EDM, the designer will find that a more efficient layout of cutting operations can be made with EDM than is possible with older diemaking techniques. In addition to this advantage, EDM places tungsten carbide at the designer's disposal as an inexpensively machined material. A third important factor is that the sequence of cutting operations can be shortened with a resultant gain in die efficiency.

As an example, Figure 15-10 shows the blanking area of two progressive strips, both of which produce identical parts. In the conventionally designed strip, the contour is blanked in two stages; in the EDM design, it is accomplished in one. The reason for two stages in the conventional design lies in the punch design (Figure 15-11). It is obviously impossible to finish grind this punch. Therefore, a total of five punches is used to cut this contour in two stages.

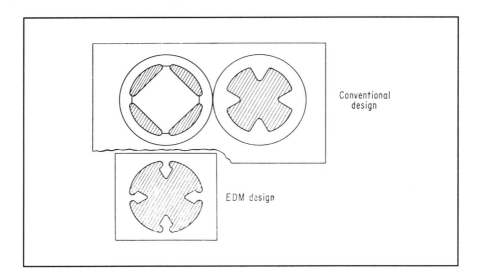

Figure 15-10. *Design simplification resulting from use of EDM.*

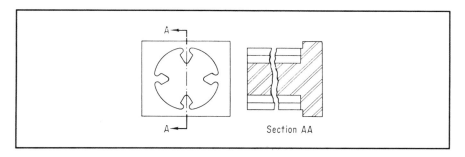

Figure 15-11. *Punch required for die design of Figure 15-10. Punches of this design cannot be machined by conventional techniques.*

Creating this contour in one stage by spark machining, however, is a simplified task performed as follows: First, a brass electrode is machined to correct size (this involves adapting an end mill to the recessed shape. In some cases it requires manufacture of a special cutter). When the brass electrode has been prepared, it is used to spark machine the required die opening in the die plate.

The die plate must be heavier than finish size because it is now used to machine the punch. The additional thickness is to provide stock for cathodic erosion. This procedure is, of course, the reverse of standard EDM practice. The polarity of the machine is reversed and the die plate which was machined as a workpiece now becomes an electrode. The punch to be machined is doweled in position, and when EDM is complete, the punch is ready for tryout. This die plate, of course, must be ground to remove the effects of electrode erosion.

Wire-Cut EDM

Production of progressive die sections by means of wire-cut electrical discharge machines (EDM) has eliminated much of the tedious surface and jig grinder work required to build most progressive dies. Figure 15-12 illustrates the principle of operation of a wire EDM machine.

Brass wire ranging from 0.004 inch to 0.012 inch (0.10 mm to 0.30 mm) in diameter is fed from a spool and through a series of pulleys which tension and guide the wire. After passing through the workpiece, several pulleys, which guide and pull the wire, feed the wire to the takeup spool.

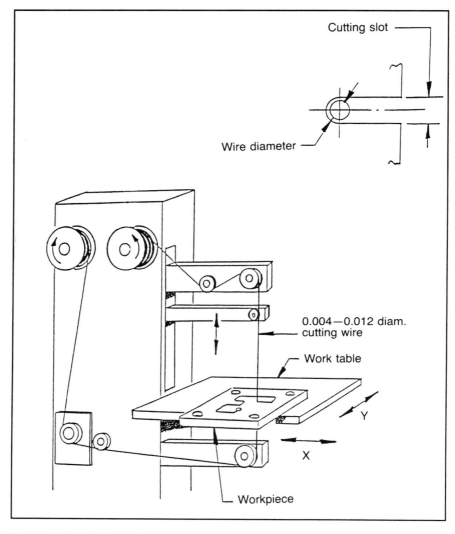

Figure 15-12. *Wire electrical discharge machining, showing machine and workpiece and cutting wire and resulting kerf.*

An electrical energy source (not shown) maintains an electrical potential difference between the wire and the workpiece. A reservoir, pump, and filter system supply dielectric fluid to the gap between the wire and the workpiece. The electrical potential difference between the wire and workpiece causes a breakdown of the dielectric resulting in multiple sparks which erode both the wire and the workpiece.

Figure 15-12 illustrates how the wire cuts a kerf in the workpiece. A 0.004-inch (0.10-mm) wire will leave a kerf approximately 0.006 inch (0.15 mm) wide. Larger diameter wire will cut more rapidly than smaller wire and leave a wider kerf. A wire 0.12 inch (3.0 mm) in diameter will leave a kerf approximately 0.018 inch (0.46 mm) in diameter.

The worktable is driven in the X and Y axis by a numerically controlled servo system. The wire path can also be tilted automatically to provide tapered relief. In addition to the wire path information needed to produce the desired die detail, the servo system also controls the rate of feed based on cutting conditions sensed by the control system.

Wire EDM Speed and Accuracy. A large die section can take in excess of eight hours to cut. The main factor is not how complicated the wire path is, but rather the length of cut and thickness of the section.

The usual practice is for one specially trained operator to set up and run a number of machines. Some shops operate these machines unattended during the night shift.

Should the wire break, some models will rethread the wire automatically, and automatic cutting of pilot holes is available on some machines. In the event of a malfunction the machine is unable to repair, an annunciator system summons the operator — by telephone or pager in some cases.

Accuracies of ±0.0002 inch (±0.005 mm) are routinely achieved, with higher accuracy attainable. A factor in the accuracy of the finished die detail is the amount of residual stress present in the blank prior to cutting. This stress is partly relieved when the blank is cut resulting in movement. This problem can be largely overcome by taking a roughing cut followed by a finishing cut.

As a general rule, tool steel types and heat-treating processes that result in the least size change during heat treatment are best suited for the production of accurate sections by wire cut EDM.

Examples of Wire-cut Construction. Figure 15-13 shows how the wire is started within the slug area (D1), and a small rectangular test cut made to check machine operation. The dowel holes are also wire-cut as shown at D2 to achieve complete interchangeability of replacement steels. Reference lines (D3) are used to establish dimensions from the die cutting path to ensure that future duplications of steels have accurate dowel locations.

Progressive Dies

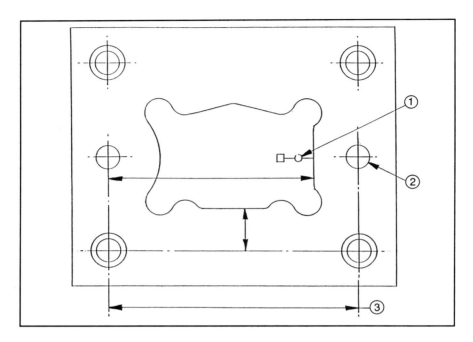

Figure 15-13. *Typical wire-cut EDM die steel with wire cut dowel holes to facilitate interchangeability.*

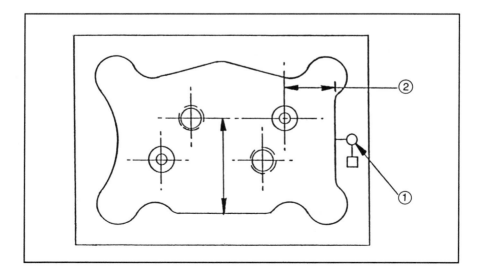

Figure 15-14. *Typical wire-cut EDM punch from rectangular block with wire cut dowel holes.*

A similar technique is used for the production of a punch steel as is illustrated in Figure 15-14. The wire is started and a test cut is made (D1). The dowels are wire cut on standard locations (D2) to provide future interchangeability. The screw holes can also be wire cut, but the counterbores will require conventional EDM machining.

Wire starting holes can either be cut by a small accessory conventional EDM electrode or drilled prior to hardening. Screw holes with proper clearances can be drilled on location prior to hardening if reasonable care is exercised.

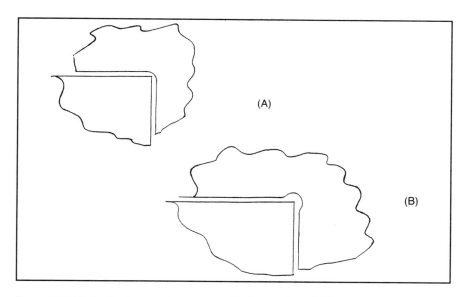

Figure 15-15. *Extra die clearance at a corner provides increased punch life.*

An advantage to EDM machining the detail completely out of a solid hardened block is that replacement parts can be produced without delays for heat treating.

Dealing with Corner Radii. Just as a perfectly square inside corner cannot be ground with a surface grinder because of normal wheel breakdown, a square corner cannot be cut with a round wire. Figure 15-15A illustrates how this radius in the die steel causes a lack of proper clearance in the corner which results in rapid breakdown of the corner of the punch.

325

One solution would be to stone a small radius on the punch steel, but this will not produce a square corner in the part.

An easy solution to the problem of obtaining a square corner is shown at Figure 15-15B. This simple wire path deviation requires very little additional time and provides the extra die clearance needed to reduce the load on the punch corner.

16

Progressive Die Mathematics

A knowledge of basic shop mathematics is as essential a tool as the micrometer in modern die shops. This is especially true in shops that specialize in progressive dies.

In addition to fundamental trigonometry, diemakers should familiarize themselves with the so-called setups peculiar to the trade. Many of these have already been described in the preceding chapters of this book. Others will be treated in this chapter, along with certain fundamentals of layout practice as it applies to individual die sections.

Trigonometric Functions

Trigonometric functions are ratios expressing a relationship between the sides of a 90-degree triangle and its angles (Figure 16-1). The most familiar one is the sine. Diemakers may or may not understand trigonometry, but they do understand and use the sine plate as an everyday tool. The sine plate is merely a mechanical expression of a trigonometric function.

The following functions are included to provide a better understanding of this tool as well as the other trigonometric functions.

$$\text{sine of angle } A = \frac{a}{c} = \frac{\text{side opposite}}{\text{hypotenuse}}$$

$$\text{cosine of angle } A = \frac{b}{c} = \frac{\text{side adjacent}}{\text{hypotenuse}}$$

$$\text{tangent of angle } A = \frac{a}{b} = \frac{\text{side opposite}}{\text{side adjacent}}$$

$$\text{cotangent of angle } A = \frac{b}{a} = \frac{\text{side adjacent}}{\text{side opposite}}$$

The definitions of these functions can be abbreviated as follows:

$$\sin A = \frac{a}{c} \quad \cos A = \frac{b}{c} \quad \tan A = \frac{a}{b} \quad \cot A = \frac{b}{a}$$

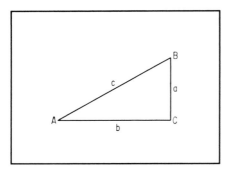

Figure 16-1. *Basic right triangle.*

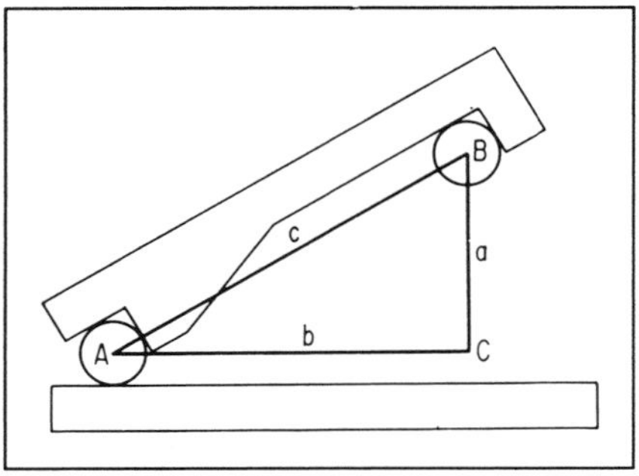

Figure 16-2. *The basic triangle applied to a sine plate.*

It is important to remember that the values of these functions cannot change: the sine of 30 degrees, for example, is always equal to 0.500. Should the sides of a triangle be measured in millimeters or miles the value of the side opposite divided by the hypotenuse remains the same. As such, 0.500 — the sine of 30 degrees — does not represent a fraction of a unit: it merely expresses a quotient. The values of the functions of all angles have been computed and can be found in the various handbooks published for the trade. Student designers and apprentice diemakers should obtain tables of natural functions for their toolboxes.

Problem 1. Using a table of natural functions, find the values* of

(a) sin 30°   (e) sin 45°   (i) sin 60°

(b) cos 30°   (f) cos 45°   (j) cos 60°

(c) tan 30°   (g) tan 45°   (k) tan 60°

(d) cot 30°   (h) cot 45°   (l) cot 60°

*See page 351 for answers.

Problem 2. From the table find the value of x in degrees when

(a) $\sin x = 0.4540$ (d) $\cot x = 1.520$

(b) $\cos x = 0.8746$ (e) $\sin x = 0.8107$

(c) $\tan x = 0.6249$ (f) $\cot x = 0.4913$

(g) If the sine of 30° equals 0.500, will the sine of 60° equal 1.000?

Problem 3. Figure 16-1 represents a 90-degree triangle whose hypotenuse equals 10. Compute the lengths of a and b when

(a) $\angle A = 30°$ (c) $\angle A = 19° 30'$

(b) $\angle A = 15°$

Hint: $\sin 30° = 0.5000 = \dfrac{a}{10}$

Sine Plates

A sine plate represents a triangle (Figure 16-2). The hinged plate to which the rolls are attached is a hypotenuse. By inserting a gage block combination under one roll, the plate can be set to a prescribed angle.

Although tables have been prepared that give the correct gageblock setting for a 5-inch (127-mm) sine plate, the diemaker should be able to set the plate from a table of natural functions. This is accomplished as follows:

$$\sin A = \frac{a}{c} = \frac{\text{gage-block setting}}{5}$$

$5 (\sin A) =$ gage-block setting

Use of this equation reduces sine-plate setting to the arithmetic task of multiplying the sine of the desired angle by the distance between the rolls.

Problem 4. A die-shop foreman selects a large piece of steel with which to make a sine plate for milling operations. He mounts the rolls 22 inches

(559 mm) apart (between contours). Assuming he wants to set the plate to an angle of 30° 15', how high must he raise the free end?

Problem 5. A diemaker is asked to duplicate a section with an angular face. To determine its angle, he places it on a 5-inch (127-mm) sine plate and indicates its face until it is level. He finds that 1.994 inches (50.65 mm) in gage blocks is required to level the section. What is the angle of the face?

Sine Setups

Diemakers are often required to make angular milling setups on large pieces of steel. These setups occur more often in line die manufacture than they do in progressive work. However, a sound knowledge of sine setups can be extremely valuable to progressive diemakers.

An example of a part requiring a sine setup is shown in Figure 16-3. This is a form pad from a heavy-duty automotive line die — a part much too large for the average shop sine plate. It is necessary, however, that the 32-degree angle be machined to close tolerance. To accomplish this, two hardened and ground sine rolls (Figure 16-4) are mounted at either end of the pad. After the rolls have been screwed to the shoes and are made parallel, they are doweled in place. The distance between roll centers is then measured and used as the hypotenuse of a right triangle. What-

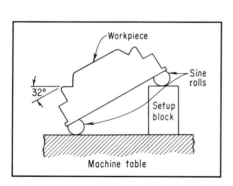

Figure 16-3. *Sine rolls applied to a large form pad.*

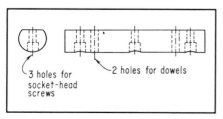

Figure 16-4. *Design of a sine roll.*

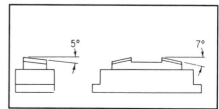

Figure 16-5. *Form pad with compound angles.*

ever this distance, it is multiplied by the sine of 32 degrees. This result is the amount one end should be raised to place the pad on a 32-degree angle.

Problem 6. Assuming that sine rolls have been mounted on a pad 25.613 inches (650.57 mm) between centers, how high must one roll be raised to machine an angle of 13° 49'?

Compound Sine Rolls

When compound angles are required on die pads and heavy sections, two compound sine rolls are often the best solution. The major drawback to a compound sine roll is that it is good for one job only. Further, the rolls are difficult to build and set. However when compound angles are required on heavy sections, compound sine rolls are often well worth the investment required in time, money, and material.

In the example shown in Figure 16-5, two surfaces angle out and forward. The outward angle is seven degrees and the forward angle is five degrees.

To mill this section, it is necessary to make the sine rolls at a five-degree angle (Figure 16-6). They are then mounted on the pad and the distance between roll centers is measured. This dimension multiplied by the sine of seven degrees provides a setting for the sine rolls.

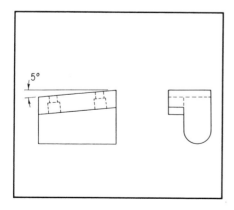

Figure 16-6. *Compound sine-roll design for machining pad in Figure 16-5.*

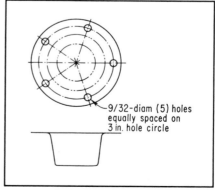

Figure 16-7. *Typical hole-circle design. Diemaker must establish hole coordinates.*

Progressive Dies

Problem 7. Compound sine rolls are required for a pad whose surfaces must be machined 13 degrees out and 3 degrees forward.

(a) If the rolls are ground on a 6-inch (152-mm) sine plate, what setting must be used?

(b) If the rolls are mounted at 40.200 inches (1021 mm) between centers, to what height must a roll be elevated to obtain the desired compound angle?

Hole Circles

Many designers use the so-called hole circle in specifying hole locations. This is convenient because product designers will often use such dimensioning devices on part drawings. An example of this is shown in Figure 16-7 in which a drawn panel receives five equally spaced 0.28-inch- (7.1-mm-) diameter holes. This is good practice, but someone — usually the diemaker — must eventually determine the exact dimensional location of each hole. In any event, the ability to translate hole-circle dimensions into coordinate dimensions is a prime requirement of diemaker and designer alike.

In Figure 16-8 the plan view of the pierce station is shown, along with the trigonometric setups for determining the dimensions. Overall block sizes are 4.5 inches by 4.5 inches (114 mm by 114 mm) with the center lines of the station coincident with the center lines of the block.

To determine hole locations, it is first necessary to find the angle between holes. In this instance there are five equally spaced holes. Because there are 360 degrees in a complete circle, the angle between holes is 360/5, or 72 degrees. Therefore, the angle β is 72 degrees. The angle between holes 2 and 1 is also 72 degrees. To determine angle α it is necessary to subtract 144 degrees (2 x 72 degrees) from 180 degrees. Thus, angle α equals 36 degrees.

Problem 8. Find the values of OA, AB, OD, and OC which will enable you to dimension this block completely.

Angular Layout

Normally, progressive dies are laid out with the axes of the panel parallel to the axes of the die. Exceptions to this type of layout occur whenever part design indicates that economies in stock can be effected by an angular

Progressive Die Mathematics

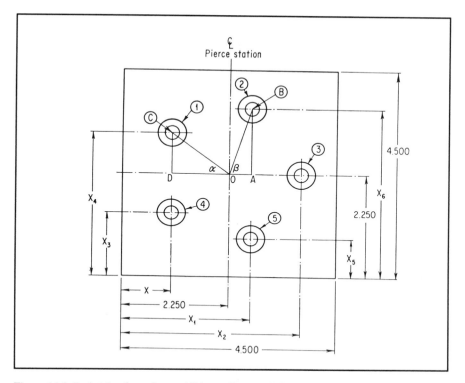

Figure 16-8. *Basic triangle used to establish coordinates of holes in preceding example.*

layout. In such cases, the die sections will generally follow an angular layout corresponding to the angular layout of the strip. (This is true if forming stages are involved. If no formation occurs in the part, die blocks will be set "square.")

Whenever a progressive die is built on the "bias," trigonometry is an indispensable aid to the diemaker. An example of this is shown in Figure 16-9 in which six sections are mounted in a pocket 8.000 inches (203 mm) wide. Sections 1, 2, and 3 are identical in size. Sections 5 and 6 are also identical, but section 4 is unique.

An interesting aspect of this layout is that all section sizes are entirely dependent upon the progression of the die. In the more standardized layouts there is often a certain amount of give and take at section lines. These can vary (except at cutting edges) as long as the progression remains undisturbed. However, such is not the case in interlocking angular designs. When diemakers establish the progression, they place a mathematical

Progressive Dies

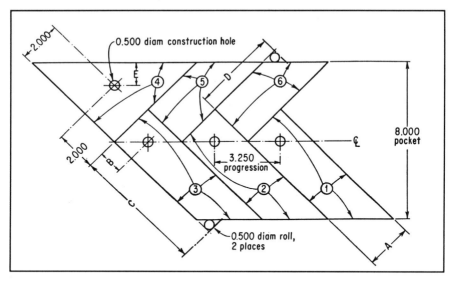

Figure 16-9. *Progressive die built to an angular layout.*

restriction on the sizes of all details — at least in the area where they interlock. The immediate concern, of course, is the calculation of their sizes.

It should be noted that the 10-times-scale layout procedure outlined in Chapter 6 does not lend itself to work of this type. Enlarged layouts are excellent for determining part sizes and development work in general. Section sizes, however, must be determined mathematically if an accumulation of error is to be avoided.

Problem 9. The progression of the die illustrated in Figure 16-9 is 3.250 inches (82.6 mm). Pilot holes lie on the center line of the die and on the center lines of their respective blocks. The milled pocket is 8.000 inches (203 mm) wide. Sections 1, 2, 3, 5, and 6 are identical in width, converging at the die center line at an angle of 45 degrees. In section 4, a 0.500-inch- (12.7-mm-) diameter construction hole is arbitrarily dimensioned as shown. Find the values of A, B, C, D, and E.

Cam Action

The subject of cam mathematics provides another application of basic trigonometry to diemaking practice. This is readily seen in the triangle shown in Fig 16-10. A driver and a cam are shown in two positions of advance. Line D represents the distance the driver has moved vertically to advance the cam a distance C horizontally. As such, cam travel is a function of the tangent of the driving angle. The driving angle is defined as the angle between the cam face and vertical. As shown in Figure 16-10, this is angle α. Therefore

$$\tan \alpha = \frac{c}{d}$$

or Tangent of driving angle = $\dfrac{\text{cam movement}}{\text{driver movement}}$

and Cam movement = driver movement × tangent of driving angle

and Driver movement = $\dfrac{\text{cam movement}}{\text{tangent of driving angle}}$

Problem 10. The driving angle of a pierce cam is 40 degrees. The stripper cams at its side have driving angles of 35 degrees. How much farther than the stripper cams will the pierce cam travel in a 1-inch (25-mm) closure of the die?

Problem 11. An engineering change in a progressive die requires widening a cam-form stage. After widening, the cam travels 0.034 inch (0.86 mm) too far. Two solutions are possible: the 40-degree driver can be either reset or reground. How much must be ground off the face of the driver to correct the 0.034-inch excessive advance?*

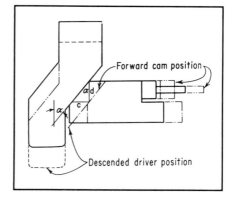

Figure 16-10. *Basic triangle used to solve problems of cam advance.*

Angular Bend Lines

An important consideration sometimes overlooked in progressive die design is the geometry of bend lines. For example, if it is necessary to form metal over an angular bend line while maintaining edge lines in two parallel planes, the edges of the blank cannot be straight lines. In other words, if

Hint: Preceding formulas do not apply. This correction is a function of the cosine.

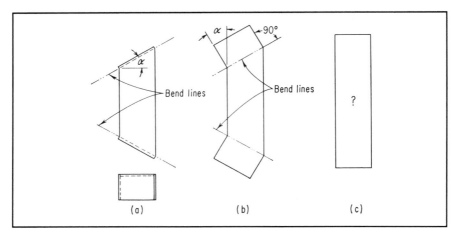

Figure 16-11. *Finished part (a) can only be made by blank (b). The blank frequently envisioned is shown at (c).*

edges after formation are to be square, the flat blank will angle at the bend line. This is seen in Figure 16-11 where the finished part is shown at *a*, the correct blank at *b*, and the incorrect blank as it is often designed at *c*. With a little study and experimentation with a piece of paper, the deviations caused by bend lines which are angular to the axis of the part will become readily evident.

The fact that the blank angles are equal to the bend line itself is illustrated at *a* and *b*, wherein α is the angle of the bend line at *a* and the angle in the blank at *b*.

In cases where an edge, after bending, is not perpendicular to the bend line (Figure 16-12), its angle must be added or subtracted from the flange angle as follows:

> To determine the angle in the flat, add the included angle to the bend angle.

Here again is a wordy rule which may be difficult to memorize. But it should be studied until its implications are obvious. This can be accomplished by further study of Figure 16-12.

Problem 12. A piece of metal is flanged over a 10-degree break line. After flanging, the included angle must be 80 degrees. What angle should be made in the developed blank to create this flange line?

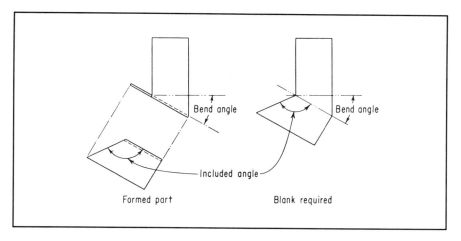

Figure 16-12. *Formed panel and blank required to make it.*

Radial Bend Lines

Parts requiring radial flange lines frequently occur in progressive die work. Two examples of this type of stamping are shown in Figure 16-13. This type of die work presents its own style of mathematical problems.

First of all, when a center is offset, problems similar to those of angular bend lines occur: the flat blank has angular sides. Again, unless the designer is aware of this, his blanking stage may be "off" far enough to require redesigning. What is worse, an error here may affect his predetermined progression, with the result that sections subsequent to the blank area are also scraped. Generally, when a designer miscalculates on angular or radial bend lines, the shoes, punches, and pilots are the only salvageable components in the entire die.

Determination of the angle of the sides of the flange is relatively simple. First, the center line of the stamping as it appears in Figure 16-13a is drawn. Another line, drawn between the center of curvature and the point where the center line intersects the line of curvature, provides an angle equal to the angles of the side. Finding this angle is another exercise in basic trigonometry.

Problem 13. If, in Figure 16-13, the radius of curvature is 1.250 inches (31.8 mm) and it is removed from the part center line by 0.250 inch (6.4 mm), what is the value of angle α which is the angle of the sides?

Progressive Dies

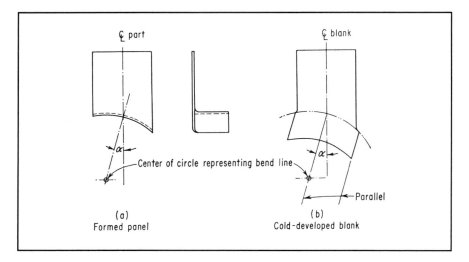

Figure 16-13. *Part flanged over a radial bend line (left). Developed blank is shown at right.*

The diemaker in most instances will develop the exact edges of the part in the blanking stage. However, it is vitally important that the designer make a reasonably accurate layout of this stage.

Whenever a progressive die is designed to form a flange over a radial bend line, two preforming stages as well as a final form stage are required. A restrike stage is required if there is a secondary flange, as shown in Figure 16-14. The first preform stage establishes the radial bend line and puts curvature in the flange.* The second forms the flange complete at a 45-degree angle. The third stage wipes the flange to a 90-degree angle.

The first form stage is difficult to build and requires a certain amount of hand barbering. The radius R is laid out and carefully prick-punched. R_1, which has the same mathematical value as R, is machined with a fly cutter or milling cutter of correct diameter. R_2 is a washed radius which runs from zero to any value the diemaker assigns it. (This value depends on the depth to which he sinks R_1 below die level. If it is 0.25 inch [6.4 mm] low, R_2 should

---

*Mathematicians may possibly question the validity of this design. Admittedly, the passage of a vertical plane through a horizontal cylinder can result only in a straight line — yet this example started with the finished curve. Because it is vitally necessary to establish first the bend line in the metal, a "nonmathematical" compromise is necessary. The curvature is finished as is its line of intersection with the horizontal plane after final formation. Between the two is a "washed radius," which is to say, an irregular curved surface.

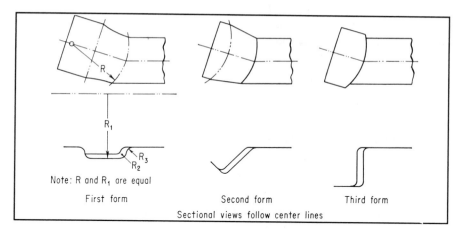

Figure 16-14. *The preforming and final form stages required when radial bend lines are used.*

equal 0.25 inch). R_3 is a part-print size which should be put on at high tolerance. Subsequent forming stages can be used to shorten it if necessary.

Locus of a Radius

A term with which diemakers and designers should become quite familiar because it occurs often in part prints — is the "locus of a radius." The locus is simply a line — straight, curved, or irregular — from which a radius is generated.

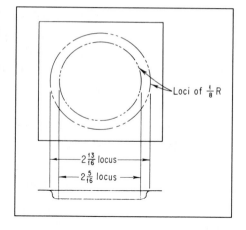

Figure 16-15. *The use of loci in showing part sizes.*

It is particularly useful to product designers in those instances where a radial design prevents dimensioning, as in Figure 16-15. In this example, the classic standards of engineering drawing make it impossible to draw the part as it actually appears. Moreover, it is impossible to dimension the part itself because there is no logical point from which to hang a dimension. For this reason, it is necessary to dimension the loci of the two curves as shown.

Problem 14. A circular drawn shell has an outside diameter that can be dimensioned. However, the designer has dimensioned from the locus of the draw radius, which is 0.125 inch (3.18 mm). Assuming a 3-inch (76-mm) locus is specified, what is the outside diameter?

Compound Angles

Experience has taught that no phase of diemaking is as perplexing as the construction of die sections having compound angles. Many top-flight diemakers have difficulty with these sections, and it is not unusual to see a section built twice, simply because of an incorrect approach to the problems of layout. Moreover, little ink is given to compounding in most books dealing with diemaking, although this subject is fundamental to the science. A few good rules are illustrated in the following examples.

Case 1. A progressive die is to be developed for the part shown in Figure 16-16. Although this part does not present a true compound angle, the forming section in the die does (Figure 16-17). However, the value of this angle is known to be 90 degrees at section *AA* because the part is formed up at right angles. The angle at which the part is flanged is 45 degrees, an easily calculated angle. Therefore, this part presents no unusual mathematical problems because all angles are known — an evenly divided 90-degree bend angling 45 degrees from the die center line. It does present a problem in grinding the V with respect to the cutoff notch, which brings to bear a basic rule of compound angles:

> Compound angles cannot be related dimensionally to any part of the die without the use of construction holes.

Following this rule does much to take the guesswork out of progressive diemaking.

From the die drawing (Figure 16-16) it is relatively simple to establish a dimension from the cutoff notch to the center of the bend line. Because the center line of the panel intersects the bend line at a point halfway between the 2-inch and 3-inch (51-mm and 76-mm) dimensions, this point of intersection is 2.5 inches (64 mm) from the end. Therefore point *A* in the die is 2.5 inches from the cutoff blade. Allowing 0.003 inch (0.076 mm) form die clearance, this dimension becomes 2.497 inches (63.4 mm) from the notch. This dimension, however, is obviously useless as a measuring tool; therefore two additional steps are necessary: (1) a construction hole must be used and (2) a roll dimension from the hole to the notch must be established.

Progressive Die Mathematics

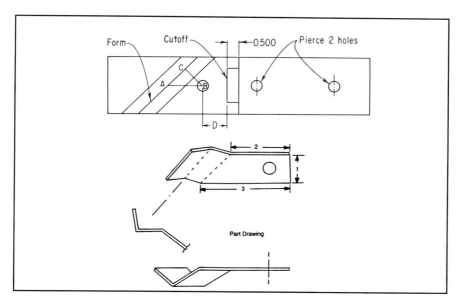

Figure 16-16. *Plan view of a typical pierce, cutoff, and form-at-angle die.*

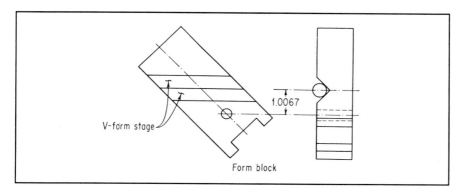

Figure 16-17. *Method of dimensioning an angular form stage.*

In doing this, it is not necessary to jig bore and jig grind the construction hole as long as ample grind stock is left in the cutoff notch and in the V. A 0.375-inch- or 0.50-inch- (9.53-mm or 12.7-mm)-diameter reamed hole serves the purpose quite well. It should be laid out, drilled, and reamed as close as possible to the center line of the block, but its transverse position is not important.

341

In dimensioning the die, the diemaker has assigned an arbitrary value of 0.500 inch (12.7 mm) to the cutoff notch. (The lateral dimension is not important as long as it is wider than the stock and the mating cutoff punch.) After the V and the notch have been shaped, the block is hardened and ready for grinding. The construction hole is lapped and the edges of the block are ground, after which the cutoff notch is ground to the predetermined 0.500-inch (12.7-mm) dimension. The distance D from the ground edge of the cutoff notch to the hole can then be measured with gage blocks. This value — assume that it measures 1.073 inch (27.25 mm) — is then subtracted from 2.497. This provides a value of 1.424 inches (36.17 mm) to the hypotenuse BA of triangle ABC. Because this is a 45-degree right triangle, the following calculation can be made:

$$\sin 45° = \frac{BC}{BA}$$

$$\sin 45° \times BA = BC$$

where $\sin 45° = 0.70711$

$BA = 1.424$

Therefore $BC = 0.70711 \times 1.424$

$= 1.006$

It is now established that the distance from the construction hole to the V is 1.0067 inches (25.57 mm). Using this dimension in a roll setup, it is comparatively simple to grind this compound angle to an exact dimension (Figure 16-17). Again this is the only way in which the V and the cutoff notch can be related dimensionally.

Problem 15. A die section (Figure 16-18) is designed with a 30-degree forming surface set at a 60-degree angle as shown. This surface must be ground so that it breaks at 2.750 inches (69.85 mm). To hold this dimension, a 0.500-inch- (12.7-mm-) diameter roll dimension must be established. It is decided to use a dowel hole for construction purposes. This hole's location is measured and found to be 0.513 inch and 0.781 inch (13.03 mm and 19.83 mm) from the edges of the block. In order to establish the roll dimension, it is necessary to find AH, the hypotenuse of a 30-degree to 60-degree triangle.

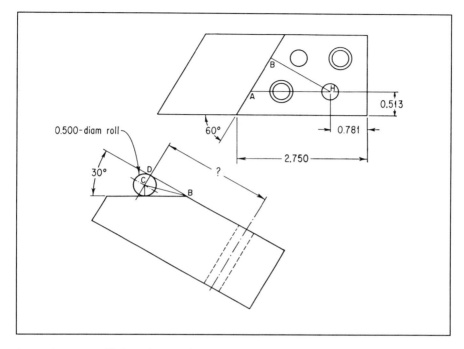

Figure 16-18. *Dowel hole used for establishing a dimension to an angular break line.*

Using *AH*, the leg *HB* is calculated. It is then necessary to evaluate *DB*. Addition of *HB* and *DB* is the desired roll dimension.

(a) Calculate *AH*.

(b) Calculate *HB*.

(c) Calculate *DB*.

(d) What is the roll dimension *HD*?

Case 2. In some instances compound angles present no unusual problems to diemakers, although they look extremely difficult. Such an example is shown in Figure 16-19. This is a preform stage — the conventional "form at 45" — mentioned often in this book. Because of the

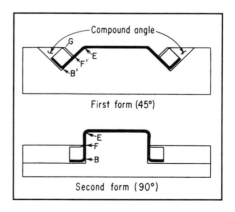

Figure 16-19. *First and second form stages. Problem: To grind the compound angle in the first form.*

343

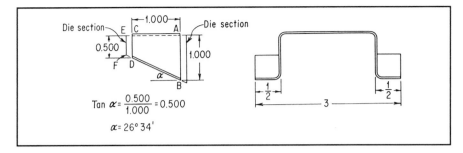

Figure 6-20. *Length of EF must be evaluated before angle can be ground.*

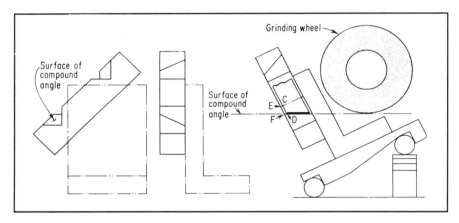

Figure 6-21. *Setup for grinding the first form compound angle.*

almost universal acceptance of this stage, many diemakers face the problem of grinding the compound angle illustrated. The best and easiest solution* to the problem of grinding the 45-degree preform stage is as follows:

1. The angle formed by the tab in its finished position is determined. (This is the angle formed by line BD with the horizontal plane (Figure 16-20).)

2. Evaluate the distance from the short edge of the flange to the end of the preform section. (This is dimension EC. Normally the diemaker determines this dimension when he dimensions the die.)

---

*A method commonly accepted by many diemakers is first to build the final form stage and then form a piece to the finished angles. This piece is then opened to the 45-degree preform angle and used to spot the preform stage. This procedure is slow and provides only an approximation at best.

3. From EC and the value of the angle formed by BD, determine the length of line EF (Figure 16-20).

4. Between the preform and final-form stages, point E does not change position relative to the die center line. However, point F in the preform stage is at point F'. Therefore line EF' is equal to EF. The contour of the notch in the preform section (at the rear face) is represented by the triangle $EF'G$.

5. The preform section is mounted on an angle plate at a 45-degree angle (Figure 16-21).

6. The angle plate is then mounted on a sine plate which is set to the angle formed by AB in step 1 (Figure 16-21).

7. The compound angle is now in the horizontal plane and the surface represented by EF' is in the vertical plane. Both surfaces can now be surface-ground in one setup.

Calculation of Compound Angles

Calculation of compound angles often presents a major difficulty in progressive die construction. Generally, the calculation of the angles is left to the diemaker when die drawings are undimensioned, but when drawings are dimensioned, they become the designer's responsibility. To aid in calculation of compound angles, several formulas have been developed in the industry and appear below for the diemaking practitioner.

Apparent Angle and Height. A compound angle often seen in progressive work is shown in Figure 16-22. In this example, two conditions are specified: the height of the part OO' and the apparent angle β. To find the true angle (which can be called α), a third condition must be known — the value of the bend angle R. Generally, it is relatively simple to calculate R and, in many cases, its value is specified.

Whenever these conditions are given — and very often they are — the following formula will yield the value of the true angle:

$$\tan \alpha = \frac{\tan \beta}{\cos R}$$

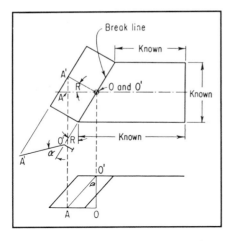 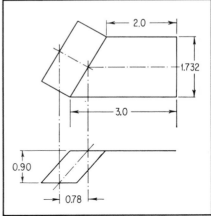

Figure 16-22. *Height and projected angle β are known. From these — and the angle of rotation — the true angle is calculated.*

Figure 16-23.

In other words, when the apparent angle and angle of rotation are known, the tangent of the true angle is equal to the tangent of the apparent angle divided by the cosine of the angle of rotation.

If, as often happens, the apparent angle β is unknown, but *OA* is known, the formula becomes

$$\tan \alpha = \frac{OA}{OO' \times \cos R}$$

In solving for β and R, the formula $\tan \alpha = \tan \beta / \cos R$ is useful in finding the angle of rotation when both the true and apparent angles are known. In such cases,

$$\cos R = \frac{\tan \beta}{\tan \alpha}$$

in addition,

$$\cos R = \frac{OA}{OO' \times \tan \alpha}$$

But a word of caution is needed: the angle of rotation is always measured between the break line and vertical, as shown in Figure 16-22. It is especially

Progressive Die Mathematics

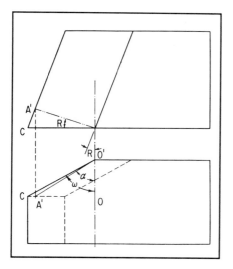
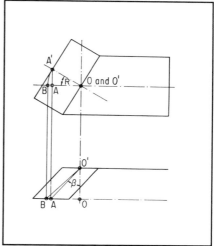

Figure 16-24. *The two variables most often given — rotation and inclination.*

Figure 16-25.

important to take care at this point if the break line is near 45 degrees because the error will not be obvious, as it will be on steeper or shallower angles.

Caution is also required in evaluating OA and β. OA is the projection (as shown by dotted lines) of $A'O$. To obtain $A'O$ it is necessary to lay out a line perpendicular to the break line. In this example, $A'O$ is parallel to the edges of the compound angle. This is a coincidence only, yet it happens often.

Problem 16. In Figure 16-23, what is the angle of the bend line? What is the true angle between the two surfaces?

Inclination and Rotation. A familiar problem in compound angles is illustrated in Figure 16-24. Found regularly in tool and die work, it is also a constantly recurring problem in the grinding of cutting tools. In this problem, two angles are given: R, the angle of rotation, and ω, the angle of inclination. (In cutting-tool work, however, the angle of inclination is measured from horizontal rather than vertical.)

Charts and tables have been developed to address this problem, but they usually are cumbersome and require a great deal of work. The following formula, however, is simple and requires little work, especially if the user is familiar with logarithms of trigonometric functions.

$$\tan \alpha = \tan \omega \times \cos R$$

where α = true angle,

ω = angle of inclination

R = angle of rotation

This problem is similar to, but not identical to, the solution of compound angles by means of the apparent angle β. ω, the angle of inclination, and β, the apparent (or projected) angle, are usually close to an equal value, particularly when a small angle of rotation is specified. There is, however, a small difference between them until R is reduced to zero value, at which time they become equal. (At this point the compound angle becomes an ordinary angle.) The reader should be most cautious in differentiating between ω and β.

Proofs

Diemakers and designers who can calculate compound angles can solve other problems of mathematics found in die manufacture as well. The following proofs serve as a means of increasing the diemaker's knowledge of compound angles.

The first proof will establish the formula

$$\tan \alpha = \frac{O}{OO' \times \cos R}$$

Proof 1. *Given:* An angle of rotation R, a height OO', and a dimension OA (Figure 16-25).

Find: Compound angle α.

1. $\cos R = \dfrac{OA}{OA'}$	1. Definition of cosine
2. $OA' = OB$	2. Radii of the same circle
3. $\cos R = \dfrac{OA}{OB}$	3. Substitution in 1
4. $OB = \dfrac{OA}{\cos R}$	4. Transposition

5. $\tan \alpha = \dfrac{OB}{OO'}$ 5. Definition of tangent

6. $\tan \alpha = \dfrac{OA/\cos R}{OO'}$ 6. Substitution of 4 into 5

7. $\tan \alpha = \dfrac{OA}{OO' \times \cos R}$ 7. Rationalization of 6

The next proof will establish that $\tan \alpha = \dfrac{\tan \beta}{\cos R}$

Proof 2. *Given:* Project angle β and angle of rotation R (Figure 16-24). *Find:* Compound angle α.

1. $\tan \beta = \dfrac{OA}{OO'}$ 1. Definition of tangent

2. $OA = OB \times \cos R$ 2. Transposition of statement 3 in Proof 1

3. $\tan \beta = \dfrac{OB \times \cos R}{OO'}$ 3. Substitution of 2 in 1

4. $OO' \times \tan \beta = OB \times \cos R$ 4. Transposition

5. $\dfrac{OO' \times \tan \beta}{\cos R} = OB$ 5. Transposition

6. $\tan \alpha = \dfrac{OB}{OO'}$ 6. Definition of tangent

7. $\tan \alpha = \dfrac{\frac{OO' \times \tan \beta}{\cos R}}{OO'}$ 7. Substitution of 5 in 6

8. $\tan \alpha = \dfrac{\tan \beta}{\cos R}$ 8. Rationalization of 7

Statement 8 can be transposed algebraically as follows:

$$\cos R = \dfrac{\tan \beta}{\tan \alpha} \quad \text{and} \quad \tan \beta = \tan \alpha \times \cos R$$

Progressive Dies

The final proof will establish that $\tan \alpha = \tan \omega \times \cos R$.

Proof 3. *Given:* Angle of inclination ω and angle of rotation R (Figure 16-25).

Find: Compound angle α.

1. $\tan \omega = \dfrac{OC}{OO'}$,   1. Definition of tangent

2. $\cos R = \dfrac{OA'}{OC}$ 2. Definition of cosine

3. $OA' = OC \times \cos R$ 3. Transposition

4. $OA' = OO' \times \tan \omega \times \cos R$ 4. Substitution of OC from 1

5. $\tan \alpha = \dfrac{OA'}{OO'}$ 5. Definition of tangent

6. $\tan \alpha = \dfrac{OO' \times \tan \omega \times \cos R}{OO'}$ 6. Substitution of 4 in 5

7. $\tan \alpha = \tan \omega \times \cos R$ 7. Rationalization of 6

Answers to Problems

1. (a) 0.5000   (g) 1.0000
   (b) 0.8660   (h) 1.0000
   (c) 0.5774   (i) 0.8660
   (d) 1.7320   (j) 0.5000
   (e) 0.7071   (k) 1.7320
   (f) 0.7071   (l) 0.5774

2. (a) 27°   (d) 33° 20'
   (b) 29°   (e) 54° 10'
   (c) 32°   (f) 63° 50'
   (g) No. Natural functions are not proportional. Sin 30° = 0.500 but sin 60° = 0.866.

3. (a) $a = 5$; $b = 8.660$
   (b) $a = 2.588$; $b = 9.659$
   (c) $a = 3.338$; $b = 9.426$

4. 11.0829 inches (281.506 mm)

5. 23° 30'

6. 6.116 inches (155.35 mm)

7. (a) 0.314 inch (7.98 mm)
   (b) 9.043 inches (229.69 mm)

8. $x = 1.0365$ inches (26.327 mm) (2.2500 - 1.2135)
   $x_1 = 2.7130$ inches (68.910 mm) (2.2500 + 0.4630)
   $x_2 = 3.750$ inches (95.25 mm) (2.250 + 1.500)
   $x_3 = 1.3684$ inches (34.757 mm) (2.2500 - 0.8816)
   $x_4 = 3.1316$ inches (79.542 mm) (2.250 + 0.8816)
   $x_5 = 0.824$ inch (20.93 mm) (2.250 - 1.426)
   $x_6 = 3.676$ inches (93.37 mm) (2.250 + 1.426)

9. $A = 2.298$ inches (58.37 mm)
   $B = 1.149$ inches (29.19 mm)

C = 6.259 inches (158.98 mm)

D = 3.961 inches (100.61 mm)

E = 1.172 inches (29.77 mm)

10. 0.139 inch (3.53 mm)
11. 0.026 inch (0.66 mm)
12. None. A flat strip bent over a 10-degree angle causes the edge of the flange to depart from vertical by 10 degrees.
13. 11° 32'
14. 3.25 inches (82.5 mm)
15. (a) 1.673 inches (42.49 mm) (2.750 — [0.781 0.513/tan 60°])

    (b) 1.4488 inches (36.80 mm) (1.673 x sin 60°)

    (c) 0.933 inch (23.70 mm) (0.250/tan 15°)

    (d) 2.382 inches (60.5 mm)
16. Bendline—30°

    True angle—45°

17

Progressive and Transfer Die Lubrication

Selection Considerations

Producing stamped components with progressive dies and various types of transfer systems is both productive and efficient. However, successful long-term, profitable, and trouble-free operation of these systems is dependent on proper lubrication of the tooling and material surfaces. Care in the selection of lubricants is critical to ensuring long tool life and part quality.

In selecting pressworking lubricants for progressive dies and transfer die systems, following proven procedures in the pre-planning stages can lead to highly cost-effective tool life. Material compatibility, easier secondary operations, and environmental issues should also be considered.

In the pre-planning stage for the selection of lubricants for progressive die, transfer press, and transfer die system operations, attention should be focused on the need for tooling interaction, material compatibility, application techniques, cleaning and finishing, and lubricant recycling.

Yet, despite all this preparation, in the end environmental factors may dictate the type of lubricant suitable for specific operations. Factors affecting this determination include air quality, hazardous waste generation, disposal cost, safety, and various restrictions imposed by local government.

Tooling Interaction

Changes in tooling philosophy as it pertains to high performance presswork are occurring at an accelerating pace. There are many new tooling materials today, more stations in presses and dies, and tooling is being integrated to perform secondary operations.

The range of tooling material available for presswork has certainly increased. Carbide, powdered metal combinations, M2, M5, S7, etc., are now commonplace and readily employed, along with high-speed tooling (D2, D4). Particle metallurgy tool steels (CPM 10V) are also gaining favor where high performance and quality piece parts are required. Understanding the differences in tooling materials can simplify the selection of a material that provides high performance. Knowledge of material properties is also important in improving part quality and length of production runs. Improved tool life — one of the keys to just-in-time inventory — reduces press downtime. In addition, upgrading tooling provides more options in the use of lubricants, especially water-soluble compound and synthetic solutions. Changing lubricant properties can be of enormous value in the finishing operation.

As presswork operations become more integrated, it may be desirable to add an assembly station to a die, or do secondary work such as tapping. Again, changes in functions may dictate changes in lubricant properties as well as the way the lubricants are applied.

In the case of carbide tooling, care and maintenance have a direct effect on tool life. In building the dies, carbide should be ground and sharpened very carefully. The carbide die areas in the pressroom should be kept clean from the residual dirt and oil that are commonly found in many stamping operations.

Also, the stock being stamped or drawn should be free of burrs and slivers; sharp or jagged material surfaces can seriously damage the surface integrity of carbide tooling.

Environment can be a major factor. Plant water used for mixing water soluble lubricants may not be suitable for use on carbide tooling. Unstable water can create a harmful acid condition.

Recirculating systems should be kept clean of metal particles, swarf, oil absorbent, and other harmful contaminants. It may be necessary to install baffles, filters, magnets, and centrifuge equipment to properly clean the coolant. In many instances, the size of the recirculating tank or coolant reservoir must be increased in capacity to provide for sufficient cleansing and filtration.

Electromotive forces (EMF) can be generated between the carbide surface and the machine tool with resulting erosion of the cobalt binder. This same condition can be caused by electrical current present around the die area in the operation of safety devices, sensing probes, and other electrical hardware.

On progressive dies, design of stock lifters is impacted by the weight of the strip material. In working extra-light stock such as "foil," it may be necessary to design the feeding mechanism with a push and pull action to provide the proper tension for operation of the progressive tooling.

In using progressive dies and transfer die systems, careful attention should be given to provide sufficient volumes of coolant or lubrication through the use of recirculating systems. This method of applying lubricant is extremely helpful in carrying away the heat of the forming process, especially on high-energy operations such as extruding, coining, punching, ironing, and embossing.

It may be necessary to install auxiliary lubricating systems to provide positive protection on tooling, especially on progressive dies. The necessary hardware needed for this type of secondary lubrication can be planned for in either the design or building stages of the die. A typical example of this type of installation using the adaptation of the airless spray application method is shown in Figure 17-1.

Often tooling is treated with special surface coatings such as chrome flashing, electrofilm, and nitrite. Specially formulated

Figure 17-1. *Auxiliary lubricating systems such as this may be necessary to provide positive protection for tooling.*

high-wetting metalforming lubricants are now available in both petroleum-based and water-soluble compounds for such applications. Many water-based solutions provide excellent wetting and penetration properties.

Consideration should be given to the benefits of washing or steam cleaning dies, followed by relubrication immediately after a production run when water-based metalforming lubricants are used. This family of lubricants is generally alkaline in nature. While they provide excellent cleaning and cooling properties for the die and tooling areas when in production, these same properties can clean out machine, guide pin, and die spring lubricants. Water and residual lubricant from water-based solutions can be trapped in die pockets, and if not cleaned out, these die areas can be attacked by rust. Rust preventives can be used to protect the die springs and other components.

Designers and operators should be mindful that just-in-time production inventory methods require short, frequent runs which may preclude frequent washing of dies.

Lubricants for transfer press operations must not only provide correct lubrication for every die but also must be compatible with the materials used to build them. Some lubricants may attack the self-lubricating graphite composition plugs pressed into guide bushings and wear plates. The possibility exists for damage to pads and seals made of rubber and elastomer products; such issues must be taken into account.

In using transfer presses and transfer die systems, care must be taken to provide sufficient volumes of cooling and/or lubrication by way of recirculating systems. This method of applying lubricant is quite effective in carrying away the heat of the forming process, particularly on high-energy operations such as extruding, coining, punching, ironing, and embossing.

Progressive Dies

Lubricant properties and application techniques for progressive tooling are even more varied than those of transfer presses and transfer die systems. Critical in progressive systems is to provide lubrication to the most severe operation in the progression. This could include extruding, coining, embossing, and in some instances, special secondary operations such as tapping or thread rolling. Progressive tooling is best lubricated with the aid of roller coaters and in some instances airless sprayers can be used in conjunction, as shown in Figure 17-1.

There can be special problems in progressive tooling where lubrication will not help much. These deficiencies should be corrected before any attempt is made to change the lubricant properties.

In progressive die work, the scrap skeleton is designed as the transfer mechanism and should be able to support its own weight. This type of design allows for even flow of material through the die without kinks and sharp drops that result in scoring and breakage caused by misalignment and insufficient lubrication.

When forming very light foil, the feeding mechanism should be able to push and pull the strip through the die, and thereby provide the proper tension. In working very light materials such as foil, it is imperative that lubricant properties be extremely light so that slug packing and jamming are prevented. Also, since this type of progressive die is used for electronic parts usually fabricated of beryllium copper, brass, silicon, stainless steel, and others, it is important to ensure that the lubricant is compatible with these surfaces.

Another area of possible trouble in progressive die operations is in stations where high energy is needed to form the piece part. Coining and embossing are cases in point. If the lubricant's viscosity is not high enough and does not have the proper physical properties, poor tool life will result.

Material Compatibility

Stamping Material Properties. The four most common categories of material surfaces are:
- Normal,
- Active,
- Inactive,
- Coated.

The pressworking lubricant must be compatible with the material to be formed.

In high-performance presswork where transfer die systems, transfer presses, and progressive dies are being used, large quantities of piece parts are produced in a relatively short period of time. For this reason, it is extremely important that the lubricant and material properties be compatible. Extra scrap and rework can result when piece parts are stained and coated stocks blister and peel.

Normal Surfaces. Most normal material surfaces have a natural affinity for retaining lubricant readily and generally do not require special wetting and polarity agents to provide sufficient lubrication.

Typical of this surface category are all bare mild steel stamping materials such as cold- and hot-rolled steel, as well as aluminum-killed and vacuum-degassed steel. Such normal surfaces naturally hold lubricant.

Problems arise even on normal surfaces when excessive amounts of contaminants such as moisture, oils, rust, scale, rust preventives, and residual pickling acid are present. These and other surface contaminants can cause problems in secondary operations such as welding, brazing, painting, plating, or cleaning. Freedom from contamination is especially important if water solubles, synthetic solubles, or chemical solutions are to be used as lubricants.

Active Surfaces. An active material surface is one in which the strength of the bond between the lubricant additive and metal atomic structure is great. Because the attractive energy of the metallic surface is high, surface chemical reactions are encouraged. This may explain why chemically active additives and wetting agents such as oleic acid, lard oil, emulsifiers, and extreme pressure agents are so effective in lubricating such materials as brass, copper, and terneplate.

However, in many instances, the use of chemically active lubricants on active material surfaces can be troublesome. Often reactive lubricants cause staining, etching, and blushing of copper and brass piece parts, especially when they are nested and stored under humid conditions. This problem is worsened if alkaline vapor or degreaser fumes are present.

Coatings can be classified as belonging to either the active or inactive family, even though the supporting base metal for these coatings is a normal cold-rolled material.

Inactive Surfaces. An inactive surface is one in which the strength of the bond between the lubricant additive and the metal atomic structure is low. In this type of surface, the attractive energy of the metallic surface is also low. For these reasons, there is much less likelihood of chemical reaction with a lubricant or its chemical additives.

Typical inactive surfaces are those of stainless steel, aluminum, and nickel. Commonly, the lubricant used with these materials has a high film strength, which actually places a mechanical barrier between the tool and the piece part to prevent metal-to-metal contact. High film strength can be obtained by using lubricants containing hydrocarbons, polymers, polar and wetting agents, as well as extreme pressure agents.

Quite often lubricants containing extreme pressure agents are used on inactive surfaces to provide a mechanical barrier rather than an actual barrier. With inactive surfaces at low temperatures, very little chemical reaction takes place with the lubricant. If the lubricant is working properly, it may simply be that it is providing a sufficient mechanical barrier for the operation.

Any lubricant formulation being used on an inactive surface must be compatible with subsequent manufacturing processes, which may include cleaning, welding, and long-term storage.

Hard-to-lubricate Surfaces. Sometimes the surface finish on the material to be formed will not retain lubricant properly. This could occur with brightly polished materials, stainless, tinplate, and bi-metals. Another culprit can be the mill oil itself. Certain coatings may keep water solubles or evaporating compounds from working properly.

Forming some steels can be exacting, especially bright or high-finished stainless steel. The surface of most stainless, for example, does not retain lubricant as readily as those of other metals. For this reason, emulsion-type lubricants with superwetting characteristics have been formulated for stainless steels. Heavy-duty evaporating compounds containing extreme pressure agents have good anti-wipe properties.

On appliance and automotive trim, felt wipers may well be required to remove any spotting from either synthetic solubles or evaporative compounds. These bright surfaces will show any residual deposits from lubricants used. The very light films may have to be removed by wipers before the parts are wrapped and packed.

Coated Surfaces. The most common coated stocks used in forming are galvanized steel and precoated materials. The zinc coating may be applied by hot-dip galvanizing or electrodeposition and is primarily intended to serve as a galvanic protective layer for the main metal substrate. Zinc-coated steel is widely used where rust protection is required.

Material pickup can occur when forming galvanized material. Special water soluble lubricants are available to keep the galvanized metal particles from building up on the tooling. Dry film and synthetic water-based solution are also effective on galvanized stock.

Corrosion of Zinc Coatings. Any metal corrosion may be aptly described as the chemical or electrochemical attack by moisture or other substances hostile to the surface. "White rust" is the destruction of the surface of galvanized steel or zinc by oxygen or other chemical elements. The reaction is accelerated by the presence of moisture.

The rate of attack is related to temperature, pH, and composition of moisture and the concentration of dissolved gases within the moisture. The higher the oxygen and carbon dioxide content and the softer the water, the greater the rate of surface breakdown. The corrosion cycle starts with the formation of zinc oxide, which in turn is converted to zinc hydroxide and then to basic zinc carbonate in the presence of carbon dioxide. The final product prevents the onset of further corrosion at the particular area of attack, providing no highly acidic or alkaline contaminant becomes involved later.

Under unventilated enclosed conditions such as blind angles and box sections, rapid attack by the available oxygen gives rise to a pitting condition as the exhausted gas is not replaced as fast as it is used. The corrosion pattern, therefore, is nonuniform and typical of "white rust" conditions.

It is essential that piece parts with pockets be drained of as much coolant as possible. Parts in process should be used as rapidly as possible to avoid stagnation of any residual fluid. Conditions of high relative humidity and temperature aggravate the condition.

When the metal surface temperature drops in the presence of high humidity, the condensed moisture on the section is rich in oxygen because the thin water film is exposed to a large volume of air. Conditions are then ideal for extensive uniform corrosion.

Zinc is rapidly attacked in acidic and alkaline aqueous conditions. Storage and manufacture within the vicinity of fumes from plating or pickling shops will accelerate metal surface breakdown.

Application Techniques

There are many ways to apply lubricants for pressworking transfer press, transfer die system, and progressive die operations. Among the most effective methods are recirculating systems, roller coaters, and airless sprays.

The use of drip application methods is not as productive and can be wasteful, unless the system is properly installed and controlled. A typical drip lubricator is commonly mounted after the stock or roll feed. The drip system is regulated by a petcock or regulating valve which can be adjusted for the flow desired. This method does not provide for automatic shutoff when the equipment stops, which results in wasted lubricant, messy parts, and housekeeping problems. If a large stock area is to be lubricated, a drip lubricator is not a wise choice.

The lower consumption of lubricant and increased productivity that can be achieved by eliminating costly drip or manual application can often pay back in a short time the cost of automatic application equipment. In laying out a new press line (space permitting), it may be advantageous to install a fully automatic coil feeder, stock straightener, and stock lubricator to obtain the maximum productivity from the press and tooling.

In most instances, transfer presses and transfer systems work well with recirculating lubrication systems. There are four advantages to be gained:

- Tooling is kept cool and clean;
- Foreign particles and metal chips are washed away and sent back to the coolant reservoir for coolant conditioning and treatment;
- Cost savings are produced through less consumption of lubricant; and
- If water soluble lubricant is used, cleaning and other downstream operations are much easier to perform.

Preliminary planning of a recirculating lubricating system for a transfer press or transfer die system should take into account many factors. The coolant system capacity should provide a five-minute lubricant cycle time if possible, with a three-minute cycle as the minimum, in order to allow for cleaning and conditioning of the coolant or lubricant. Sizing the system's storage tank at 5 to 10 times the rated capacity of the coolant delivery pump is a good rule.

Roller Coating. This is one of the most popular methods of applying lubricant. Several types are available, but regardless of the roller coater system chosen, there are certain design features to consider.

First, the location of the roller coater is important. The preferred position is between the fabricating equipment and the feeding mechanism. Placing the coater before the coil feed can cause the lubricant to be mechanically worked off the metal surface and can result in slippage in the coil feeding mechanism.

There should be positive control of the lubricant being applied. The lubricant flow should be stopped when the press is not operating.

The unpowered roller coater is the most widely used, wherein the coil feeding mechanism mechanically pushes the stock through the roller coaters. With this type of roller coater, either one or both sides of the material can be lubricated; the stock is coated with lubricant as it passes through the applicator rollers. Any excess lubricant is then squeezed off by wiper rollers and returned to the recirculating reservoir where it is filtered and available for reapplication. A typical roller coater unit of this type is shown in Figure 17-2. An important feature of a roller coater is the ability to coat the top, bottom, or both sides of the stock.

The weight of the lubricant being applied by roller coater has a bearing on the pump components involved. When light oils or water-base fluids are involved, a centrifugal pump is adequate for recirculation of the lubricant. When heavy oils are used, a gear-type pump may be required.

Common roll materials are steel, neoprene, felt, and polyurethane. The surface of the roller can be textured to retain lubricant.

When working nonferrous material or specially coated stock, a soft roller made of either felt, neoprene, or polyurethane should be used so that the material surface will not be scratched or marred.

Airless Sprayers. An airless spray is a mechanical method of producing a finely divided spray of lubricant. Pressure is applied by means of an intensifier and then carried via a high-pressure hose to a tiny orifice in the nozzle, where the lubricant is expelled as a fine spray.

A typical airless setup consists of an intensifier assembly (piston and barrel), a check valve, a lubricant reservoir of the required size, the necessary

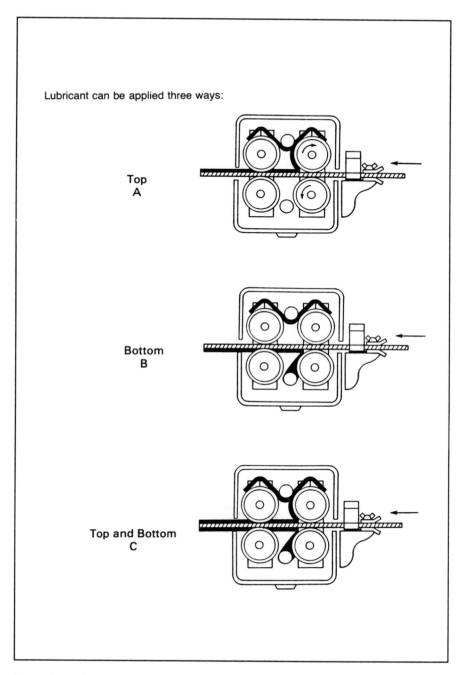

Figure 17-2. *Roller coaters provide lubrication coverage and flexibility not available in many conventional lubrication systems.*

number of nozzles needed to cover a particular piece of operations, and an air valve to activate the unit on each stroke of the press.

The spray pattern obtained with the use of airless spray can be either round or fan-shaped. Owing to the various spray patterns available, it is an excellent method for spot lubrication, either lubricating the stock before it enters the die or in the die itself.

A modern airless spray system does not produce mist or fog that results in overspray problems, can be precisely directed at a target area in the die, and is timed to operate in conjunction with the equipment cycle.

Cleaning and Finishing Requirements

In most cases, the cleaning and finishing processes being used determine the type of pressworking lubricant. For example, low-temperature cleaning lines generally are not capable of removing animal fats, residual oils, and certain extreme pressure agents. For low-temperature cleaning, specially formulated water solubles are generally recommended.

Hot alkaline wash systems can clean heavy residual oils and certain drawing compounds but require careful maintenance. The waste disposal costs for alkaline systems can be quite high if skimmers and clarification equipment are not used to lengthen the life of the cleaner.

Solvent vapor degreasing is being phased out by many facilities because of the air quality and environmental considerations. If vapor degreasing is used, water-based lubricants such as water soluble oils, soaps, pastes, and dry films should not be used to form the part. A lubricant compatible with a vapor system should be selected.

Special Considerations. There are certain restrictive or governing considerations that must be met by the pressworking lubricant for compatibility with secondary operations. Suitability for use with finishing operations such as painting, powder coating, assembly, or joining is of prime importance.

The use of some types of water extendable pressworking lubricants (emulsions and chemical solutions) may allow welding without smoke and enhance weld integrity. Painting can be performed without prior cleaning in some cases.

The need for long-term storage and/or long-term rust prevention may be another controlling factor.

Many of the major manufacturers have established manufacturing standards for the type of pressworking lubricants to be used on their respective components. The ultimate goal is to improve part quality, secondary operations, and ease of finishing.

Microbiology of Pressworking Lubricants

Pressworking lubricants containing water are apt to be affected by any number of the microbes that occur in the environment. Lubricants containing water are affected by contamination, allowing microbes to grow. In the case of emulsions containing mineral oil, bacteria are the principal microbe. In water-based solutions, mold and yeast are the principal microbiotic organisms.

Lubricant constituents are the food that feeds these microbes. When growth occurs by attacking a constituent, in a short period of time (often as rapidly as one day) a major constituent within the lubricant can be changed chemically by the microbe's feeding, no longer allowing the lubricant to perform as intended.

An emulsifier may be unable to promote stable emulsification, rust-preventive additives may no longer prevent rust formation, or a lubricant additive may no longer provide an effective lubricating film.

Anaerobic bacteria are active in the absence of oxygen, often causing the breakdown of sulfur compounds within the fluid. This can result in the strong rotten egg odor called the "Monday morning smell."

Aerobic bacteria growing in the presence of oxygen are a more aggressive microbe. They can be responsible for the overall chemical breakdown of the lubricant.

Mold and yeast can cause a different type of problem by blocking filters as they continue to grow. They also can cause a slime buildup on tools and dies.

The problem of bacteria, mold, and yeast within the lubricants can be controlled by a balanced formulation of the lubricant. The use of effective biocides together with proper clarification and filtration, are good procedures to follow. Also effective are clean work habits, including the elimination of contaminants such as plain dirt, food, shop rags, and other forms of contamination.

Lubricant Clarification, Recycling, and Disposal

The cost of oil, chemicals, and the disposal of pressworking lubricants has increased rapidly. For this reason, it is desirable to conserve and recycle pressworking lubricants. New rules and regulations to protect the environment have also made lubricant disposal much more costly.

Extending the useful life of pressworking lubricants depends largely on understanding the source of contaminants that can affect its useful life.

The type and choice of clarification, recycling, and disposal technique depend upon the type of lubricant and its contaminants. Selection can be determined by analysis of the lubricant. If the lubricant is oil-based, some of the chemical components of the lubricants may, with time, precipitate and circulate within the fluid, or they may be deposited on surfaces wetted by the fluid. If solid lubricants are used in suspension, these too may be deposited. If emulsions and soluble oils are used, the stability of the emulsion can be reduced and the particle size of the discontinuous phase can increase. These changes influence the choice of the clarification, recycling, and disposal process and equipment. If water-based solutions are used, changes in pH may cause precipitation. In any pressworking operation there is a potential of biological degradation and resulting particulate contamination, which then must be removed to maximize the effectiveness of the lubricant.

The particulates that can contaminate lubricants can include particles from the material that break loose during the pressworking operation. Typical are mill scale, aluminum oxides, and galvanized particulates. The clarification units can remove these particles from the lubricant reservoir by magnetic action, filters, etc.

Pressworking lubricants to be recycled may contain a wide variety of solid and liquid contaminants other than those directly resulting from the process. The contaminants present in the fluid may come from varied sources; cleaning fluids, oil-absorbent material from the floors, food, and various other random contamination sources may end up in recirculating systems. Obviously, these random occurrences are difficult to predict on a systematic basis, but should be taken into account when process and equipment selections are made.

Water. The presence of calcium and magnesium promotes the formation of insoluble precipitates, particularly when soaps are present in the emulsion or water-based solution. During recirculation of the pressworking lubricant or the process of clarification or recycling, crystalline precipitates may be produced since their solubility in water is very low.

Understanding these in-process changes of the pressworking lubricant can improve any decision in selecting the proper clarification equipment and the design of the recycling system.

Clarification. As suggested, many factors encourage extending the effective life of a given metalworking lubricant. A significant contribution to extending the life of a given lubricating system is the installation of appropriate clarification equipment.

Settling. Particulates in many metalworking lubricant systems are often adequately removed by installation of a simple gravity-settling tank. To facilitate removal of the settled debris, fines, and mold, a drag conveyor is often installed that serves to remove the buildup from the bottom of the tank. Separation of the tank into separate sections by partitioning walls over which the fluid flows increases the effectiveness of the settling system by progressively removing the contaminant that may flow from one section to the other. Several baffles can be installed at the top of the tank. The first baffle can prevent tramp oil from flowing into the tank, while the second baffle drops heavy residuals and metal fines.

In light of their low cost and simplicity, gravity settling systems are often used as the first stage in a total system which might include cyclone separators, filtration devices, or a centrifuge. This approach reduces the operating cost of more sophisticated equipment, and the productivity of the system is frequently improved.

Flexible piping should be used only if there is no possibility of buildup of microbial contamination on the walls of the pipe that cannot be readily removed by flushing. All pipes and their attendant valves should contain as few bends as possible to facilitate cleaning.

Reclamation of metalworking lubricants should be carried out on an ongoing basis so that dead storage of used lubricants does not occur, again to eliminate excessive microbial growth and deterioration of the fluid.

Structure of Pressworking Lubricants

Pressworking lubricants fall into three major categories: fluids, paste-soaps, and dry films. Each of these families has varying combinations of chemical and physical constituents (Table 17-1).

Fluids, the most widely used pressworking lubricants, fall into two major classes: solutions and emulsions.

Solutions. Three types of solutions are available to the pressworker: oil-based, water-based, and synthetic.

A solution is a fluid in which all of the ingredients are mutually soluble. A typical oil-based solution may consist of a mineral oil base, a wetting agent, a rust inhibitor, and an extreme pressure agent. These are all oil-soluble constituents that can be properly balanced and compounded into a solution.

Oil-Based Solutions. These solutions employ mineral oil as a base, and provide good fluid integrity. Generally safe from biological attack, oil-based solutions can be recycled and clarified. They provide their own natural lubricant and have good wetting properties. Oil-based solutions have some inherent degree of corrosion protection and are available in a wide range of viscosities.

Water-based Solutions. Water-based solutions are true solutions consisting of water surfactants, soluble esters, soluble rust preventives, and in some cases extreme pressure agents.

These solutions differ greatly from oil-based solutions or emulsions in that they need special handling and some additional operating procedures, especially during changeover and startup.

Synthetic Solutions. The synthetics are man-made compounded pressworking lubricants combining excellent high-temperature properties and good boundary lubrication. The main ingredients of synthetics are: synthesized hydrocarbons (polyalphaolefins) and polybutane derivatives. The polyalphaolefins are much like oil-based solutions in their physical characteristics, but have a higher degree of resistance to oxidation. The polyglycols, polyesters, and dibasic acid esters have superior high-temperature stability. Synthetic solutions, though effective, are higher in cost than oil-based solutions.

Emulsions. An emulsion system is based on one immiscible fluid suspended in another. The mixture with formed droplets is an emulsion. In pressworking lubricants the continuous phase is water, the suspended phase is oil or a synthetic solution. The continuous phase is critical. Stable emulsions require surface active ingredients as well as special mixing devices and techniques.

Table 17-1
Comparison of Pressworking Lubricants

Function	Oil-based Solution	Water-based Solution	Synthetic Solution	Emulsion
Reduce friction between tool and workpiece	1	1	1	2
Reduce heat caused by plastic deformation transferring to the tool	5	1	5	2
Reduce wear and gulling between tool and workpiece due to chemical surface activity	4	1	3	3
Flushing action to prevent buildup of dirt on tooling	5	2	5	3
Minimize subsequent processing costs such as welding and painting	5	1	5	2
Provide lubrication at high pressure boundary conditions	1	4	1	2
Provide a cushion between the workpiece and tool to reduce adhesion and pickup	1	4	1	2
Nonstaining characteristics to protect surface finish	5	1	4	2
Minimize environmental problems with air contamination and disposal problems	5	1	5	2

NOTE: 1 = most effective   5 = least effective. *(Tower Oil and Technology Company)*

There are two families of conventional emulsions, multipurpose and heavy-duty. Multipurpose emulsions generally consist of an emulsifier, oil, wetting agent, and a rust inhibitor. Heavy-duty emulsions contain emulsifiers, oils, wetting agents, extreme pressure agents, fats or polymers, and nonmisting additives.

In invert emulsion formulation, the oil is the continuous phase and the water is the discontinuous phase. This type of emulsion is not effective in such operations as drawing, where a good physical barrier is required.

Pastes, Coatings, and Suspensions. When severe pressworking operations are encountered, such as drawing or cold forming, high film strength lubricants are necessary. For these operations, pastes, suspensions, and conversion coatings are recommended.

Pastes can be made several ways, formulated with an oil base or water base, and can be pigmented or nonpigmented.

The pigments used in pastes are much like the pigments in paints. They are fine particles of solids which are insoluble in water, fat, or oil. Some of the common pigments are talc, china clay, and lithophone. Pigmented pastes come in one of two forms. *Emulsion compounds* are pastes composed of fats and fatty oils and sometimes mineral oil, pigment, emulsifier, soap, and water. They can be used as supplied or diluted with water or oil for easier handling and application.

Oil compounds are pastes consisting of pigments dispersed in fatty oils and/or mineral oils which may be treated with extreme pressure agents (chlorine, sulfur, or phosphate). These compounds are generally used straight or diluted with mineral oil. Dried-on lubricants and pigmented coatings can be used in sheet metal work, especially in heavy forming and drawing.

Nonpigmented pastes come in four different categories:

Emulsion drawing compounds, which are pastes composed of fats and fatty oils and their fatty acids, sometimes containing free mineral oil, various emulsifiers, and water. These products are diluted with water before use.

Fats are fatty oils and fatty acids sometimes used straight, but usually mixed with mineral oil for use.

Mineral oil and greases can be used straight when necessary.

Emulsions are usually diluted with water.

Dry-processed Coatings. These coatings are rapidly coming into use because of their easy and economical application, clean and tidy characteristics, and ease of handling and cleaning.

For dry soap films, stock or parts should be cleaned in degreasing solutions at about 190°F (88°C) and rinsed at about 160°F to 180°F (71°C to 82°C). Water-soluble soap solution should be applied at temperatures ranging from 180°F to 200°F (82°C to 93°C). For low production requirements, dip coating may be most economical. For high production, roller coating is preferred for sheet and coil stock.

Wax or wax-fatty coatings for light to medium drawing, especially of nonferrous stock, may be applied by hot dipping at 120°F to 150°F (49°C to 66°C), by spraying the material hot, or by cold application in a solvent vehicle. In the cold method, the vehicle evaporates, leaving a dry coating.

Phosphate coatings are properly chemical immersion coatings that provide a measure of lubricity.

Graphite coatings are useful under high-temperature and heavy-unit-load conditions where it is not feasible to use water-base, oil-base, or other solid lubricants. Graphite has the disadvantage of difficult removal and is consequently used for drawing only when strictly necessary. In fact, the unauthorized use of graphite in automotive body panel drawing dies has caused subsequent paint adhesion problems that have resulted in customer dissatisfaction.

Subsequent Operations. Some of the greatest benefits of using water-based solutions properly in pressworking application are the substantial savings that can be realized in the operations that must be performed after the metalforming operation has been completed.

Some typical examples are: possible elimination of degreasing altogether, painting over the lubricant without prior cleaning, easier cleaning and elimination of smoke in the heat-treating operation, as well as welding. Not only may smoke be reduced or eliminated, the actual quality of the welding operation and heat treatment may be improved.

Other advantages of water-based solutions are reduced maintenance of equipment and machines because of cleaner operating conditions, and cleaner and safer work areas.

Troubleshooting Die and Lubricant Problems

Scoring. Often, scoring occurs due to lack of lubrication to the tooling area caused by any of a number of reasons — the heat of deformation may be too great for the physical and chemical properties of the lubricant, the lubricant may thin excessively on hot die surfaces. Matching the requirements for lubricity with lubricant characteristics at the actual operating temperature can be helpful. Water cooling of the die is another useful technique. The use of water-based lubricants is a good way to reduce die temperatures by evaporative cooling.

Another cause of scoring is poor lubricant application technique — the lubricant must reach the tooling area. Poor penetration can be still another cause for scoring — the lubricant may be too heavy.

In analyzing scoring problems, a good place to start is with the stock condition. Make sure that the stock is free of burrs, protective coatings, excess mill oil, and foreign particles. The stock thickness may be too great for the clearances provided in the tooling.

In forming hard materials, more force and energy are required to form the part than is the case with soft materials. The lubricant must be able to withstand these extra forces and still not wipe or squeeze out, especially in the areas surrounding a tight form radius.

Softer material surfaces such as clad aluminum alloys and deep-drawing grades of electro-galvanized steel can contribute to scoring. Metal pickup on the tooling surfaces can generate fine metal particles during drawing and forming operations. These particles must be flushed out by the lubricant and not allowed to accumulate. Material particles such as those derived from beryllium copper can be quite abrasive, resulting in accelerated tool wear.

Tool Wear. Rapid tool wear can be caused by several conditions, including the lack of lubricant. The most common cause is poor application technique. On high-speed applications, lubricant must be present at all forming stations to provide good tool life. On very severe operations such as coining, extruding, or tapping, a secondary lubricant may have to be used and applied by some auxiliary means such as an airless sprayer.

The characteristics of the lubricant used can also contribute to tool wear. Large amounts of uninhibited extreme pressure agents can, in some instances, promote tool wear. When using carbide tooling, care should be taken to ensure that the lubricant is indeed compatible with the tooling surface and the chemical properties are safe for use on carbide tooling.

Metal particles generated during the forming process can bring on rapid wear if they are not properly flushed out by the forming lubricant. This problem can become acute when working materials that generate metal fines on a large scale, such as aluminum, hot-rolled steel, aluminum killed stock, and even stainless steel.

Slug Pulling. Slug pulling is caused by the vacuum created by a lubricant seal between the slug and the face of the punch. It is an unwanted event and can be resolved by the use of vented or spring ejector punches, tighter clearances, and even air or vacuum ejection. Lubrication can also have a negative influence on the ease of slug shedding: too much lubricant may be applied, causing sticking, and the lubricant may be too thick.

In some instances, the physical properties of the lubricant are too adhesive in nature and, therefore, not satisfactory for high-speed perforating. Often just providing correct die clearances will eliminate slug pulling.

Tooling Problems in Drawing. The two most common problems in drawing are wrinkling and fracturing or tearing.

Wrinkling can have a number of causes. If the draw die is known to produce good parts and the onset of wrinkling occurs at the start of a production run, the binder adjustment is the first thing to check. Generally, wrinkling is a sign that the blankholder is too loose. In severe cases, wrinkling and fracturing may occur simultaneously. This is because the wrinkled stock cannot pass over the draw radius and the metal flow is locked out.

Worn draw beads are a frequent cause of wrinkling in localized areas. If wrinkling occurs only on one side of the die, a slug or other foreign object may be lodged under the die. Press out-of-level conditions can also cause wrinkling on one side of the die and splits or fractures on the other side.

Incorrect material specifications and excessive amounts of lubricants, as well as excessive lubricant viscosity, can be a contributing factor in wrinkling. The lubricant may need to be thinned or, possibly, applied more evenly.

The most frequent cause of splitting or fracturing is excessive blankholder pressure. Again, splits on one side of the die can be due to a foreign object under the die or a press out-of-level condition. Insufficient clearance between the punch and die can be the cause for fracturing or tearing. The percentage of reduction or depth of draw may be excessive for the type of stock being drawn. The material may be too thin or age hardened. Circle grid analysis of the area of the fracture and checking of tensile samples against stock specifications are good troubleshooting techniques.

The lubricant must be correct for the working temperature of the die. Insufficient lubricant due to improper application or poor wetting or penetration of lubricant can cause fracture. The lubricant properties must provide sufficient barrier or boundary lubrication.

Finishing Operations. A stamped part or component is not really completed until all other subsequent operations have been performed. If there are problems with cleaning, painting, brazing, welding, or performing other secondary operations, the lubricant could very easily be at fault. Substantial savings can be realized by understanding what part lubricant properties actually play in secondary operations.

Finishing is very important, where integrated forming and stamping operations are involved. The correct lubricant can reduce costs and improve quality. This is especially true when special types of cleaning have to be performed, or annealing, stress relieving, or deburring is required.

Annealing. In many instances, annealing operations are performed on a metal formed part before or during the finishing process. The compatibility of the forming lubricant with the annealing operation must not be overlooked.

Annealing is generally performed at temperatures of 1400°F to 1700°F (760°C to 925°C) in a controlled atmosphere. The forming lubricant must burn off clean to obtain a bright anneal. The ingredients in the forming lube must not contaminate the part or the furnace atmosphere. Forming lubricants that contain sulfur, chlorine, animal fats, and pigments can cause adverse surface reactions during annealing, and also contaminate the furnace atmosphere. Before performing any annealing operation, it is a good practice to first check the forming lubricant for overall compatibility both in the furnace atmosphere and on the part's surface.

Sometimes it may be necessary to first clean the formed part of any undesirable substances before the annealing can be done successfully.

Stress Relieving. Stress relieving is another finishing process that requires a careful choice of the forming lubricant properties. Here again, the ultimate goal may be to achieve a clean part before relieving.

The temperatures required for stress relieving are usually in the range of 600°F to 800°F (315°C to 425°C). This is substantially lower than the temperature required for annealing. The lower temperature means that lubricant residues do not completely burn off and can cause contamination problems. This can include staining of parts, furnace atmosphere problems, and eventual damage to the furnace. Among the problem lubricants are sulfurized and chlorinated oils, heavy residual oils, animal fats, graphite, pigments, and soaps.

Cleaning Stamped Parts. One of the most important operations in preparing a metal formed part for finishing operations, such as painting or plating, is cleaning. The need for the forming lubricant to be compatible with the cleaning system cannot be overemphasized.

Low-temperature Cleaning. These systems generally operate at temperatures of approximately 100°F (38°C) and can be quite effective in removing light oils, soluble oils, synthetic lubes, and some specially formulated oils designed for low-temperature cleaning. Low-temperature cleaning systems conserve valuable energy, but they also have their limitations. For example, this type of system will not completely remove pigments, pastes, soaps, animal fats, heavy residual oils, and heavily compounded forming lubricants.

The performance of a low-temperature cleaning system can often be improved, either by using a different type of forming lubricant more compatible with the system, or modifying the cleaner to improve its cleaning ability.

Vapor Degreasing. This cleaning method uses hot vapors of chlorinated hydrocarbon solvents (generally trichlorethylene or trichlorethane) to remove many types of forming lubricants, generally petroleum-based.

Vapor degreasing can also remove mill oils and rust preventives, along with some cleaning residues. Vapor degreasing can be a costly operation when too much oil is left on fabricated components before cleaning. This excessive amount of lubricant can result in a buildup of residual oils and other compounds on the boiling or vapor chambers, which can cause foaming or a reduction in the evaporation efficiency. It is not uncommon for a manufacturing plant to increase the life of its chlorinated solvent simply by changing the type of application technique for its forming lubricant, or using a different type of lubricant for the piece part or component in question.

There are some types of lubricants that should not be removed by vapor degreasing. Metalforming lubricants which contain free fatty acids, chlorine, or sulfur can upset the inhibitors in the chlorinated solvent, resulting in hydrochloric acid formation in the degreaser. This can result in corrosion of the heating elements and other working surfaces.

Another family of forming lubricants that can create problems are emulsions. Many of these lubricants are compounded with fatty acids and also leave some water on piece parts to be cleaned. Even though the chlorinated degreasing solvents are especially inhibited against the effects of hydrochloric acid formation in the presence of water, excess moisture should not enter the degreaser. Another side effect of attempting to remove water soluble fabricating compounds by vapor degreasing is the white powder deposit left on the piece part after vapor degreasing. These deposits may be due to components of the soap, or may be present in the original fabricating compound.

The operator must be concerned with the following problems when operating a vapor degreaser:

- Depletion of the stabilizer in the chlorinated solvent,
- Formation of acids,
- Deposits on piece parts,
- Buildup of oily residuals in the solvent.

Alkaline Cleaning. One common goal in cleaning is to maintain the cleaner properly to extend its life as long as possible. This fundamental rule certainly applies to alkaline cleaning.

Fabricated parts should be drained of excess forming or stamping lubricant whenever possible. Parts can also be stacked in such a way as to maximize draining of the lubricant. Another technique is to blow off the lube before cleaning. Vibratory conveyors can also aid in keeping the amount of lube entering the cleaning tank to a minimum. The recovered lubricants can often be recycled, increasing the savings.

Water-based Pressworking Lubricants

Many benefits can be derived from the use of water-based lubricant solutions where possible. When compared to mineral oil-based lubricants the advantages include lower initial cost, more efficient secondary operations, and elimination or reduction of cleaning.

In some instances portions of the production process require modification to permit the use of water-based lubricants. For example, additional clarification and contamination control equipment may have to be installed to maintain lubricant stability and product quality.

Tooling Problems. The maintenance and operation of tooling when working with water-based solutions requires several changes in operating procedure. Most synthetics are alkaline in nature and act as detergent soaps. Several changes may be required in tooling maintenance procedures. These solutions remove most conventional greases and machine oils. It may become necessary to protect the tooling and related die components with rust preventives especially when the tooling is not in operation during extended periods of time. All that may be required are a simple floor tank with a few baffles added to drop out metal fines, and a skimmer to drag off the accumulations of tramp oil and greases. These simple measures will lengthen lubricant life, provide better finish to formed parts, and reduce the cost of secondary operations.

There are, of course, exceptions. Some systems may generate large amounts of metal fines. If this is the case, magnetic separators or cyclonic units may be installed to remove the large amount of metallic particles. Media filters are also an effective means of removing foreign particles.

System Startup. It is possible for contamination to occur during the initial operation startup. It is therefore important to make the required adjustments for contamination control. There are three considerations pertaining to systems performance that need to receive careful attention when starting up a pressworking operation using water-based solutions:

- Overall system cleanliness,
- Machine-lubricant compatibility,
- Water source analysis.

The importance of starting up the pressworking system in a clean condition cannot be overemphasized. Clean formed parts cannot be obtained unless the system itself is clean, including machinery, tooling, lubricant reservoirs, and any piping that carries lubricant to the forming areas. Generally, the lubricant supplier will specify the type of cleaner recommended for use in this preparation period, as well as any special cleaning, especially when the tooling is not in operation during extended periods of time.

Material Considerations. When specifying material for use with water-based solutions, the material should be ordered clean, dry, and as free as possible of mill oil and rust preventives.

Paper-clad and plastic-film protected material finishes can be formed with water-based solutions without damage to the protective coverings by simply applying the lubricant with roller coaters or spray units.

An area of importance when considering a water-based solution is the interaction between the solution and the material surface. Tests should be conducted to determine that the water-based solution will not react with materials such as galvanized steel and aluminum in an adverse manner, resulting in corrosion or staining. The proper dilution strength should be carefully noted. Operating with the proper concentration can be the difference between success and failure. The same surface tests should also be performed on coated stocks.

Contamination Control. While system cleanliness is extremely important, it only goes halfway in assuring success. The overall forming system is also of great importance and should be mechanically designed, modified, or altered to operate optimally with the water-based lubricant solution.

In the majority of cases, the contamination control devices required to keep the lubricants clean and effective are quite minor. Again, that may be required are a simple floor tank with a few baffles added to drop out metal fines, and a skimmer to drag off the accumulations of tramp oil and greases.

Subsequent Operations. Some of the greatest benefits of using water-based solutions properly in pressworking applications can be realized in the operations that must be performed after the metalforming operation has been completed.

Some typical examples are possible elimination of degreasing altogether, painting over the lubricant without prior cleaning, easier cleaning, and elimination of smoke in heat treating or welding operations. Not only may smoke be reduced or eliminated, the actual quality of the welding operation and/or the heat treatment may be improved. Other advantages of water-based solutions are reduced maintenance of equipment and machines because of cleaner operating conditions, and cleaner and safer work areas.

Incorrect material specifications and excessive amounts of lubricant, as well as excessive lubricant viscosity, can be factors in wrinkling. The lubricant may need to be thinned. The lubricant must also be applied evenly, and must be correct for the working temperature of the die. Insufficient lubricant, due to improper application, poor wetting, or poor penetration of lubricant, can cause fracture. The lubricant properties must provide sufficient barrier or boundary lubrication.

Special Considerations

Disposal of spent lubricants may require that your lubricant be biodegradable. Because of the newer OSHA regulations, toxicity, safety, and nonmisting properties may hold special interest. Long-term outdoor storage of lubricants is often overlooked as a special consideration in certain areas of the country.

For long-range planning, it is advisable to consider an alternate source of procurement, especially if and when an energy squeeze re-emerges.

Certain areas of the country now have restrictions on pollution that involve lubricants containing lead, pigments, and other physical or chemical properties. A review of current lubricants can ensure compliance with the latest environmental regulations.

A host of intangible variables in lubricants make a difference in overall system performance and profitability. An in-depth analysis of the manufacturing process as it pertains to lubricant compatibility can be a productive tool for the metalformer. Good lubrication expertise is not an expense, it is a profitable tool whose importance continues to increase as stamping and forming systems become more integrated and performance-sensitive.

18

Electronic Sensors and Die Protection

Introduction

Dies and their development represent a sizable investment for the manufacturer, in terms of both money and manpower. It is no surprise, then, that protecting that investment is the focus of increasing attention as the design of dies has become more sophisticated and their manufacture more complex.

All die protection systems (Figure 18-1) have a similar broad set of major components: *sensors* mounted in and around the die to monitor specific events; a *control* to perform logic functions, provide decision-making capability, and interface with the press stop circuits; a *timing device* to allow the control to perform logic functions based on the press crankshaft angle.

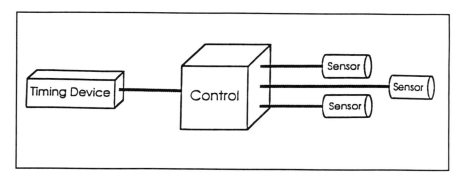

Figure 18-1. *Die protection system components.*

The control is the "brains" of the die protection system. Its job is to interpret the signals from the sensors and decide whether to let the press run. Recent advances in die protection control technology have spawned so many new controls with so many features that selecting a control can be confusing. Diemakers must be able to look beyond all the "bells and whistles" and select a control with only the features needed. Some of the latest controls, although seemingly advanced, in reality lack certain features that can make the difference between smashing and protecting your dies.

History

The earliest die protection systems were not much more than normally closed switches connected in series with the press stop circuit. When one of these switches opened, the press would stop. These switches, when operated by pilot punches, were called *hot pilots* — hot because they operated on 120 volts. With this system, a sensor actuation involved the switching of 120 volts. The possibility of electrical shock was high in dies equipped with hot pilots, so stampers started looking for something a little safer.

In the late 1950s, a metal stamper named George Wintriss designed a safer way to provide die protection. His system used a lower (and much safer) DC voltage to run the sensors. A sensor actuation in this type of system was simply a closure to ground. With this arrangement, the sensors and the press stop circuits operated at different voltages so the sensors could no longer connect directly to the press' stop circuits. It became necessary to have some sort of controller to provide DC power for the sensors and to interface with the press stop circuit(s).

The next major advance in die protection was the introduction of external timing signals to the controller. Timing signals enable the control to monitor functions that happen at certain angles during the press stroke, such as part ejection and full feed. Most controls accepted one timing signal input. Typically these controls had interchangeable modules that allowed the user to choose different modules for different sensor operating characteristics. Die protection became increasingly popular with the introduction of these controls.

Today's die protection controls are programmable. Programmability allows the user to change operating characteristics by pressing a button instead of changing a module. These controls still operate on the same basic principles as the first modular controls: a sensor actuation is still a closure to ground, sensors still operate on low-voltage DC, and the control is still connected to the stop circuit(s). However, modern controls have many features. They may accept input from as many as 16 sensors and have internally generated timing, setup memory, built-in counters, plus more.

Protection Monitoring

Examples of events that die protection systems can monitor include part ejection, misfeed, material buckle, and end of stock. However, not all events are monitored the same way. Those events that die protection systems can monitor fall into one of two categories: static or cyclic.

The term "static event" describes a function that *should not* happen unless there is a fault, material buckle being a typical example. Cyclic events are functions that should happen every stroke. If a cyclic event fails to occur once per stroke, there is a problem. Part ejection characterizes a cyclic event.

The early modular die protection controls had different modules for different event types. Pressroom personnel had to choose the correct module for the type of event they wanted to monitor. Since most pressroom personnel were not familiar with electrical or electronic terms, the event types were named after colors to make module selection easier, as shown in Figure 18-2. The sensors that monitor static events are called *red* (if the sensor is normally closed to ground) and *yellow* (if the sensor is normally open to ground). The sensors that monitor cyclic events are called *green* sensors. These names stuck and are still used today. A new sensor type called *green-special* is available with some controls. Green-special sensors can be used to detect slugs stacking up in a hole.

Static Events

A sensor that monitors a static event or condition needs no timing input. It constantly monitors the event throughout the entire stroke, regardless of crankshaft position. A static sensor does not change state unless there is a fault. An example of static event monitoring is a pilot-operated misfeed sensor. The sensor itself is a normally closed switch mounted behind a spring-loaded nonworking pilot. If the material feeds properly, the pilot will slide into a prepunched pilot hole. If the material is misfed, the pilot will miss the pilot hole, be pushed up, and open the switch. A word of caution: the more sensors connected to a single input, the more difficult it is to identify exactly which sensor signaled the fault. If the control has enough inputs, it is good practice to connect only one sensor to each input.

Progressive Dies

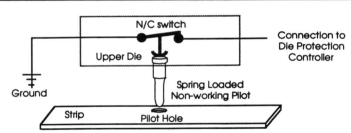

This switch is normally closed, so this is a "red" sensor. If the switch were normally open (as below) this would be a "yellow" sensor.

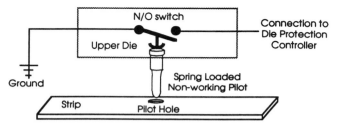

More than one static event sensor can be connected to a single die protection input. Red sensors can be connected in series to a single input (see diagram below); however, since this is a series connection, the sensors must be two-tiered or switch-type sensors.

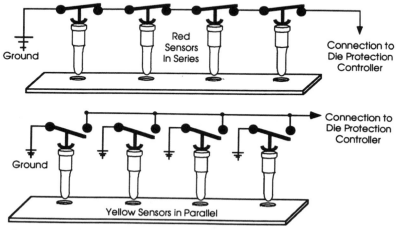

Yellow sensors of any type can be connected in parallel to a single die protection input. A word of caution: the more sensors connected to a single input, the more difficult it is to identify which sensor caused a fault. If the control has enough inputs, it is recommended that only one sensor be connected to each input.

Figure 18-2. *Red and yellow sensor configurations.*

Cyclic Events

Cyclic or green sensors monitor events that happen with every stroke, though they are not constantly monitored throughout the entire stroke like the static sensors. Typical of an event monitored by a green sensor is part ejection. The controller needs a timing input to tell it when to look for the green sensors to happen. Called a "ready signal," this input is the timing window within which the control must see a green sensor actuate. Static sensors will not work for part ejection because a static sensor must constantly be on, and the part is not constantly coming out of the die. Rather, the part comes out of the die once per stroke, usually on the upstroke, and the part activates the sensor only momentarily as it passes it or bounces off it. This momentary actuation must happen during the ready signal. The ready signal for part ejection would probably begin after bottom and end somewhere before top dead center (TDC). If the part ejection sensor does not actuate during the ready signal, the die protection control will signal the press to stop. The control decides whether or not to stop the press at the end of the ready signal. This is called the *decision point* (see Figure 18-3).

Some die protection controls allow only one ready signal for all of the green sensor inputs. This ready signal has to be long enough to allow *all* cyclic events to occur. Having such a long ready signal can significantly delay the decision point. For example, to monitor feed as well as part ejection, the ready signal shown in Figure 18-3 would have to stay on until the feed finishes. Depending on the type of feeder used, this could delay the stopping decision such that the die partially closes on a non-ejected part. Ideally, there should be a separate ready signal for each cyclic sensor. Many newer controls can provide multiple ready signals, Figure 18-4.

Some controls perform an additional sensor check often called the "sensor fault." Sensor fault is a term traditionally used to describe a control's ability to check if a sensor has

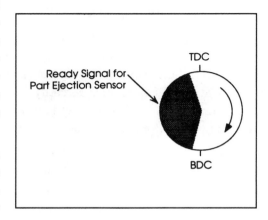

Figure 18-3. *Part ejection cyclic event monitored by a green sensor.*

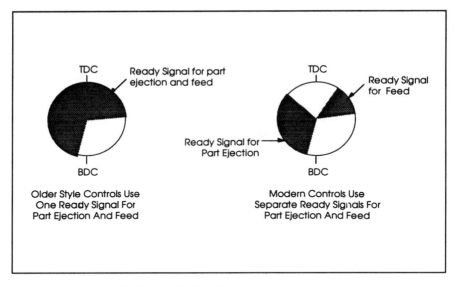

Figure 18-4. *Sensor monitoring of multiple cyclic events.*

failed by shorting to ground. It is not meant to imply that a control can verify failsafe operation of the sensor. The fault check requires green sensors to turn off at some point while the ready signal is off. This ensures that the sensor is not shorted to ground or stuck on or actuated. Without this check, a part could stick to the part ejection sensor and the control would always be satisfied because the sensor is on during the ready signal (because the part is stuck to it). Later, if a part failed to eject, the control would not know, so it would not stop the press.

For years, the only type of cyclic monitoring that die protection controls were capable of was that done by green sensors as described above. For most applications, simply detecting a momentary sensor actuation during the ready signal was enough to verify proper operation. However, as die protection gained in popularity and was applied to more and more complex stamping operations, it became evident that the standard green sensor needed additional capabilities.

The *green constant* is a modification to the standard green sensor that adds such capabilities. Instead of a momentary actuation at some point during the ready signal, the control requires the sensor to be constantly actuated during the entire ready signal, Figure 18-5. Like a standard green, the green constant must turn off somewhere outside the ready signal to ensure that it

Electronic Sensors and Die Protection

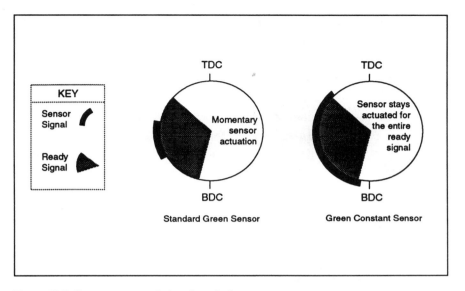

Figure 18-5. *Green constant variation of standard green sensor.*

is not stuck. Green constant was traditionally called "red bypass" because originally a red sensor was bypassed using an external timing input. This bypass signal shunts the red sensor input to ground. The bypass timing turns on shortly before the sensor is supposed to turn off. It turns off shortly after the sensor is supposed to have turned on. This way, either the sensor or the bypass is on at all times. The bypass and sensor are both connected to the die protection control's red input. One drawback of this method (red input with a bypass signal) is that the control may need an additional external timing input. Another is that the control has no way of checking the sensor to see if it is stuck or shorted. The green constant function checks to make sure the sensor is not stuck.

A typical application for the green constant sensor is shortfeed monitoring. A green constant sensor first verifies that the material is fully fed and then continues to monitor the sensor to see if the material stays in place when the feeder ungrips to allow the pilots to align the material. If a green constant shortfeed sensor detects a fault, it will stop the press and not allow it to run again unless the material is fully fed, even if the control is reset. This prevents someone from inadvertently resetting and running the press before the problem has been fixed.

Another popular use for the green constant function is transfer monitoring. The sensors monitor the transfer fingers to ensure that the part is gripped at all times during the transfer. If the transfer drops one of the parts at any time, the press can stop immediately instead of waiting until the end of the transfer.

Another type of green sensor is the "green quick-check," shown in Figure 18-6. This function could also be called "sensor outside ready not allowed." A green quick-check, like a standard green must turn on at least momentarily during the ready signal. What makes it different is that the green quick check cannot be on at any time outside the ready signal. A standard green sensor can actuate before the ready signal starts, stay on during the ready signal, and turn off after the ready signal ends. The control will allow the press to run since all the requirements of the standard green have been met. A green quick-check sensor must actuate after the ready signal starts and turn off before the ready signal ends, or the press will stop.

Green quick-check sensors should be used when it is necessary to ensure that a green sensor is off by a particular time. An example is when checking ejection of a very long part. The leading edge of a long part may actuate the part ejection sensor while the trailing edge of the part is still in the die. If the part hangs up after the sensor is actuated, the sensor will still be actuated

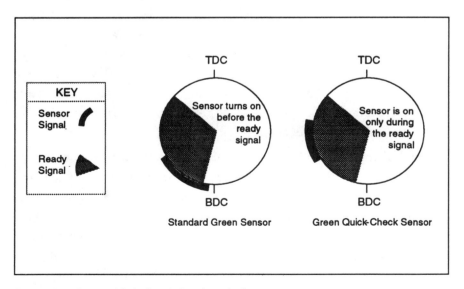

Figure 18-6. *Green quick-check variation of standard green sensor.*

after the end of the ready signal. A green quick-check will stop the press. A standard green will let the press run, possibly hitting the exiting part.

Semicyclic Events

Certain events are best described as "semicyclic," by definition, neither static nor cyclic. An example of such an event is the detection of slugs stacking up in a hole. When slugs are ejected through the bottom of the die, they do not necessarily come out on every stroke. Several strokes may go by with the slugs sticking in the hole before they finally break free and fall out of the bottom of the die. This is normal. If the slugs jam, and too many stack up in the hole, the die can be damaged. Slug stacking is impossible to monitor with a green sensor because the slugs do not come out with every stroke. A static sensor is equally ineffective because the slugs must come out at some time and will cause the sensor to change state. To effectively monitor slug ejection a special sensor had to be developed. This sensor type is called "green special." Instead of once per stroke like a cyclic sensor, a green special sensor must actuate at least once during a preset number of strokes and instead of setting a ready signal, the operator sets the maximum number of strokes the press can run without this sensor actuating. For example, if the press is set to clear all the slugs out of particular hole about every four strokes and no slugs come out in six strokes, something is wrong. Then, if the green special stroke counter was set at six, and no slugs are ejected after six strokes, the press will stop. Almost all die protection controls allow red, yellow, and standard green sensor types. Only more recent controls have the green constant, green quick check, and green special sensor functions.

Ready Signals

As stated, one of the biggest advances in die protection was the introduction of controls that could accept timing inputs, thus relating die protection events to specific angles in the press crankshaft rotation. The timing signals, called ready signals, can be generated in a number of ways. The ready signal itself is simply a closure to ground for a specified number of degrees of press rotation. Ready signals can come either from external switches or be generated internally by the control itself.

Older controls require that ready signals come from an external device. Some of the common ways to generate ready signals for these controls are rotary limit switches, ram-mounted limit switches, programmable cams, and Candy®, switches. A Candy switch is a single-station rotary limit switch with knobs to adjust the on angle and dwell. Older controls that allow multiple ready signals require a separate external switch for each ready signal. External ready signals are difficult or impossible to adjust when the press is running and often must be reset each time a die is changed.

Controls with internally generated ready signals are more flexible, more accurate, and easier to use than the older style controls that require external ready signals. Two methods are commonly employed by die protection controls to internally generate ready signals: "time-based" and "angle-based."

Less sophisticated controls use the time-based method. Time-based controls typically have one or two sensors mounted on the crankshaft of the press. These sensors send synchronization signals to the control. The control calculates the crankshaft angle based on the time interval between these signals. Because they are less sophisticated, time-based controls are inaccurate. Also, they do not work for single-stroke operations, or for the first few strokes of continuous operations. Nor do they react well to variations in press speed.

Greater accuracy is obtained through angle-based controls which use a resolver or encoder driven one-to-one with the press crankshaft. Angle-based die protection controls always know the exact angle of the press. They are very accurate, work well in single-stroke mode, start monitoring as soon as the press starts, and are unaffected by speed variations. Where there is a choice, angle-based controls are preferable. One angle-based control feature to check is the *resolution*, which is the smallest increment of rotation the control is capable of monitoring. Most controls have a 1-degree resolution; accurate enough for most monitoring. However, to cut costs, some manufacturers offer controls with resolutions as coarse as 10 degrees. These controls are not accurate enough for most stamping applications.

Ready Signal Verification

Another point to consider is whether the die protection system has a way of indicating whether the press is running. Although seemingly trivial, this is a very important point. How can the control know when to start looking for the green sensors if it doesn't know when the press is running? In systems with external ready signal inputs, the only way the control can tell if the press is running is if the ready signal is turning on and off. If something happens that causes the ready signal to fail, the control could allow the press to run virtually unchecked. Fortunately, most controls have what is called "ready fault check" circuitry to prevent this from happening. The ready fault check circuit consists of an internal timer connected to the press stop circuit. The timer is set to stay on slightly longer than it takes for the press to make one stroke. Each time the ready signal turns on, it retriggers this timer; so while the press is running the timer never turns off. If the ready signal fails, it will not retrigger the failsafe timer, and after one stroke the timer will turn off. This causes the press to stop. During normal operation when the press stops, the timer will turn off and must be reset before the press can run again. Die protection controls that must be reset immediately before the press is run likely have a fault timer.

Systems with internally generated ready signals usually monitor the solenoid of the clutch/brake valve, so they always know when the press should be running. Fault check circuitry is built into these systems. They do not have to be reset before the press is run.

Stop Outputs

Most presses have two stop circuits, the emergency or immediate stop circuit and the delayed or top stop circuit. Die protection controls can have either one or two stop circuit outputs. The systems that have two outputs can be tied into both stop circuits and typically allow the user to select which faults will cause the machine to top stop and which faults will cause the machine to emergency stop. When selecting the stop type for an event, a good rule to follow is to always use emergency stop unless:

1. The event being monitored is "end of material." This is not a die-threatening fault. The press can run back up to top before stopping.

2. The event being monitored has a decision point that comes after the critical angle (the critical angle is the latest angle in the crankshaft rotation where the press can stop before die closure). If a fault occurs and the press tries to emergency stop, it is possible that the die will still smash and the press may stick on bottom.

The big drawback of controls with only one stop output is that if any of the events happen after the critical angle, there is danger of sticking the press on bottom when the control signals an emergency stop.

Productivity Enhancing Features

The features described thus far are directed specifically toward protecting the dies. Other control features exist that are best described as productivity enhancements. Controls with productivity enhancing features do not necessarily protect dies better than controls without them. However, since it is not desirable to sacrifice productivity to protect the dies, these features should be considered in the selection process.

Setup Memory. Setup memory describes a control's ability to store job setup information. Information like ready timing, sensor type (red, yellow, or green) and stop type (top stop or emergency stop) is likely to differ from one job to the next. Without setup memory, the control would have to be reprogrammed every die change. With setup memory, the operator simply recalls previously programmed information for each die. Controls typically store this information by program number; however, some controls allow setup information to be stored using a company's own tool numbers.

Sensor Timing Display. Most programmable die protection controls will display the ready timing for each sensor. Some also display the actual on/off timing of the sensor itself. This feature can be quite valuable in the initial debugging stages of sensor implementation. The first time a die is set up with die protection, ready timing for each sensor is a matter of best guess. A trial is then needed to adjust the timing as necessary to get the job to run. With the actual sensor timing displayed along with the ready timing quick adjustment and much more accurate ready timing is possible.

On-the-fly Adjustability. It is inevitable that periodic minor adjustments to the ready timing for most dies will have to be made. Without on-the-fly adjustability the press must be stopped, the control adjusted, and the press run again to check the adjustment. This is a tedious, frustrating, and time-consuming process. There are controls, however, that allow adjustment while the press is running. Typically, these adjustments are automatically stored with the setup information.

Fault Identification. How does the control react when it detects a fault? First it will send a stop signal to the press. It will also give some sort of visual signal to the operator to indicate that a fault condition exists. If the control simply trips an indicator light, it is up to the operator to find the fault, what caused the fault, and when it occurred. Some controls provide detailed error messages that indicate which sensor detected the fault and describe the exact nature of the fault and the angle at which it occurred. The control may also tell if an event occurred too late, or not at all. Good descriptive error messages are important if the control has many sensors connected to it.

Ease of Use. The ease with which the operator uses the control is the most difficult feature to quantify. Virtually all manufacturers claim that their control is easy to use. Unfortunately, user friendliness is not gaged by reading a brochure or listening to salespersons. The only certain way to know is to use it — there is no substitute for hands-on experience. If the control supplier has a demonstrator model, ask to program it or at least ask for a live demonstration. Answer these questions:

- Am I able to program the control without using the manual?
- Is the user manual complete and easy to understand?
- If I have questions, is there somebody I can call to get answers?

Control Checklist

The features described are not standard on all controls. In selecting a control, decide which features best suit the application. If the press already has a control, check which features it has and which it does not have. This can help ensure maximum protection of dies. As an aid in evaluating controls, the control checklist on page 394 cites many of the control features available and the page numbers on which their description appears in this book.

Using Electronic Sensors in Die Protection

Die protection sensors fall into two broad categories: electromechanical and electronic. Until the early 1980s, virtually all die protection was done with electromechanical sensors, the most popular of which were spring probes or cat whiskers. As the name implies, an electromechanical sensor requires some kind of mechanical contact to send an electrical signal to the control. Often this simply meant that the electrically grounded material would contact the end of a spring.

Die Protection Control Checklist

_____Multiple Ready Signals..Page 385
Separate ready signals for each green sensor allows the control to stop the press as soon as a fault is detected.

_____Sensor Check..Page 385
Control checks that green sensors are not "stuck" or shorted to ground.

_____Green Constant Sensor Type...Page 386
More effective shortfeed and transfer monitoring.

_____Green Quick-check Sensor Type......................................Page 388
Control can verify sensor operation immediately upon completion of the ready signal.

_____Semicyclic Sensor Type..Page 389
Control can check events like slug ejection even if they do not occur on every stroke.

_____Internally Generated Ready Signals..............................Page 389
Eliminates expensive and hard-to-adjust external timing devices.

_____Angle-based Ready Signals..Page 390
More accurate than time-based controls. Starts protecting the die as soon as press starts.

_____Ready Failsafe Circuitry..Page 391
Verifies proper operation of the ready signal.

_____Selectable Stop-type for Sensors.....................................Page 391
Allows user to select stop-type for each sensor individually. Helps to ensure that the press will not stick on bottom in the event of a die protection fault.

_____Setup Memory...Page 392
Allows quick recall settings for each die.

_____Sensor Timing Display...Page 392
Allows more efficient debugging and more accurate ready signals.

_____On-the-fly Adjustability...Page 392
Allows control adjustment without stopping the press.

_____Fault Identification...Page 392
Does the control provide detailed error messages?

_____Ease of Use..Page 393
Can I program the control without having to use the manual? Is the manual complete and easy to understand? Will the press operators be able to use the control?

These sensors, however, have drawbacks. Because of their mechanical contact, they can wear out. They also require frequent minor adjustments that can add to setup time. And because of "contact bounce," they do not work well in high-speed applications. Electromechanical sensors also are often too large and cumbersome to mount in the die.

This is not to imply, though, that electromechanical sensors should not be used. On the contrary, electromechanical sensors have been part of die protection for more than 30 years and will continue to be important for a long time to come. In applications such as material buckle detection, where the sensor does not have to change state on every stroke, electromechanical sensors are still the device of choice.

For detecting events that occur on every stroke, electronic sensors work best. Electronic sensors solve many of the problems of their electromechanical predecessors and are a must for applications that require improved accuracy, smaller size, and high speed. They eliminate contact bounce and greatly reduce adjustment time. Electronic sensors do not make contact with the material being sensed, so they are also called non-contact sensors. They work extremely well with painted or precoated materials and have been successfully used in other industries. Today, they are slowly making their way into metal-stamping operations.

Though "electronic sensor" can be used to describe thousands of different devices using hundreds of different technologies, most electronic sensors are not suitable for use in die protection because of the extremely harsh environment. The two types of electronic sensors commonly and successfully used for die protection are inductive proximity sensors and photoelectric sensors. Inductive proximity sensors are widely used to check for full-feed progression and photoelectric sensors to detect part ejection, though by no means are they limited to these applications.

Proximity coil sensors are often used to detect small, fast-moving parts. Versions of photoelectric sensors (particularly fiberoptic) are well suited for checking feed progression.

Inductive Proximity Sensors. Proximity sensors are popular for a number of reasons. They are small enough to mount in the die, effectively becoming part of the tool; they are the lowest-cost electronic sensor; they are immune to nonmetallic contaminants like dirt and oil; proximity sensors are virtually maintenance free; and they are very accurate and come in sizes suitable for many applications. Proximity sensors, moreover, do not require contact with the material and have no moving parts, so they will not wear out. One drawback of proximity sensors is their short ranges. If mounted farther than 0.38 inch (9.7 mm) from the target object, a proximity sensor probably will not work.

An inductive proximity sensor uses an internally generated magnetic sensing field to detect metallic objects. Its primary components are a coil of wire inside a ferrite core, an oscillator circuit, a detector circuit, an output amplifier and an indicator light (see Figure 18-7).

An oscillator is an electronic circuit that generates a high-frequency electrical signal. This signal is fed to the coil, causing the coil to emit a magnetic field. The ferrite core around the coil helps to "bundle" or direct the field toward the front of the sensor. The magnetic field is also called the sensing field. When a metallic object (called the "target") enters the sensing field, it absorbs some of the field's energy. The detector circuit detects the energy loss and activates the output. When the target object leaves the sensing field, the field regenerates and the sensor turns off.

Proximity Sensors for Full-feed Detection. A proximity sensor is a good choice for checking full-feed progression. It is extremely accurate and small enough to be embedded in the die and is not affected by a buildup of oil and grime, making it truly maintenance free. However, care must be taken when installing these sensors to ensure that the highest level of accuracy is realized.

A notion prevails that for feed detection, a proximity sensor can simply be installed in the bottom die so that it detects a hole or edge of the material. This simplistic approach may work in some dies, but most of the time it results in poor accuracy. The low accuracy level stems from the shape of the proximity sensor's sensing field and the target material's tendency to move up and down in the die. The sensing field of a shielded proximity sensor is shaped like a slightly rounded cone.

The sensing field narrows as the distance from the sensor face increases. The amount of the sensor face that must be covered by the target material (the strip stock) is determined by the distance between the target and the sensor (see Figure 18-8).

Therefore, any up and down motion of the strip within the die will throw off the actuation point, thus affecting the accuracy of the proximity sensor. To achieve the level of accuracy required for full-feed detection, the material has to be kept at a constant distance from the sensor. Unfortunately, in most cases the only time the material is at a known distance from the bottom die is when the stripper plate is holding the material down. By the time the stripper plate is holding the material, it is too late to stop the press if there is a misfeed.

However, this is not to say that proximity sensors cannot be used for detecting feed progression. Following are descriptions of four specific applications that use inductive proximity sensors to check feed progression.

The first uses stock guides in the die that limit the up-and-down motion of the strip. By confining the strip, the sensor-to-strip distance is kept constant enough to allow for accurate feed detection. However, confining the strip this way can increase the possibility of the material hanging up and buckling during the feed. This application uses two proximity sensors; one to detect overfeed and one to detect shortfeed.

In the second application, the sensor is brought to the strip rather than bringing the strip to the sensor. The sensor is installed in a spring-loaded bushing that is in constant contact with the strip. If the strip moves up and down, the bushing and sensor will move with it. The sensor is recessed slightly in the bushing so that it does not wear as the strip moves across the

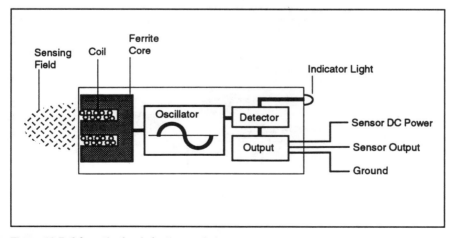

Figure 18-7. *Schematic of an inductive proximity sensor.*

bushing. This method works well when the up-and-down motion of the strip is less than 0.25 inch (6.4 mm). If the strip is on lifters, this method is not recommended. The sensor's cable has a tendency to get pinched in the spring if the sensor has to move more than 0.25 inch.

In the last two applications, the sensors are installed at the end of the die, and the material feeds up against them. In Application No. 3, with a short feed length, the material simply feeds toward the sensor. When the material is fully fed, the end of the material enters the sensing field and actuates the sensor.

In Application No. 4, with a longer feed length, more material is hanging over the end of the die, which can cause the strip to sometimes sag below the sensor. To reliably detect the material, the sensor is installed in a plunger assembly. The strip feeds up against the plunger, and the plunger actuates the sensor. The plunger has a larger diameter than the sensor and gives the strip a bigger target to feed against.

Application 1 — Using Proximity Sensors to Detect Shortfeed and Overfeed

The Application. The part being stamped in this application is a very large 24-inch (610-mm) motor lamination made from 0.025-inch (0.60 mm) cold rolled steel. An air feed is used and the feed progression is 24 inches.

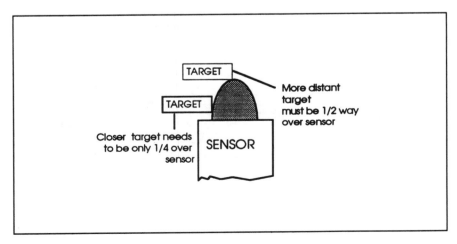

Figure 18-8. *Proximity sensor showing cone-shaped sensing field.*

Electronic Sensors and Die Protection

The die has working pilots that can align the material as long as it is within 0.030 inch (0.76 mm) of full progression. To verify that the material is in place, both shortfeed and overfeed are monitored. The press speed is 25 stampings per minute (SPM).

The Sensors. The sensors used for this application are 0.12-inch (3.0 mm), fixed sensitivity proximity sensors. They have a sensing field about 0.040 inch (1.02 mm) in diameter with a detecting distance of about 0.023 inch (0.58 mm) from the face of the sensor. These are fixed sensitivity sensors, so this distance cannot be adjusted; the only adjustment option is changing the distance to the target.

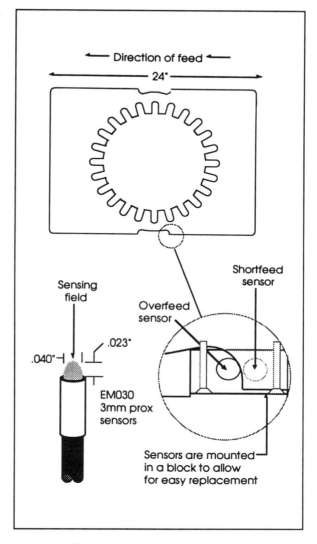

Figure A-1. *Shortfeed and overfeed detection.*

The sensors are mounted next to each other in a block in the bottom die focused at a notch in the side of the material. Although they come from the same manufacturer, these sensors are available with different operating frequencies so they can be mounted next to each other without interference. The stock guides along the side of the material in the area of the sensors are designed to be more restrictive to confine the material at a more or less fixed distance from the sensor.

Control and Timing. Because of the slow press speed and a quick airfeed, the material is fully fed by 300 degrees. The shortfeed sensor is a green sensor type. The ready signal for this sensor turns on at 280 degrees and off at 300 degrees. If the material at the end of the notch doesn't cover the sensor by 300 degrees, the press will emergency stop. The overfeed sensor is a green quick check type which stops the press if it is on anywhere outside of the ready signal. The ready signal for the overfeed sensor turns on at 100 degrees and off at 300 degrees. If the material overfeeds, it will cover the overfeed sensor at 300 degrees. The sensor will be on outside of the ready signal and the press will emergency stop.

Application 2 — Using a Proximity Sensor in the Lower Die to Detect Feed Progression

The Application. This application involves sensing full-feed progression for a progressive die running 0.075-inch (1.90-mm) steel. This die has three pilots that can align the material if it is within ±0.015 inch (0.38 mm) of full progression. This die originally had four pilots. To accommodate the sensor, one of the pilots was removed and its entry hole in the lower die was enlarged. A bushing was installed in the enlarged hole. A proximity sensor was installed in this bushing, and a spring placed below the bushing to keep it in contact with the strip at all times. The proximity sensor is slightly recessed in the bushing so that the face of the sensor will not wear. Since the bushing is always in contact with the strip, the proximity sensor is always at a fixed distance from the strip. The proximity sensor detects the presence of the pilot hole in the strip. The sensor and bushing are installed in one of the first stations in the tool (after the pilot holes are pierced, but before any forming occurs).

The Sensor. The sensor is a 0.48-inch (12.1-mm) inductive proximity sensor. It is connected to an adjustable amplifier, which allows the sensor to be precisely tuned to the size of the hole being sensed and the material composition.

Adjustment Procedure. Turn the amplifier's sensitivity adjustment all the way down. Install the strip in the die so that it is 0.015 inch (0.38 mm) underfed (the maximum amount the pilots can accommodate to align the strip). Slowly increase sensitivity adjustment on the amplifier (turn the screw clockwise) until the sensor barely turns on. Move the strip ahead

0.015 inch (0.38 mm) (to its fully fed position) and verify that the sensor turns off. Continue to move the strip until it is 0.015 inch overfed and verify that the sensor turns on again.

Control and Timing. This input's sensor type is a green constant. The feed finishes at 5 degrees. The ready signal is set to turn on at 10 degrees and off at 170 degrees. The sensor is set to emergency stop the press if the material is improperly fed anytime between 10 degrees and 170 degrees.

Application 3 — Using An Adjustable Proximity Sensor For Feed Detection

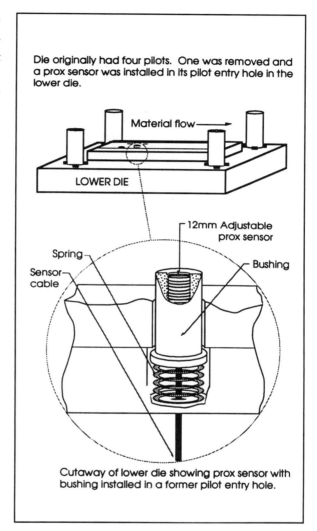

Figure A-2. *Feed progression detection in lower die.*

The Application. This application involves the detection of feed progression for a 9-station, 2-up progressive die. The material is 0.038-inch (0.97-mm) steel-backed bearing material. The die has working pilots that can align the material if it is within 0.020 inch (0.51 mm) of full progression. The press speed is 80 SPM, and a light film of lubricant is applied to the strip and die surfaces. The sensor is installed at the end of the die and detects the scrap at the end of the strip. This scrap is cut off and falls through an opening in

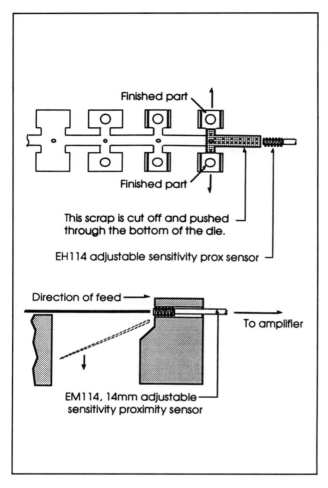

Figure A-3. *Feed progression detection with an adjustable proximity sensor.*

the die. The sensor also detects a buildup of scrap if the scrap fails to fall through the bottom of the die

The Sensor. The sensor is a 0.56-inch (14-mm) adjustable sensitivity proximity sensor with an amplifier that allows the sensing field to be adjusted. This sensor's maximum detecting distance is about 0.196 inch (5 mm) and can be adjusted down to 0.079 inch (2 mm). This sensor was chosen for its long detecting distance and adjustable sensitivity. A long range sensor is needed because of the small profile (only 0.038 inch (0.97 mm)) of the edge being detected. The adjustable sensitivity allows for the detection of materials with different compositions.

Adjustment Procedure. With the material fully fed, increase the sensitivity adjustment of the amplifier slowly until the sensor turns on. Verify the adjustment by backing the material out 0.020 inch (0.51 mm) and seeing if the sensor turns off.

Control and Timing. This sensor is a green sensor type. The ready signal turns on at 330 degrees and off at 30 degrees. The control signals the press to emergency stop if the material is not fed by 30 degrees. A green sensor must turn off at some point outside of the ready signal. If the scrap that is cut off does not fall through the die properly, it will build up in the die and eventually stay in front of the sensor. This will cause the sensor to stay on for an entire stroke. When this happens the sensor will signal an emergency stop.

Application 4 — Checking Feed at the End of the Die with a Proximity Sensor

The Application. It is sometimes difficult to install a feed sensor inside the die. In such cases, installing it at the end of the die is recommended. If the material is properly fed at the end of the die, it must be fed everywhere else. In this application, the material is fed over a cutoff station past the end of the die where it actuates the feed sensor. At bottom the part is cut off the strip and drops onto a ramp.

The Sensors. The progression is checked with a 0.32-inch (8-mm) adjustable sensitivity proximity sensor. For this application, the sensor was incorporated into a homemade metal case with a spring-loaded plunger. The strip feeds up against the plunger and moves it back toward the sensor. When the plunger gets close enough, the sensor actuates. This installation is optimum for a couple of reasons. First, since the edge of the material is somewhat thin, it may not reliably actuate the sensor. Second, this sensor must detect the edge of a part that is past the end of the die, yet there is nothing past the end of the die to support the part. The material does not always feed directly at the center of the sensor, and the end of the strip may sag below the centerline of the sensor or ride higher than the centerline. By having the strip feed up against the plunger and designing the plunger to operate the sensor, the sensor's target remains consistent and the negative effects are minimized.

Control and Timing. The sensor type is set to green constant. The material feeds up against the sensor at about 0 degrees and stays against the sensor until the part is cut off. The ready signal for this sensor turns on at 5 degrees and off at 170 degrees. If the material is not fully fed by 5 degrees, or if it pulls back off of the sensor anytime between 5 degrees and 170 degrees, an emergency stop signal will be sent to the press.

Photoelectric Sensors. Sometimes it is impossible or undesirable to install a sensor close to the object that it will sense. When this is the case, a proximity sensor will not work. As a rule of thumb, if the sensor cannot be

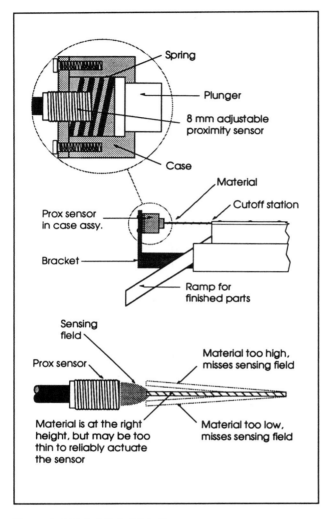

Figure A-4. *End-of-die feed detection.*

installed within 0.38 inch (3.7 mm) of the target, a proximity sensor cannot be used. That being the case, the best choice is a photoelectric sensor, which has a longer range than most proximity sensors. Often the range is long enough to install the sensor away from the target, sometimes even outside the die looking in. Photoelectric sensors are also quite versatile. They can be used to sense most everything from end of material to part ejection.

As the name implies, photoelectric sensors use light and electricity to accomplish the sensing task. A transmitter in the sensor (also called the emitter) converts electrical energy into light energy using devices called light-emitting diodes (LED), a semiconductor that emits a small amount of light when current flows through it. The receiver of a photoelectric sensor converts light energy into an electrical signal using various devices that are sensitive to the amount of incoming light. The receiver may also have circuitry to allow adjustment of the amount of light required to actuate the sensor.

Photoelectric sensors use either visible or infrared light. Infrared light has a longer wavelength than visible light and is invisible to the naked eye. Sensors with infrared light are less affected by droplets of oil in the sensing area because infrared light is not reflected by oil; rather, it burns though it. Visible light sensors are easier to align than infrared because the spot of light is clearly visible. However, visible light sensors can be affected by oil in the detecting area.

Photoelectric sensors have two operating modes called "light-on" and "dark-on," Figure 18-9. If a sensor operates in the light-on mode, it actuates (turns on) when the light from the transmitter is allowed to reach the receiver. If a sensor is operating in the dark-on mode, it actuates when the light beam between the transmitter and receiver is interrupted (thus causing the receiver to become "dark"). Some photoelectric sensors have a switch to select light-on or dark-on. Switching a photoelectric sensor from light-on to dark-on is similar to changing it from normally open to normally closed.

Opposed-mode Sensors. Application 5 (page 410) uses opposed-mode photoelectric sensors (also known as through-beam or beam-break sensors) to detect part ejection. Each of the opposed-mode sensors has a separate

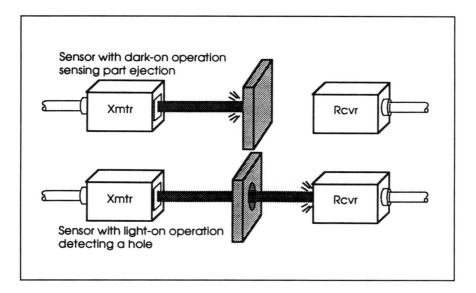

Figure 18-9. *Dark-on and light-on modes of photosensors.*

Progressive Dies

transmitter and receiver that usually operate in the dark-on mode for part detection, Figure 18-10. The transmitter targets a beam of light directly at the receiver. These sensors have an extremely long range; the receiver and transmitter of the model used in this application could be separated by a maximum of 27 inches (686 mm). However, when the sensors are installed in this die, they are not more than 4 inches (102 mm) apart. The excess gain of the sensor (the difference between 4 inches and 27 inches) acts in effect as a "reserve" to burn through the film of grime that will inevitably build up on the sensor.

In Application 6 (page 412), another type of photosensor is used, called a diffuse reflective photosensor (Figure 18-11). Diffuse reflective sensors

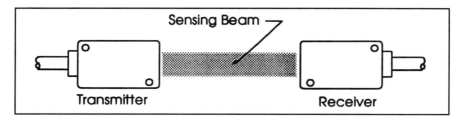

Figure 18-10. *Components of beam-break photosensors.*

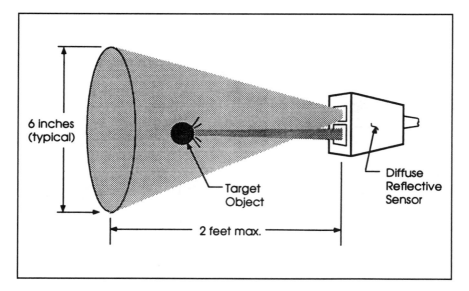

Figure 18-11. *Diffuse reflective photosensor.*

(also called diffuse proximity sensors) house the receiver and transmitter in a common unit. The beam of light from the transmitter strikes the part and is scattered — or diffused — in all directions. The sensor actuates when a small amount of this light reflects off the part and bounces back to the receiver (light on).

The sensing field of a diffuse sensor spreads out from the face of the sensor, becoming wider as the distance from the sensor increases. The sensing field also weakens as the distance increases, requiring larger and larger targets to reflect enough light back to the sensor to actuate it. Finally, about two feet (0.6 m) from the sensor, the field gets so weak that any target, regardless of size, will fail to actuate the sensor.

Application 6 needed a large sensing area so several diffuse reflective sensors were "stacked" and connected to one input on the control to make a diffuse light curtain.

The diffuse proximity sensors used in this application operate on infrared light and are not prone to sending false signals reflecting from oil droplets in the sensing field. A buildup of grime on the sensor face may cause it to operate erratically, but a quick cleaning is usually all that is needed to return it to smooth operation. The main disadvantage of diffuse sensors is their inability to detect objects near a background. They must be installed to focus into empty space, so that the background will not interfere with the sensing field.

Applications 7 and 8 (pages 413 and 415) use fiberoptic sensors to detect ejected parts. A fiberoptic sensor consists of the light source containing a transmitter and receiver and fiber or "light pipe" (Figure 18-12). The sensor's light source is similar to a reflective photosensor in that the transmitter and receiver are located side-by-side in the same housing. Most fiberoptic sensor light sources have adjustable sensitivity and a switch to select light-on or dark-on operation. The fibers are typically attached to the light source by screws or a twist or cam-lock mechanism.

A fiberoptic light source can use visible or infrared light, the range of which is determined by the type of fiber used. Plastic fibers have a longer range when used with visible light. Glass fibers work better with infrared light.

The fiber, or light pipe, for the fiberoptic sensor (see Figure 18-13) "conducts" light between the light source and the sensing area. A fiberoptic fiber consists of a core with a high refractive index surrounded by cladding

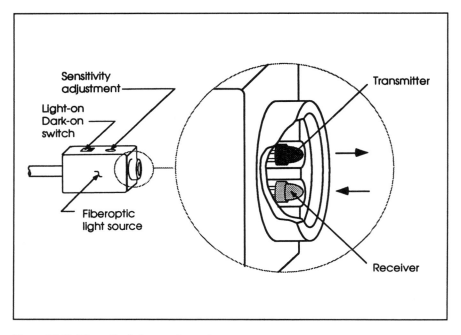

Figure 18-12. *Fiberoptic photosensor transceiver.*

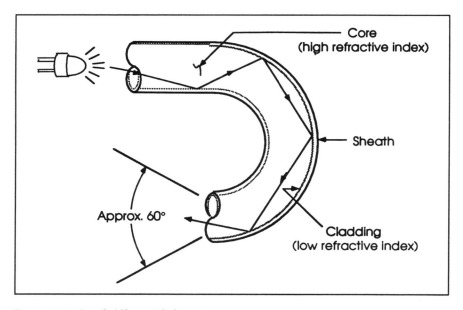

Figure 18-13. *Detail of fiberoptic light pipe.*

Electronic Sensors and Die Protection

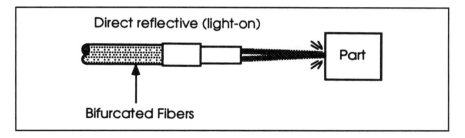

Figure 18-14. *Direct reflective sensing with bifurcated fiberoptics.*

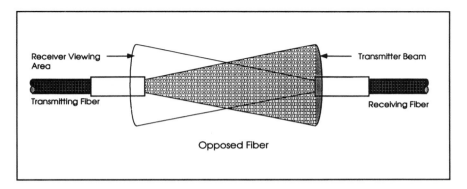

Figure 18-15. *Opposed-fiber sensing.*

having a low refractive index. The core allows light energy to pass through very quickly. The cladding does not allow light energy to pass; rather, it "reflects" the energy back into the core. The light travels through the fiber and exits out of the opposite end. Light is typically emitted from the end of the fiber in a 60-degree cone shape. Lenses (like the one used in Application 8) are available that can focus the light into about a 10-degree cone.

Application 7 (page 413) uses reflective fibers, Figure 18-14. A reflective (or bifurcated) fiber is a fiberoptic assembly with the fibers from the emitter and receiver combined into a single unit. Reflective fibers are typically used for direct reflective sensing in which the target object reflects the emitted light back to the fiber (light-on mode).

Application 8 (page 415) uses an opposed fiber. Opposed fibers consist of two identical but separate fibers (Figure 18-15). One fiber carries light from the transmitter to the sensing area, the other carries light from the sensing area back to the receiver. Opposed fibers usually operate in the dark-on mode when sensing part ejection.

Application 5 — Using Photosensors to Detect Part Ejection for a Four-up Die

The Application. This application involves sensing part ejection for a four-up die. The parts are cut off in the last station and drop through four cutouts in the lower die onto a conveyor. The parts are made from 0.125-inch (3.18-mm) mild steel and are about 1.5 inches x 1 inch (38.1 mm by 25.4 mm) in size. The press speed is 60 SPM. The sensors are mounted in a phenolic (plastic) plate installed under the cutoff station. The sensors' wires are routed through slots cut into the mounting plate and joined in a junction box at the end of the die.

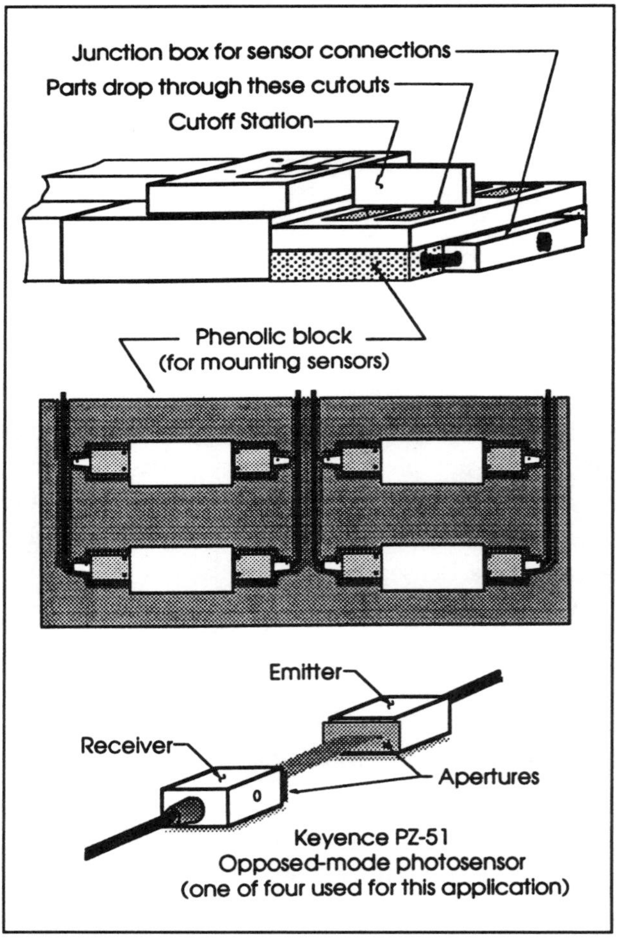

The Sensors. The sensors are opposed-mode photosensors. Apertures are installed on the sensors to limit their field of view. Without the apertures, the conveyor and die could interfere with the sensors' operation, causing false signals to be sent to the control. Opposed-mode photosensors are used instead of fiberoptics because their longer range gives these sensors enough extra power (or "ex-

Figure A-5. *Photosensing part ejection in a four-up die.*

cess gain") to burn through the film of oil and grime that is often deposited on the face of a sensor over time. The sensors' transmitters and receivers are installed about 4 inches (102 mm) apart. The PZ-51 has both light-on and dark-on outputs. For this application, the sensors operate in the dark-on mode so the dark-on outputs are connected to the control. When a part drops through the die, it breaks the beam between the transmitter and receiver, and actuates the sensor.

Adjustment Procedure. These sensors require minimal adjustment. The sensitivity needs to be set only once: Simply turn the sensors' sensitivity adjustments to the maximum. Drop parts through the cutouts in the die and verify that the sensor actuates when the part passes by it. If the sensor misses the part, gradually decrease the sensitivity until it detects the part.

Control and Timing. The control input's sensor type is a standard green. The ready signal turns on at 180 degrees and off at 270 degrees. The control is set up so that it will send an emergency stop signal to the press if the part is not detected by 270 degrees.

Application 6 — Using a Diffuse Reflective Sensor "Screen" to Detect Part Ejection

The Application. This part is produced in a compound die running at 60 SPM. The part is carried up with the upper die, knocked out at the top of the stroke and is ejected out the back of the press by a blast of air. The part is hat-shaped and is about 4 inches (102 mm) in diameter by 2 inches (51 mm) high.

The Sensors. The sensors for this application are diffuse reflective photosensors. Four sensors are stacked on top of one another to make a diffuse reflective "screen." The sensors are separated by 2 inches (51 mm) so the screen is approximately 10 inches (254 mm) high. The sensors are mounted on a metal plate and offset slightly to allow access to each sensor's adjustment screw. This light screen can detect large parts to a distance of about 2 feet (0.6 m). If the parts are smaller than about 1 inch by 1 inch by 1 inch (25 mm by 25 mm by 25 mm), a diffuse light screen might not detect them. The sensors are wired in parallel, also called OR mode, to one die protection input. If one or more sensors detects the part, the input will be activated.

Progressive Dies

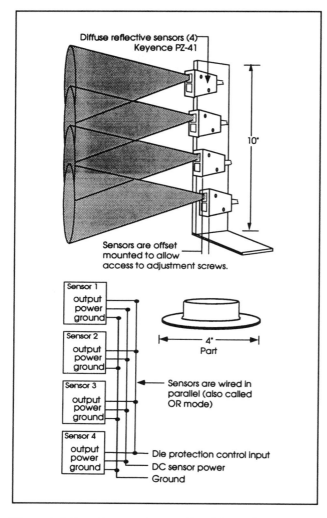

Figure A-6. *Detecting part ejection with a diffuse reflective sensing screen.*

Adjustment Procedure. The main concern when using diffuse reflective sensors is interference from background objects. The adjustment procedure will ensure that the sensors will reliably detect the parts and not the background; therefore, increase the sensitivity on each sensor to the maximum level. Install the screen so that it looks across the path of the flying parts as close to the die as possible. If possible, install the screen so that it looks off into space. If any of the sensors are on (because they detect objects in the background), slowly decrease their sensitivity until they turn off.

Control and Timing. All four sensors are wired to a single die protection input. This input's sensor type is a green. The ready signal for this sensor turns on at 350 degrees and off at 70 degrees. If a part is not detected by 70 degrees, the control signals the press to emergency stop.

Application 7 — Using Reflective Fiberoptics to Check Part Ejection

The Application. Two identical parts are produced each stroke. The final station of the die is a cutoff station. When the parts are cut off, they slide down 45-degree ramps on either side of the die and drop through square cutouts about 1.25 inch by 1.25 inch (31.8 mm by 31.8 mm). The parts are small, about 0.38 inch by 0.63 inch by 0.5 inch (9.7 mm by 12.7 mm by 31.8 mm). The parts follow a very repeatable path as they slide down the ramps. Press speed is about 80 SPM, a light lubricant is applied to the die and strip.

The Sensors. The sensors are reflective fiberoptics with a visible red light source. These sensors have a range of 3 inches (76 mm). The fibers are installed in blocks about 2 inches (51 mm) from the ramps looking a-cross the cutout. They are aimed so they detect the part at the very end of the ramp as it is falling through the cutout.

Adjustment Procedure. With no parts in place, increase the light

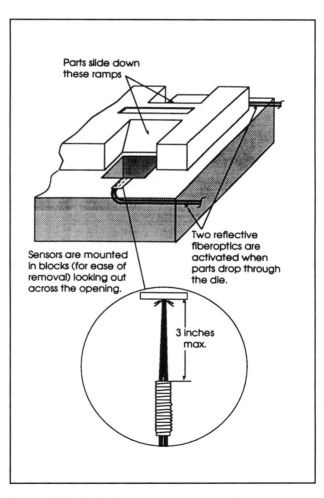

Figure A-7. *Reflective fiberoptics monitoring part ejection.*

source's sensitivity until the sensors detect the ramp that the parts slide down. Then, back off the sensitivity slightly until the ramp is no longer detected. This gives the maximum sensitivity necessary to detect small parts.

Control and Timing. These sensors are connected to separate inputs on the die protection control. They operate in the light-on mode and the part reflects light back to the receiving fiber. They are green sensors and send a momentary signal to the control when the part passes by. The ready signal for each of these sensors is the same: on at 235 degrees and off at zero degrees. The control is set to emergency stop in the event of a missed part.

Maintenance. There is a film of light lubricant on the strip and die surfaces. Some of this will deposit onto the end of the fiberoptics. A periodic wiping of the end of the fiber will prevent nuisance stops.

Application 8 — Part Ejection Detection in a Tight-clearance Application

The Application. The part is a 1-inch- (24-mm-) diameter rimmed cup made from 0.045-inch (1.14-mm) mild steel. It is made in an 8-station progressive die at 65 SPM. The strip is on lifters and the part is pushed out of the bottom of the strip in the last station. An air blowoff blasts the part out of the die shortly after bottom dead center. Sometimes the part "hovers" in place during the air blast and stays in the die. When this happens, the press must be stopped.

The Sensors. The sensor is a through-beam fiberoptic sensor fitted with a lens. Fiberoptic sensors can be used because the parts follow a repeatable path when they leave the die. The fibers are about 10 inches (254 mm) apart. The addition of a lens is very important in this application. The lens is not used to increase the range; the sensor has a range of 12.5 inches (318 mm); instead the lens is needed to focus the beam. The part is being ejected under the strip and above the lower die, with only about 1.5 inches (38 mm) of clearance. Without a lens, the transmitting fiber sends light out in about a 60-degree cone. Enough light would reflect off of the lower die and the bottom of the strip so that the sensor would miss the parts. By putting a lens on the transmitting fiber, the light is focused into about a 10-degree cone.

This makes the beam narrow enough so that the sensor can only be actuated by the part and not the strip or lower die.

Adjustment Procedure. Turn the sensitivity adjustment to the maximum. Single stroke the press while gradually decreasing the sensitivity. Mark the point at which the sensor first misses a part. Set the sensitivity midway between this point and the maximum sensitivity.

Control and Timing. This input's sensor type is a green. The part is ejected just after bottom dead center, or BDC. The ready signal turns on at 185 degrees and off at 300 degrees. The sensor is set to emergency stop the press if a part is not ejected by 300 degrees.

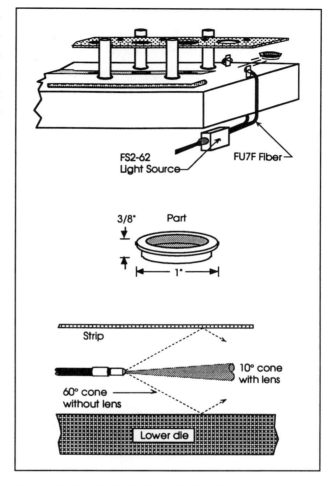

Figure A-8. *Beam-break detection for limited-clearance situations.*

||||19||||

Quick Die Change

The use of quick die change methods and equipment can increase the stamping shop capacity without increasing the size of the facility and purchasing more presses. By increasing press up-time from 50 percent to 90 percent, shop capacity goes up 80 percent. Quick die change can also improve quality and reduce inventory. Moreover, Just-in-Time manufacturing can be elusive without quick die change.

Advantages

Increased Capacity. The time spent setting up a stamping press is essentially idle press time. Setup time reduction is the most cost-effective means to increase the capacity of a stamping plant. The required capital expenditure is low compared to the cost of additional presses.

Improved Quality. Quick die change techniques require the exact duplication of a standardized setup. Examples of this are positive die location and common shut heights. The elimination of trial-and-error techniques practically guarantees repeatability.

Scrap Reduction. The goal of quick die change is to quickly change from one standardized setup to another and produce top-quality parts when production resumes. The setup repeatability that quick die change requires ensures that the first hit produces a good part. Eliminating trial and error adjustments eliminates scrap.

Enhanced Job Security. Metal stamping is a highly competitive business, and survival depends upon making a profit. World class means the ability to meet the challenge of any shop in the world in terms of deliverable quality, actual productivity, and real profitability. Backing away from the challenge simply means that ultimate defeat and insolvency has been accepted.

Increased Safety. Conversion to quick die change provides an opportunity to improve die clamping methods. For example, if strap clamps are currently being used in conjunction with poorly made setup blocks, the quick die change conversion provides the opportunity to adopt more secure clamping methods.

Reduced Die and Press Maintenance Costs. A large amount of total die and press damage occurs when diesetting. Shut-height errors are a persistent source of difficulty. However, if a common shut-height is established, these errors can be eliminated. Quick die change requires secure die handling methods, by ensuring that dies are not accidentally dropped.

Reduced Inventory. The increased press up-time and productivity of quick die change methods can cause the warehouse to overflow rather quickly. The reduced setup cost and increased setup confidence make possible low economic order quantities (EOQ) and short production runs, making JIT a realistic goal.

Grouping Presses and Dies for Quick Die Change

Several good arguments can be made for designating a home or primary press for each die. In fact, many stampers do so on the part-process sheets. Always using a die in the same press improves process repeatability from one run to another.

The amount that a press deflects per ton varies from press to press. It is especially true when comparing presses made by different manufacturers. This can be a critical factor when maintaining dies at a common shut height.

A great many other factors can limit which two or more presses are co-primary for a die. For example, a straightside press in good condition remains level if symmetrically loaded. A gap-frame press has an unavoidable angular deflection of approximately 0.0015 inch (0.038 mm) per 1.0 inch (25.4 mm) of front-to-back bed distance at full load.

A large cutting die with close clearances may run satisfactorily in the straightside press and be damaged each time an attempt is made to run it in a gap-frame press, even though the gap press had plenty of tonnage and bed size.

Some attempts have been made to reduce the angular deflection inherent in gap-frame machines by fitting the open side of the machine with tie rods or keyed tie bars. This will reduce the angular deflection, but not completely eliminate it. The combined cross-sectional area of the rods or bars is small compared to that of the frame on the opposite side of the die space. In addition, the rods or bars will not permit access to the full press opening for the large workpieces.

Dealing With a Mix of Equipment

A Typical Contract Stamper's Dilemma. Many contract stampers started business years ago by running service parts and other low-volume jobs. These were jobs that large automotive stamping plants as well as high-volume contract shops couldn't afford to run on automated presses. It simply was not profitable to use high-volume automated transfer presses and tandem lines for short-run hand-loaded work.

Mix of Work. The mix of work tends to be diverse. To be successful, contract stampers must establish a "can do" reputation. No work within the shop's specialty is turned down, provided there is sufficient plant capacity and a profit can be made on the job.

In most cases, the customer retains ownership of the tooling and does not enter into a long-term agreement with the stamper. A minimum commitment on the part of the customer means that the stamper is reluctant to invest in such items as permanently mounted parallels and subplates needed for quick die change adaptation.

Types of Presses Used. The type of press chosen by a small contract stamper is often determined by what is available at an affordable price when additional stamping capacity is needed. Often, used machinery is purchased for this reason. This sometimes results in presses with a great diversity of tonnage capacities, shut heights, and bolster sizes.

Used Presses Are Often a Problem. Every new press is built with a specific category of work in mind. When buying a used press, the selection criteria often do not go beyond satisfying the following considerations:

- Is the tonnage capacity sufficient for the job?
- Is the bed size large enough?
- Are repairs required?
- How much does the press cost?

Presses Have Personalities. Presses deflect differently under load. Manufacturing flexibility is the main reason for setting the same die in more than one press and if dies are operated at a common shut height, the differences in the way presses deflect is an important factor when grouping presses.

A good progressive die blanking press should be of very robust construction to resist deflection and hence limit the severity of snap-through energy release. A blanking press should have a short stroke to keep the actual metal-shearing velocity to reasonable speeds.

Presses designed for progressive deep drawing often have longer stroke lengths than a blanker of comparable size. The greater stroke length permits deep drawing to be performed and still have a large enough opening to permit ease of the progression strip movement.

In addition, feeding equipment may vary from press to press, alignment is not the same on all presses, and speeds may vary.

Planning

Determining which press to use should be based on a predetermined plan. It should not be a decision largely determined by which presses are available at any given moment.

Grouping Presses. Once a decision is made to group presses for flexible scheduling, a plan should be drawn up. Usually the large straightside presses are a logical place to begin the evaluation. There are several reasons why this makes sense:
- These presses often run high-volume jobs,
- These presses are the most expensive,
- These presses have the best feeders and auxiliary equipment, and
- These presses offer the greatest opportunity for a quick payback.

Build a Press Database. Most shops maintain records of all of their presses. This information is needed to perform preventive maintenance and order spare parts. Press specifications are also needed by process engineers in order to determine process feasibility. In the event that accurate information is not available, a database should be developed.

The best place to store the information is in a computerized database. A number of database software packages are available that make doing this quite easy. While a mainframe system can be used for this purpose, desktop personal computers are powerful enough to easily store all of the information for the equipment in a large pressroom.

Gathering the Data. Figure 19-1 illustrates a form developed for surveying presses on the shop floor when starting such a database. The information concerns only the parameters required for grouping presses for quick die change and manufacturing flexibility.

If press maintenance information is being gathered, data on motor horsepower, frame size, motor speed, type and number of drive belts, etc. will be needed. A common database program may also be configured to do the job.

Common Press Factors

Maximum Pass Height. Unless the press bolster has an opening to discharge scrap, any scrap generated will have to be removed by either gravity or mechanically assisted means. Where feasible, gravity chutes are often the easiest and simplest means for discharging both scrap and finished parts.

If dies are operated on permanently attached parallels, it is usually possible to avoid adjusting the stock feeder, a factor that can easily save five or more minutes per dieset.

Available Shut Height. Another important factor to consider is the amount of shut height and range of adjustment of each press. Not only does this determine the upper die buildup requirements, it is required for common shut height decisions.

Ram Risers. It may be found that the shut height adjustment at one press will not permit interchanging dies between presses without the need for parallels on top of the dies. It is possible that that press was designed for a deep drawing application and is now being used exclusively for progressive die operation. Not only does the placement of heavy parallels on top of the die delay diesetting, it can result in back strain.

A permanently attached ram riser can be attached to the upper slide, if there is a rare need for shut height capability. This riser can even incorporate built-in hydraulic die clamps if desired.

Ram risers are either fabricated from steel plate or cast from grey iron. The major cost in either case is for the required machine work. Usually, it is desirable to have standard JIC T-slots in the riser.

PRESS DATA WORKSHEET

PRESS NUMBER	DATE
TYPE	

MAXIMUM PASS HEIGHT
MAXIMUM SHUT HEIGHT
WIDTH OF BOLSTER (L-R)
DEPTH OF BOLSTER (F-B)
SCRAP REMOVAL METHOD
TONNAGE
STROKE
STROKES PER MINUTE MAX   MIN
TYPE OF FEEDER
TYPE OF DECOILER
TYPE OF STRAIGHTENER
TYPE OF LUBRICATOR
MAXIMUM THICKNESS
MAXIMUM WIDTH
MAXIMUM PITCH

Figure 19-1. *Worksheet for gathering data for a press grouping feasibility study.*

To improve operations within a limited budget, adapting a ram riser from a used equipment dealer may prove worthwhile. Generally, they cost about one half that of new equipment.

An allowance for re-machining should be included when estimating cost. Used risers are normally badly rusted and have a lot of "battle scars" that should be cleaned up.

It is important to make certain that there is enough press counterbalance capacity to support the extra weight of the riser and the heaviest upper die to be used. Used bolster plates can be used for ram risers. If weight is a problem, investigate milling lightening slots. Figure 19-2 illustrates a ram riser made from a press bolster by machining lightening slots.

A simple cost study will reveal where the best payback is to be obtained. The payback can be achieved either by avoiding the cost of permanently installing parallels on the dies or in the setup time reduction. This will avoid the need to place parallels on top of the die in the press.

A hidden savings is that the riser will provide better support for the upper die shoe than parallels. This can improve quality and reduce die maintenance in some operations.

Bolster Size. When matching dies to available presses, bolster size is a very important factor. Not only does it determine the maximum die size, it can also determine scrap shedding feasibility.

Tonnage. Available tonnage is certainly an important factor. Not only must the press not be overloaded at any time, the load must be centered and the die shoes must properly cover the bolster.

There must also be enough flywheel energy to do the desired work throughout the press stroke. Even though the press may not be overloaded from a peak tonnage standpoint, the flywheel may slow excessively, particularly when deep drawing.

Press Speed. The actual metalworking speed is also important in stamping. The actual speed is a product of several factors including stroke length, strokes per minute (SPM), distance from the bottom of stroke, and press motion curve.

Just as the speed of a machine tool such as a lathe is a major factor in determining tool bit life and surface finish, the actual forming and cutting

Progressive Dies

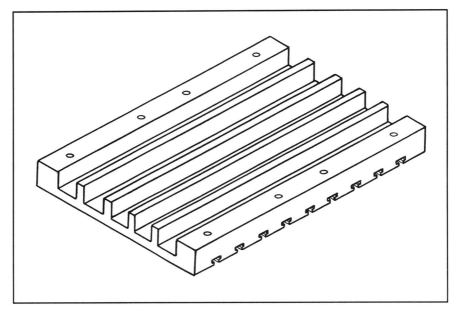

Figure 19-2. *A surplus press bolster converted into a ram riser for reducing excess press shut height by milling lightening slots.*

speed is a press personality factor that must be considered along with stamping flexibility.

Deflection. Deflection in any machine is unavoidable. Knowing when press deflection will be a problem is an important success factor in pressworking management. Deflection is also a factor that affects the success of the stamping process in straightside machines.

If dies are interchanged between presses that deflect differently for a given amount of tonnage, the ability to operate dies at a common shut height will be limited.

A further consideration is the effect of deflection on snap-through energy in blanking operations. A progressive die that does a lot of cutting work that can be used successfully in a very stiff machine may shake the building when operated in a press fabricated from lightweight steel plate.

Press Condition. Problems such as a slight twist in a crankshaft can result in irregular motion of the press slide. In the case of a straightside having two or more slide connections, both sides of the slide may not reach bottom dead center simultaneously. Dies having close clearances may wear rapidly in such a press.

Jobs such as flattening small heavy details and setting noncritical embossments can often be run satisfactorily, provided full press tonnage is not required. The size of the workpiece in relationship to press size is a factor in the successful use of a worn or misaligned machine.

Using Existing Records

Existing records can form a basis for determining the best press for a given job. Records should include:
- Production efficiency,
- Die maintenance costs,
- Press maintenance costs,
- Labor cost per part,
- Scrap rates,
- Overall labor costs, and
- Quality issues.

Determining the best presses for a given job will depend to a large extent upon how extensive and accessible the data is. For example, if all of the data listed above is accessible from a common computerized source by press, job, part, and die number, it probably can be sorted in a number of useful ways and reports generated easily. This will permit decision-making based upon good factual data.

If, on the other hand, the data is in the form of handwritten time card and repair-part invoices, the job will be difficult.

Important Decision Factors. The main consideration in assigning a job to a primary or home press will tend to be where the job ran the best. This is based on the maximum number of pieces per hour produced. This has a pitfall, however; nearly all jobs will be loaded in the best presses, which can have a drastic effect on plant capacity.

There are other factors to consider such as:
- In which press or press line was the job consistently achieving the highest percentage of theoretical machine capacity?
- In which operation did the machine or line produce the lowest die maintenance costs?

- What operation (as determined by SPC data) verifies the most consistent quality?

Running Jobs in the Home Press

The proven benefits of the primary or home press concept in improving and stabilizing quality are beyond dispute, but it is a difficult goal to achieve in many stamping shops. Some critical factors are:

- Advanced planning of production,
- Better teamwork in many shops,
- An effective preventive maintenance program in place to avoid scheduling problems due to breakdowns,
- An accurate knowledge of the capacity to do work.

Success Factors. To be successful in improving quality and machine utilization while meeting delivery requirements, there should be:

- A backup plan in case of press downtime,
- A backup plan in the event of die problems,
- Procedures to accommodate normal stock variation such as the normal range of stock camber,
- A plan for engineering cost studies to be conducted on a routine basis to ensure that the most cost-effective solution to any problem is followed,
- Implementation of those quick die change methods that offer a reasonable payback,
- An ongoing emphasis on maximization of human resources.

Equipping Presses for Quick Die Change

One of the central features of modern tandem and transfer press die exchanging systems is the moving press bolster system. This arrangement permits a second moving bolster outside of the press to be prestaged for the next dieset. The same dual bolster system can be used for prestaging large progressive dies.

An Ideal Arrangement. The most rapid exchanging of dies will occur if both bolsters can move simultaneously in one side of the press and out the other. Danly QDC tandem lines, that are over 20 years old, routinely change over and produce a different part in under 10 minutes.

Custom-built Quick Die Change Systems

Although a wide variety of methods are used for die change systems, several basic concepts are normally utilized.

T-slot Lifters. First, instead of dragging or skidding the dies out of the press like a caveman pulling his load, modern die change systems have reapplied the same caveman's invention: the wheel. With these die change systems, the dies move on rollers out of the press. This eliminates skidding and reduces frictional forces so that the die is moved smoothly and efficiently. To accomplish this rolling feature, most modern systems use hydraulic or pneumatic T-slot lifters, such as the one shown in Figure 19-3.

These lifters raise the die or die subplate from the press bolster and place it on wheels, thus reducing the force required to move the die to approximately 1 percent of its total weight. When equipment is designed for this concept, a safety factor should be applied, and total thrust should be 10 percent of the weight to be moved. In contrast, when the skidding technique

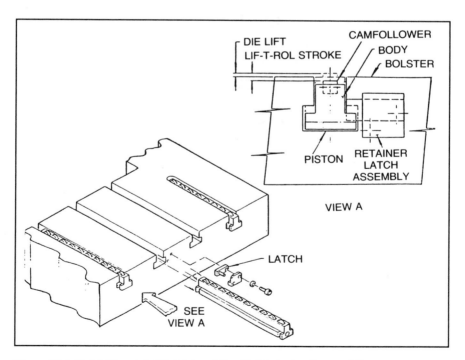

Figure 19-3. *T-slot lifters retrofitted into existing T-slots to permit ease of die movement.*

Progressive Dies

is used, the equipment must be designed for 70 percent of the weight. These higher loads mean that the equipment is seven times heavier, with higher initial and potential maintenance costs.

Secondly, many systems use subplates. A subplate is normally the same size as the press bed. Most subplates have tapped hole patterns to mount a number of different dies, location holes to locate the various dies accurately on the plate, and hold-down slots or holes to attach the plate to the bolster. Figure 19-4 shows a subplate with two dies attached.

In certain cases involving only a small quantity of dies, the dies are permanently attached to the subplates. Some thicker plates have T-slots and keyways. The only exception is in cases where dies on a particular job are designed and built with outside dimensions identical to the press and the die press mechanism. Several attributes make the subplate important.

- A smooth surface provides an unbroken surface for the hydraulic T-slot lifters to roll against, rather than the various holes and cutouts normally found on the bottom surface of dies.
- The tool holder allows mounting of any size die in the press, similar to a tool holder in a machining center.

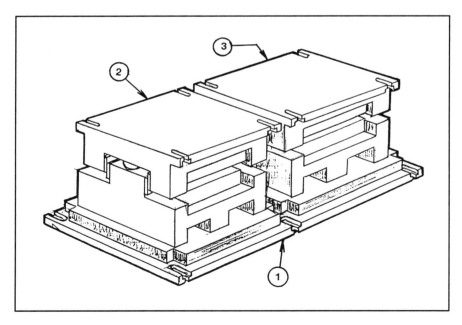

Figure 19-4. *Two dies mounted on a common subplate: (1) subplate with slots for automatic die clamps. (2) chassis base die. (3) chassis top die.*

- Prestaging provides the opportunity to prestage the die on the subplate during production, thereby keeping the die change time and work stoppage on the press to a minimum.
- Location provides consistent location of the die in the press through the use of a keyway or side guides. This assists in realigning automation and attachment to the press and ram. No prying or jacking the die into position is required.
- Consistent location in the press simplifies attachment. Bolt-down slots are provided in the same location on each die change, allowing the same-length bolt to be used time after time.
- Cleanliness keeps the bolster and hydraulic T-slot lifters clean since the only time the press is exposed is during a die change.

Types of Die Change Systems

Systems to change dies range from simple hydraulic T-slot lifters and bolster extensions on a single press, to automatic die cart systems for large multi-press stamping lines or rolling bolsters. The most efficient system for a plant should be designed around the specific requirements at the location in question. Figure 19-5 lists the types of systems that could be used.

The simplest system uses bolster extensions (fixed, pivoted, mobile, or removable), mounted to the front and/or back of the press as illustrated in Figure 19-6. These extensions support a new die ready to install, or an old die when it is removed from the press. Hydraulic T-slot lifters are installed in the bolster T-slots to reduce the force required to move the die into and out of the press. The dies are rolled manually if they weigh less than 5000 pounds (2275 k). To accurately locate the die in the press, side guides or keyways are used on the press and bolster extensions.

The advantages of this type of system are low cost and ease of installation. The disadvantages are that dies cannot be prestaged without interfering with either the front or rear of the press. In other words, the die change cannot start until the press area is clear of the operator and production has stopped.

Die Change Tables. The next component in die change equipment is a die table. As noted in the previous summary of die change systems, there are a number of types of die tables that can be used. The simplest is a fixed-height table that remains in place on one side or on the end of the press. This type is illustrated in Figure 19-7. Automation can be built around or over

A. BOLSTER EXTENSIONS
   1. Removable Bolster Extensions,
   2. Swing Away Bolster Extensions,
   3. Fixed Bolster Extensions, or
   4. Mobile Bolster Extensions.
B. DIE TABLES
   1. Fixed Height Table.
   2. Adjustable Height Table.

   Options:
      a. Portable,
      b. Tilting,
      c. Disappearing,
      d. Lift Die, or
      e. Rotary Type.
   3. Skid Type–Fixed Height.
   4. Skid Type–Adjustable Height.
   5. Tee Table.
      a. Powered Push/Pull Type.
      b. Unpowered.
   6. Self-propelled Mobile Die Tables.
C. DIE CARTS
   1. Transfer Carts,
   2. Single Station Carts, or
   3. Dual Station Carts.

   Options:
      a. Unpowered,
      b. Air Powered,
      c. Electric Powered,
      d. Air-Hydraulic Powered,
      e. Electric/Hydraulic Powered,
      f. Die Storage or Prestaging Racks,
      g. Height Variation,
      h. Long Reach Type,
      i. Rotary Type,
      j. Load Rails or Rollers,
      k. Drop Deck Design,
      l. Vertically Actuated Grippers,
      m. Light Duty,
      n. Medium Duty, or
      o. Heavy Duty.
D. ROLLING BOLSTERS
   1. Single Axis, or
   2. Dual Axis.

Figure 19-5. *Types of die change systems.*

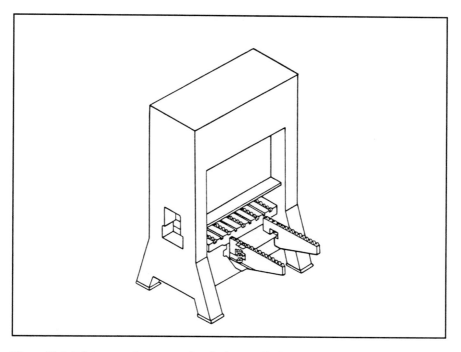

Figure 19-6. *Bolster extensions mounted to the front and/or back of the press to support dies being installed or removed from the press.*

this table as required. Hydraulic T-slot lifters reduce the forces required and assist in accurate placement of the subplate/die in the press.

An adjustable-height die table such as in Figure 19-8 could be used for an installation servicing two adjacent presses of different heights or as a portable table servicing several different presses with varying heights. The portable table would normally be transported from place to place by forklift or crane.

There are two advantages to die tables. The first is low cost. Secondly, they provide a way to accomplish die change when the bolster-to-floor height is below 12 inches (305 mm) or when there is not enough room to move carts out into the aisle area. The disadvantage of a die table is that prestaging cannot be accomplished prior to shut-down of production.

T-tables. Another approach to die change is the T-table. This concept uses hydraulic T-slot lifters in the press bolster plus side guides or a keyway to control left-to-right location. As in the previous systems, dies are mounted

Progressive Dies

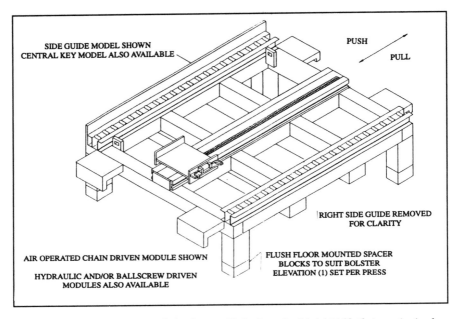

Figure 19-7. *An example of a simple die change table having a fixed-height table that remains in place on one side or end of the press.*

to subplates so that all dies appear to be the same size in the die change system. A three-station powered roller conveyor (Figure 19-9) is mounted behind and parallel to the bed of the press. Pivoted or fixed bolster extensions equipped with rollers are mounted to this table and pin to the bed of the press when the die is ready to be moved. The pivoting feature permits ease of access to the press. The cam-follower rollers on the extensions provide support and side guiding for the die and subplate as it is being removed from the press. The hydraulic T-slot lifters are raised after the die is unclamped from the bolster.

Heavier dies are rolled out using a powered push-pull module. A new die is located on one end of the three-station powered conveyor. After the die is removed from the press, it is lowered onto the powered conveyor, and both the old die and the new die are moved simultaneously to place the new die in the center of the conveyor and in line with the press. The new die is then rolled into the press to a fixed stop. The hydraulic T-slot lifters are lowered and the die is bolted or hydraulically clamped to the bolster.

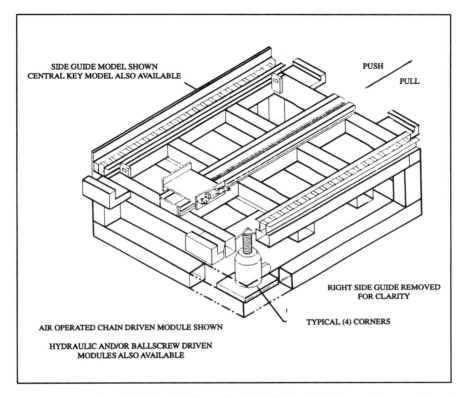

Figure 19-8. *An adjustable-height die table can also be used for an installation servicing two adjacent presses of different heights.*

For lightweight die installations, a manual system uses a two-station linear actuator. In this type of system, the die is pulled out of the press onto one linear actuator carriage equipped with nonpowered rollers. This carriage is pushed to one side and the second carriage, with the new die, is pushed into the center position. The new die is pushed into the press and clamped in final position.

The only disadvantage of the T-table system is that it takes up a fixed area behind the press. The important advantages of the system are low cost and quick changes without the use of forklift trucks or cranes. In addition, this system does not affect either side or the front of the press, and can be used for two presses back-to-back with the addition of another set of bolster extensions on the other side of the table.

Progressive Dies

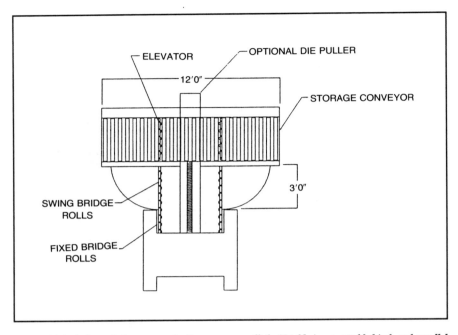

Figure 19-9. *A three-station powered roller conveyor called a T-table is mounted behind, and parallel to, the bed of the press to aid quick die change.*

Self-propelled Die Tables. In certain cases, a mobile self-propelled die table may be required where space is at a premium or the pressroom layout precludes any other options. This system uses a storage battery-powered walk-behind device with a die table attached to the top of the power unit. The portable unit illustrated in Figure 19-10 allows removal or insertion of the subplate and die from presses or high-density die storage racks in the range of 20 inches (0.5 m) to 80 inches (2.0 m) from floor level. A maximum of 12,000 pounds (5455 kg) can be handled with this unit. The front corners of the platform attach to the press bolster or the die rack and are leveled with the power unit for easy removal or insertion. Additional units are available for heavier dies but they do not have the lift-stroke capacity to store dies in multilevel storage racks.

Because of the various press sizes that may be considered, a center key is used for alignment, which in turn requires a thicker subplate. A minimum of two subplates is required for each press to be serviced by this system.

The main advantage of this system is versatility. A number of presses, regardless of their relationship to one another, can be serviced by the same

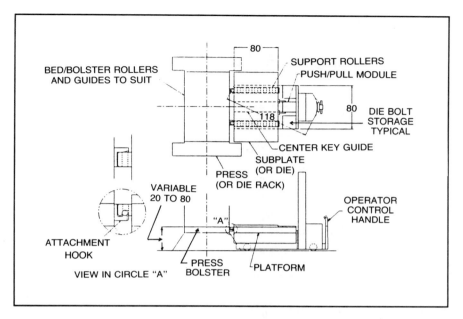

Figure 19-10. *A mobile self-propelled die table utilizes a storage battery-powered walk-behind device with a die table attached to the top of the power unit.*

unit. The main disadvantage is the increased time needed for the die change due to the various steps required. Also there is a greater dependency on operator skills to transfer the new die to the press correctly, level, and make the change in a smooth and efficient manner, and then unhook and transfer the old die to a transfer station.

Die Carts. Another approach to quick die change is the use of die carts. These devices come in a number of designs and configurations depending upon the situations at the plant location. These systems also utilize subplates.

One possibility is a double-station die cart system illustrated in Figure 19-11. This is a good approach when the front or back of the press is congested with automation or other equipment. Flush floor tracks are installed on one side of the press. In this case, the new subplate and die are prestaged on one side of the cart while production is being run in the press.

When production is complete, the die cart is moved into position and the old subplate and die are rolled onto the empty station. The cart is then moved so that the station with the new subplate and die is in line with the press. The subplate and die are then rolled into the press, and the die cart is

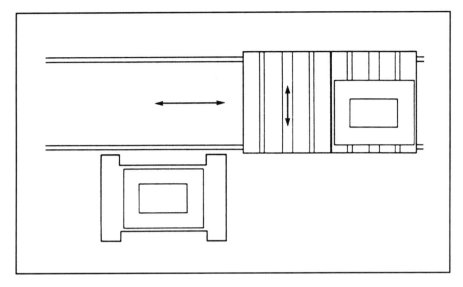

Figure 19-11. *A double-station die-cart system is a good approach when the front or back of the press is congested with automation or other equipment.*

moved to its storage area away from the press. At this point, the automation can be rolled back into position and production can restart in a manner of minutes.

Die Carts With Storage Racks. Another possibility is to use a single cart with storage racks. This system allows dies to be stored adjacent to the press at bolster heights and uses a smaller single-station cart for making the die change (Figure 19-12). Flush floor rails are installed on one side of the press. In this case, the new subplate and die would be on the racks and at least one rack must be empty. When production is complete, the die cart is moved into position in front of the press, and the subplate and die are rolled onto the cart. The die cart is then moved to the storage rack to store the old die and get the new one to place in the press. Finally, the die cart is stored out of the press area.

This system is less expensive than the double-station cart and only increases the die change time by a few minutes. Flexibility is an important advantage of this system because additional storage racks can be added at any time as space permits. This system does not require the use of forklift trucks and cranes once the dies are placed on the racks.

Quick Die Change

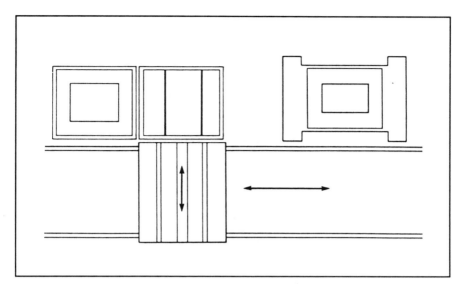

Figure 19-12. *A single cart used with storage racks allows dies to be stored adjacent to the press at bolster heights.*

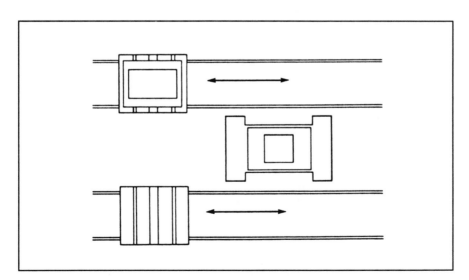

Figure 19-13. *A two-die cart system: one cart is on the front of the press while the other is located at the back.*

Two-die Cart System. Another option is a two-die cart system such as the one illustrated in Figure 19-13. One cart is on the front of the press while the other is located at the back. Flush floor tracks are installed on both sides of

Progressive Dies

the press. A new die is prestaged on a subplate on one cart while both carts are in the storage position and production is running. When production is completed, the automation is moved out of the way and the carts are moved into position. The old subplate and die are rolled onto the empty cart and the new subplate and die are rolled into the press from the second cart. When the die change is finished, the carts are moved back to the storage position.

The principal advantage of this system is that it can handle very large dies and it allows the fastest possible die change. The one disadvantage is higher cost due to the two carts and a double track.

Proper Applications. Any of the previous examples could be used for press access, if the press considered has windows. This is often required if the front and back of the press are tied up with equipment. An example of this sort of situation would be a blanking press with coil handling equipment and stacking systems that are located on the front and back of the press.

Multiple press lines can easily utilize die carts. Figure 19-14 illustrates a press line with three presses. Four carts are used for the die change. During production, die carts are prestaged with the new dies on subplates. When production is complete, the automation is moved out of the way, and the die carts are moved into position in line with the presses. The top cart, which is

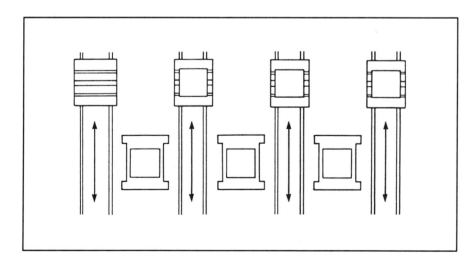

Figure 19-14. *A three-press line using four carts for the die change.*

empty, pulls the old die out of press number one. The next cart pulls the old die out of press number two while pushing the new die into press number one. The third and fourth carts function the same as the first two carts, completely changing the balance of dies. The carts are moved out of the way and the automation is reinstalled, often utilizing the same flush floor tracks. Production resumes in a matter of minutes since automation is easily set up with the accurate die location obtained with the carts.

The advantage of this system is that all dies for the next run can be stored or prestaged on the carts. No storage racks are necessary, unless storage for the third run is desired. This is the quickest die change possible for a complete line since two dies are removed or installed at the same time. The main disadvantage is cost, since one more cart than the number of presses is required. Another disadvantage is space, since the carts must go into the aisle for storage and prestaging.

With the proper layout of equipment, this die cart system can be used for a second or third line of presses. This, of course, reduces the cost of the die change system on a per-press basis and generates a much better return on investment.

In some pressrooms, the spacing from one press to another and/or building restrictions require that an additional cart be installed. Large variations in bolster height can require the use of an additional cart. Minor variations in bolster height of 1 to 3 inches (25 to 76 mm) can be handled with an optional cart design.

Ideally, the press location should be designed to best utilize a die cart system. However, this is usually impractical on existing facilities. Most die cart systems must be tailored to the conditions that exist at the customer's plant.

Certain factors are important when designing for an application.
- Size of press bed.
- Bolster height from floor.
- Weight of dies to be moved.
- Size of dies to be moved.
- Frequency of change.
- Availability of floor space.
- Automation used.

- Availability of crane or fork lift.
- Is there enough shut height adjustment to allow a subplate under dies?
- Location of die storage.
- Number of dies to be changed.
- Use of cushion pins.
- T-slots in bolster? Size?

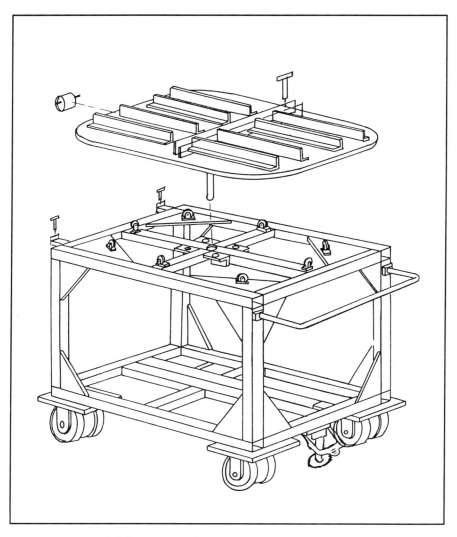

Figure 19-15. *An exploded view of a simple light-duty die cart.*

Simple Light-duty Die Cart

Figure 19-15 illustrates a simple locally fabricated die cart made of common structural and hardware components. The wheels are made of phenolic and have a groove in the center to permit them to be guided by a track made of strap-iron affixed to the floor. The table rotates on eight small wheels attached to the framework. The exploded view shows only one of the rollers supported by the angle irons welded to the table top. A means is provided to prevent the die from falling off the cart while in transit. A typical application would be transporting dies from a storage or a prestage area to and from the press(es).

Glossary

Air draw: A draw operation performed in a single-action press with the blankholder pressure supplied by an air cushion.

Annealing: A process involving the heating and cooling of a metal, commonly used to induce softening. The term refers to treatments intended to alter mechanical or physical properties or to produce a definite microstructure.

Bead: A narrow ridge in a sheet-metal workpiece or part, commonly formed for reinforcement.

Bead, draw: (a) A bead used for controlling metal flow; (b) rib-like projections on draw-ring or hold-down surfaces for controlling metal flow.

Bend allowance: The developed arc length along the neutral axis of bent metal.

Bending: The straining of material, usually flat sheet or strip metal, by moving it around a straight axis which lies in the neutral plane. Metal flow takes place within the plastic range of the metal, so that the bent part retains a permanent set after removal of the applied stress. The cross section of the bend inward from the neutral plane is in compression; the rest of the bend is in tension.

Bend radius: (a) The inside radius at the bend in the work; (b) the corresponding radius on the punch or on the die.

Blank: A precut metal shape, ready for a subsequent press operation.

Blank development: (a) The technique of determining the size and shape of a blank; (b) the resultant flat pattern.

Blankholder: The part of a drawing or forming die which holds the workpiece against the draw ring to control metal flow.

Blankholder, multi-slide: A mechanical device either spring or cam actuated for firmly clamping a part in place prior to severing from the carrying strip and retaining in place for forming.

Blanking: The operation of cutting or shearing a piece out of stock to a predetermined contour.

Bulging: The process of expanding the walls of a cup, shell, or tube with an internally expanding segmental punch or a punch composed of air, liquids, or semi-liquids, such as waxes or tallow, or of rubber and other elastomers.

Burnishing: The process of smoothing or plastically smearing a metal surface to improve its finish.

Bushing, guidepost: A replaceable insert usually fitted in the upper shoe to provide better alignment.

Cam action: A motion at an angle to the direction of an applied force, achieved by a wedge or cam.

Carburizing: A process that introduces carbon into a solid ferrous alloy by heating the metal in contact with a carbonaceous material — solid, liquid, or gas — to a temperature above the transformation range and holding it at that temperature.

Case hardening: Any process of hardening a ferrous alloy so that the case or surface is substantially harder than the interior or core.

Clearance, die: The space, per side, between the punch and die.

Coining: A closed-die squeezing operation in which all surfaces or the work are confined or restrained.

Cold heading: The process of upsetting the ends of bar, wire, or tube stock while cold.

Cold working: Working of a metal, such as by bending or drawing, to plastically deform it and produce strain hardening.

Complementary output sensor: A solid state sensor having both a normally open output and a normally closed output.

Crimping: A forming operation used to set down or close in a seam.

Cryogenic treatment: A low-temperature steel treatment process used to improve the toughness of hardened tool steel by continuing the transformation of retained austenite into the more desirable martensite, at temperatures ranging from -120° to -300°F (-84° to -184°C).

Cup: Any shallow cylindrical part or shell closed at one end.

Cupping: An operation that produces a cup-shaped part.

Curling: Forming an edge or circular cross section along a sheet or at the end of a shell or tube. (See also Wiring.)

Cylinder, nitrogen: A cylinder containing high-pressure nitrogen gas used as a die spring in pressure pad, draw ring, and cam return applications. These are available in individual piped, manifold, and self-contained styles.

Deep drawing: The drawing of deeply recessed parts from sheet material through plastic flow of the material, when the depth of the recess equals or exceeds the minimum part width.

Deflection: The deviation of a body from a straight line or plane when a force is applied to it.

Dial feed: (a) A press feed which conveys the work to the dies by a circular motion; (b) a mechanism which moves dies under punches by a circular motion and into definite indexed positions.

Glossary

Die, assembling: A die which assembles and fastens parts together by riveting, press fitting, folding, staking, curling, hemming, crimping, seaming, or wiring.

Die, bending: A die which permanently deforms sheet or strip metal along a straight axis.

Die, blanking: A die for cutting blanks by shearing.

Die, burnishing: A die which improves surface or size by plastically smearing the metal surface of the part.

Die, cam: A die in which the direction of moving elements is at an angle to the direction of forces supplied by a press.

Die, combination: A die in which a cutting operation and a noncutting operation on a part are accomplished in one stroke of the press. The most common type of combination die blanks and draws a part.

Die, compound: A die in which two cutting operations are accomplished in one press stroke. The most common type of compound die blanks and pierces a part.

Die, compound-combination: A die in which a part is blanked, drawn, and pierced in one stroke of the press.

Die, curling: A forming die in which the edge of the work is bent into a loop or circle along a straight or curved axis.

Die, dimpling: A forming die which produces a conical flange (stretch flange) encircling a hole in one or more sheets of metal.

Die, dinking: A die which consists of a press or hand-operated hollow punch with knife-edges for cutting blanks from soft sheet metals and nonmetallic materials.

Die, double-action: A die in which pressure is first applied to a blank through the blankholder and is then applied to the punch.

Die, embossing: A die set which is relatively heavy and rigid for producing shallow or raised indentations with little or no change in metal thickness.

Die, expanding: A die in which a part is stretched, bulged, or expanded by water, oil, rubber, tallow, or an expanding metal punch.

Die, extrusion: A die in which a punch forces metal to plastically flow through a die orifice so that the metal assumes the contour and cross-sectional area of the orifice.

Die, floating (or punch): A die (or punch) so designed that its mounting provides for a slight amount of motion, usually laterally.

Die, forming: A die in which the shape of the punch and die is directly reproduced in the metal with little or no metal flow.

Die, heading: (a) A die used in a forging machine or press for upsetting the heads of bolts, rivets, and similar parts; (b) a die used in a horizontal heading machine for upsetting the flanged heads on cartridges and similar shells.

Die, hemming: A die which folds the edge of the part back over on itself; the edge may or may not be completely flattened to form a closed hem.

Die, inverted: A die in which the conventional positions of the male and female members are reversed.

Die, joggle: A die which forms an offset in a flanged section.

Die, lancing: A die which cuts along a line in the workpiece without producing a separation in the workpiece and without yielding a slug.

Die, multiple: A die used for producing two or more identical parts in one press stroke.

Die, perforating: A die in which a number of holes are pierced or punched simultaneously or progressively in a single stroke of the press.

Die, piercing: A die which cuts out a slug (which is usually scrap) in sheet or plate material.

Die, progressive: A die in which two or more sequential operations are performed at two or more stations upon the work, which is moved from station to station.

Die, riveting: A die that joins two or more parts by riveting.

Die, sectional (segmental): A die punch, or form block, which is made up of pieces, sections, laminations, segments, or sectors.

Die, shaving: A die usually having square cutting edges, negligible punch and die clearance, and no shear on either punch or die.

Die, shimmy (Brehm trimming die): A cam-driven die which cuts laterally through the walls of shells in directions determined by the position of cams.

Die, single-action: A drawing die that has no blankholder action, since it is used with a single-action press without the use of a draw cushion.

Die, swaging: A die in which part of the metal under compression plastically flows into contours of the die; the remaining metal is unconfined and flows generally at an angle to the direction of applied pressure.

Die, trimming: A die that cuts or shears surplus material from stock or workpieces.

Die, triple-action: A die in which a third force is applied to a lower punch in addition to forces applied to the blankholder and the punch fastened to the inner slide.

Die, two-step: A drawing or reducing die in which the reduction is made in two stages or levels, one above the other, in a single stroke of the press.

Die, waffle: A type of flattening die that sets a waffle or crisscross design in the blank or workpiece without deforming it.

Die block: (a) A block or plate out of which the die proper is cut; (b) the block or plate to which sections or parts of the die proper are secured.

Die cushion: A press accessory located beneath or within a bolster or die block, to provide an additional motion or pressure for stamping operations; actuated by air, oil, rubber, or springs, or a combination thereof.

Die height (shut height): The distance from the finished top face of the upper shoe to the finished bottom face of the lower shoe, immediately after the die operation and with the work in the die.

Die holder: A plate or block upon which the die block is mounted.

Die pad: A movable plate or pad in a female die, usually for part ejection by mechanical means, springs, or fluid cushions.

Die set: A standardized unit consisting of a lower shoe, an upper shoe, and guide pins or posts.

Die shoe: A plate or block, upon which a die holder and in which guide posts are mounted.

Die space: The maximum space within a press bounded by the top of the bed (bolster), the bottom of the slide, and any other press parts.

Dimpling: Localized indent forming of sheet metal, so as to permit the head of a rivet or a bolt to fasten down flush with the surface of the sheet.

Dishing: Forming a large-radiused concave surface in a part.

Double seaming: The process of joining metal edges, each edge being flanged, curled, and crimped.

Downloading: The process of transferring digitalized data from a host computer or CAD system to another digitally controlled device such as a CNC machine tool.

Draft: The taper given to a die so as to allow the part to fall through the die or be removed.

Drawability: (a) A measure of the feasible deformation of a blank during a drawing process; (b) percentage of reduction in diameter of a blank when it is drawn to a shell of maximum practical depth.

Drawing: A process in which a punch causes flat metal to flow into a die cavity to assume the shape of a seamless hollow vessel.

Draw line: A surface imperfection in a drawn part caused by the initial punch contact of a character line on the blank that is subsequently drawn or stretched to a plane surface where it becomes a visible defect.

Draw radius: The radius at the edge of a die or punch over which the work is drawn.

Draw ring: A ring-shaped die part (either the die ring itself or a separate ring), over the inner edge of which the metal is drawn by the punch.

Ductility: The property of a material that permits it to sustain permanent deformation in tension without rupture.

Dwell: The time interval in a press cycle during which there is no movement of a press member.

Eccentric: A machine element that converts rotary motion to straight-line motion.

Elastic limit: The maximum stress to which a material can be subjected, and yet return to its original shape and dimensions on removal of the stress.

Embossing: A process that produces relatively shallow indentations or raised designs with theoretically no change in metal thickness.

Extrusion: The plastic flow of a metal through a die orifice.

Eyelet machine: A multiple-slide press, usually employing a cut-and-carry or a transfer feed for sequential operations in successive stations.

False wiring: Curling the edge of a sheet, shell, or tube without inserting a wire or rod inside the curl.

Feed: A device that moves or delivers stock or workpieces to a die.

Feed, grip (slide, hitch): A type of feed mechanism employing a set of jaws to grip strip stock and feed it to the die.

Flaring: (a) The process of forming an outward flange on a tubular part; (b) forming a flange on a head.

Fluting: (a) The forming of longitudinal recesses in a cylindrical part; (b) a surface imperfection in formed or drawn parts; See Luders' lines.

Formability: The ability of a material to undergo plastic deformation without fracture; a quantitative expression of formability, applicable to explosive forming, it is the true strain at fracture.

Form block: A punch or die used in the rubber-pad process to form materials.

Form block, mechanical: A special die used in rubber-pad forming to perform operations which cannot be made with the simpler, regular form blocks.

Forming: Making any change in the shape of a metal piece which does not intentionally reduce the metal thickness.

Forming, high-velocity: Shaping of a workpiece at forming velocities on the order of hundreds of feet per second. This usually requires the use of electrical, magnetic, high-speed-mechanical, or explosive energy sources.

Forming, pneumatic-mechanical: Shaping of a workpiece with the force generated by the impact of a ram accelerated by the release of compressed gas.

Four slide machine: A type of multi-slide forming machine having four synchronous rotating shafts on a horizontal plane. The shafts carry cams that power four or more forming slides plus auxiliary motions for feeding materials, piercing and trimming strip stock, severing the blank, and transferring or ejecting the finished part.

Front slide machines: The forming slides on a four slide or vertical slide machine and their respective positions.

Fulcrum slide: A self-contained slide with a fulcrum lever used in a multi-slide forming machine for added power at the sacrifice of stroke length or conversely added stroke length at the sacrifice of power.

Gag: A metal spacer to be inserted so as to render a floating tool or punch inoperative.

Gage: A device used to position work in a die accurately.

Gage, finger: A manually operated device to limit the linear travel of material.

Galling: The friction-induced roughness of two metal surfaces in direct sliding contact.

Gibs, adjustable: Guides or shoes designed to ensure the proper sliding fit between two machine parts.

Guerin process: A forming method in which a pliable rubber pad attached to the press slide is forced by pressure to become a mating die for a punch or punches which have been placed on the press bed.

Guide posts (guide pins, leader pins): Pins or posts usually fixed in the lower shoe and accurately fitted to bushings in the upper shoe to ensure precise alignment of the two members of a die set.

Guide, stock: A device used to direct strip or sheet material to the die.

Heel block: A block or plate usually mounted on or attached to a lower die, and serving to prevent or minimize deflection of punches or cams.

Hole flanging: Turning up or drawing out a flange around a hole; also called "extruding."

Hysteresis: In sensor terminology, it is the difference in percentage of effective sensor range or distance between the on and off point when the target is moving away from the sensor face.

Inching: A control process in which the motion of the working members is precisely controlled in short increments.

Incremental deflection factor: The amount that a press deflects due to a given unit of loading. It is usually expressed as the amount of deflection that occurs as a result of one ton per corner (four tons total) increase in tonnage in a straightside press.

Ironing: An operation in which the thickness of the shell wall is reduced and its surface smoothed.

Knockout: A mechanism for ejecting blanks or other work from a die. Commonly located on the slide, but may be located under the bolster.

Lancing: Cutting along a line in the workpiece without producing a detached slug from the workpiece.

LED: A light emitting diode. LEDs are widely used as indicators of sensor status.

Lightening hole: A hole punched in a part for the purpose of saving weight.

Limit switch: A type of electric switch used to control the operations of a machine automatically.

Luders' lines: Depressed elongated markings parallel to the direction of maximum shear stress, or elevated elongated markings appearing on the surface of some materials, particularly on iron and low-carbon steel, when deformation is beyond the yield point in tension or compression, respectively.

Manifold: A system of piping for connecting several air, fluid, or nitrogen operated devices to a common source. A cross-drilled steel plate fitted with nitrogen cylinders.

Mainframe: A large, often centrally located computer capable of doing a number of tasks at once.

Motion diagram: A graph showing the relative motions of the moving members of a machine.

Necking (necking in): Reducing the diameter of a portion of the length of a cylindrical shell.

Normalizing: A process in which a ferrous alloy is heated to a suitable temperature above the critical range and then cooled in air at room temperature.

Nosing: Forming a curved portion, with reduced diameters, at the end of a tubular part.

Notching: The cutting out of various shapes from the edge of a strip, blank, or part.

Offal: Scrap metal trimmed from stamping operations that is used to produce small stampings in recovery dies.

Oil canning (canning): The distortion of a flat or nearly flat surface by finger pressure, and its reversion to normal.

Olsen ductility test: A test for indicating the ductility of sheet metals by forcing a hemispherical-shaped punch or hardened ball into the metal and measuring the depth at which fracture occurs.

Overcrown: Added crowning of a curved surface in a drawing die for large irregular shapes to compensate for springback.

Overbending: Bending metal to a greater amount than that called for in the finished piece, so as to compensate for springback.

Pad: The general term used for that part of a die which delivers holding pressure to the metal being worked.

Parting: An operation usually performed to produce two or more parts from one common stamping.

Perforating: The piercing or punching of many holes, usually identical and arranged in a regular pattern.

Pilot: A pin or projection provided for locating work in subsequent operations from a previously punched or drilled hole.

Plastic flow: The phenomenon which takes place when a substance is deformed permanently without rupture.

Plasticity: The property of a substance that permits it to undergo a permanent change in shape without rupture.

Plastic working: The processing of a substance by causing a permanent change in its shape without rupture.

Platen: The sliding member or ram of a power press.

Powdered metallurgy process: The production of tool steel from finely divided steel powder through fusion at welding temperature in an evacuated steel canister. A much more uniform product is obtained than can be produced from conventional cast ingots.

Press, C-frame: A press having uprights or housing resembling the form of the letter C.

Press, double-action mechanical: A press having one slide within the other, the outer slide usually being toggle or cam-operated, resulting in independent parallel slide movements.

Press, geared: A press whose main crank or eccentric shaft is connected to the drive shaft or flywheel shaft by one or more sets of gears.

Press, hydraulic: A press actuated by a hydraulic cylinder and piston.

Press, inclinable: A press whose main frame may be tilted backward, usually up to 45 degrees, to facilitate ejection of parts by gravity through an open back.

Press, knuckle-joint: A heavy powerful short-stroke press in which the slide is actuated by a toggle (knuckle) joint.

Press, multiple-slide: (a) A press having individual slides built into the main slide, or (b) a press of more than on slide in which each slide has its own connections to the main shaft.

Press, open-back inclinable: An inclinable press in which the opening at the back between the uprights is usually slightly more than the left-to-right dimension of the slide flange. (See also Press, inclinable.)

Press, punch: (a) Most commonly, an end-wheel gap press of the fixed-bed type; (b) a name loosely used to designate any mechanical press.

Press, rack-and-pinion reducing: A long-stroke reducing press actuated by a rack and pinion.

Press, reducing: A long-stroke, single-crank press used for redrawing (reducing) operations.

Press, rubber-pad: Any single-action hydraulic press with its slide equipped with a rubber pad for rubber pad forming.

Press, single-action: Any press with a single slide; usually considered to be without any other motion or pressure device.

Press, straightside: An upright press open at front and back with the columns (uprights) at the ends of the bed.

Press, tapering: A press designed to permit placing a blank in a die without the need for a slide plate, and to deliver an exceptionally long stroke.

Press, toggle: (a) Any mechanical press in which a slide, or slides, is actuated by one or more toggle joints; (b) a term applied to double-action and triple-action presses.

Press, toggle drawing: A press in which the outer or blankholder slide is actuated by a series of toggle joints and the inner slide by the crankshaft or eccentrics.

Press, trimming: A special-purpose mechanical press in which shearing and trimming operations are usually done on forgings.

Press, triple-action: A press having three slides with three motions synchronized for such operations as drawing, redrawing, and forming, where the third action is opposite in direction to the first two.

Press, tryout (spotting): A press used in the final finishing of dies to locate inaccuracies of mating parts.

Press, twin-drive: A press having two main gears on the crankshaft meshing with two main pinions on the first intermediate shaft.

Press, two-point: A mechanical press in which the slide is operated by two connections to the crankshaft.

Press, underdrive: A press in which the driving mechanism is located within or under the bed or below the floor line.

Press-brake (bending brake): An open-frame press for bending, cutting, and forming; usually handling relatively long work in strips.

Puckering: A wavy condition in the walls of a deep drawn part.

Punch: (a) The male die part, usually the upper member and mounted on the slide; (b) to die-cut a hole in sheet or plate material; (c) a general term for the press operation of producing holes of various sizes in sheets, plates or rolled shapes.

Punch holder: The plate or part of the die which holds the punch.

Quill-type punch: A frail or small-sized punch mounted in a shouldered sleeve or quill.

Rack-and-pinion drive: A drive incorporating a rack and pinion and commonly used to actuate roll feeds.

Redrawing: Second and following drawing operations in which cuplike shells are deepened and reduced in cross-sectional dimensions.

Reducing: Any operation that decreases the cross-sectional dimensions of a shell or tubular part; includes drawing, ironing, necking, tapering, and redrawing.

Residual stress: Stresses left within a metal as the result of nonuniform plastic deformation or by drastic gradients of temperature from quenching or welding.

Restriking: A sizing operation in which compressive strains are introduced in the stamping to counteract tensile strains set up by previous operations.

Riser block: A plate inserted between the top of the bed and the bolster to decrease the height of the die space.

Roll straightener: A mechanism equipped with rolls to straighten sheet or strip stock, usually used with a feed mechanism for pressworking.

Rotary slide machines: A vertical multi-slide forming machine whose cam driven slides are powered by a large sun gear rather than shafts. There can be as many slides as space permits and they can be positioned virtually anywhere around the sun gear and at any angular orientation affording virtually limitless flexibility.

Rubber pad (blanket): A flat piece of rubber used as an auxiliary tool for rubber forming.

Rubber pad forming: (See Guerin process.)

Scoring: (a) The scratching of a part as it slides over a die; (b) reducing the thickness of a material along a line to weaken it purposely along that line.

Scrap cutter: A shear or cutter operated by the press or built into a die for cutting scrap into sizes for convenient removal from the die or for disposal.

Seam: (a) The fold or ridge formed at the juncture of two pieces of sheet material; (b) on the surface of a metal, a crack that has been closed but not welded; usually produced by some defect either in casting or in working, such as blowholes that have become oxidized, or folds and laps that have formed during working.

Seaming: The process of joining two edges of sheet metal by multiple bending.

Seizing: Welding of metal from the workpiece to a die member under the combined action of pressure and sliding friction.

Sensing range: The distance away from the sensing face at which a nominal target will be detected.

Sensor, analog: A sensor that provides a voltage output that is proportional to the distance from the target.

Sensor, capacitive: An electronic sensor that is triggered by the effect a change in the surrounding dielectric causes to an electrostatic field emanating from the sensor.

Sensor, digital output: A sensor that provides an output that is either on or off with no intermediate states.

Sensor, electronic: A device used in die and press automation to sense position by means of a change in an electromagnetic (inductive) or electrostatic (capacitive) field.

Sensor, optical: A device used in die and press automation to sense position by means of reflection or interruption of a light beam.

Setback: The distance from the intersection of two corresponding mold lines to the bend line.

Shank, punch-holder: The stem or projection from the upper shoe which enters the slide flange recess and is clamped to the slide.

Shaving: A secondary shearing or cutting operation in which the surface of a previously cut edge is finished or smoothed.

Shear: (a) A tool for cutting metal and other material by the closing motion of two sharp, closely adjoining edges; (b) to cut by shearing dies or blades; (c) an inclination between two cutting edges.

Shedder: A pin, rod, ring, or plate, operated by mechanical means, air, or a rubber cushion, that either ejects blanks, parts, or adhering scrap from a die, or releases them from punch, die, or pad surfaces.

Shoe: (a) A metal block used in bending processes to form or support the part being processed; (b) the upper or lower component of a die set.

Shut height of a press: The distance from the top of the bed to the bottom of the slide with the stroke down and adjustment up. In general, the shut height of a press is the maximum die height that can be accommodated for normal operation, taking the bolster into consideration.

Sizing: A secondary pressworking operation to obtain dimensional accuracy by metal flow. The final forming in a die of a workpiece which has previously been formed (for example, by free forming) to a shape approximately equal to the shape of the die.

Slide: The main reciprocating press member; also called the ram, the plunger, or the platen.

Slitting: Cutting or shearing along single lines; used either to cut strips from a sheet or to cut along lines of a given length or contour in a sheet or part.

Spotting: The fitting of one part of a die to another, by applying an oil color to the surface of the finished part and bringing it against the surface of the intended mating part, the high spots being marked by the transferred color.

Springback: The extent to which metal tends to return to its original shape or position after undergoing a forming operation.

Staking: The process of permanently fastening two parts together by recessing one part within the other and then causing plastic flow of the material at the joint.

Steel rule die: A metal-cutting die employing a thin strip of steel (printer's rule) formed to the outline of a part and a thin steel punch mounted to a suitable die set. A flat metal plate or block of wood is substituted for the punch when cutting nonmetallic materials and soft metals.

Stock oiler: A device, generally consisting of felt-wick wipers or rolls, which spreads oil over the faces of sheet or strip stock.

Stop pin: A device for positioning stock or parts in a die.

Straight slide: A slide used in multi-slide forming machines whose cam produces direct motion equaling the throw of the cam.

Straightener rolls: (See Roll straightener.)

Strain: The deformation, or change in size or shape of a body, produced by stress in that body. Unit strain is the amount of deformation per unit length. Deformation is either elastic or plastic. In the literature on forming, the term "strain" is frequently used for the plastic deformation, specifically the quantitative expression for a component of the plastic strain. A strain component may be a normal strain, associated with change of length, either elongation or compression, or it may be a shear (shearing) strain, associated with change of angle between intersecting planes.

Stress: The internal force or forces set up within a body by outside applied forces or loads. Unit stress is the amount of load per unit area.

Stress relief (relieving): A heat treatment which is done primarily for reducing residual stresses.

Stretch (stretcher) forming: The shaping or forming of a sheet by stretching it over a formed shape.

Stripper: A device for removing the workpiece or part from the punch.

Stripper plate: A plate (solid or movable) used to strip the workpiece or part from the punch; it may also guide the stock.

Stripping: The operation of removing the workpiece or part from the punch.

Surge tank: A pressure vessel attached to a die and used in conjunction with a pneumatic or high-pressure nitrogen die pressure system to prevent an excessive pressure buildup during press cycling.

Swaging: A squeezing operation in which part of the metal under compression plastically flows into contours of the die; the remaining metal is unconfined and flows generally at an angle to the direction of applied pressure.

Tapering: A swaging or reducing operation, in which the metal is elongated in compression, for producing conical surfaces on tubular parts.

Tempering (drawing): A heat-treating process for removing internal stresses in metal at temperatures above those for stress relieving, but in no case above the lower critical temperature.

Tensile strength: The ultimate strength of a material, measured in pounds per square inch in tension on the original cross section tested, which, if exceeded, causes sectional deformation leading to ultimate rupture.

Transition temperature: The temperature at which there is a transition between ductile to brittle behavior in the fracture of material.

Trimming: Trimming is the term applied to the operation of cutting scrap off a partially or fully shaped part to an established trim line.

Trimming, pinch: Trimming the edge of a tubular part by pinching or pushing the flange or lip of the part over the cutting edge of a draw or stationary punch.

True strain and stress: Terms used in stress and strain analysis, indicating that the strain and stress are calculated on the basis of actual and instantaneous values of length and cross-sectional area. Also known as "natural strain and stress." True strain is also known as "logarithmic strain."

Ultimate strength: The maximum stress which a material can withstand before or at rupture.

Upsetting: A squeezing or compressing operation in which a larger cross section is formed on the part by gathering material in such a way as to reduce the length.

Vent: A small hole in a punch or die for admitting air to avoid suction holding, or to relieve pockets of trapped air which would prevent proper die closure or action.

Yield point: The stress at which a pronounced increase in strain is shown without an increase in stress.

||||Bibliography||||

Ivaska, Joseph, Jr., *Lubricants -- Productive Tool in the Metal Stamping Process.* Society of Manufacturing Engineers Technical Paper TE77-499. Dearborn, Mich.: SME, 1977.

Ivaska, Joseph, Jr., *How Metal Forming Lubricants Affect the Finishing Process.* Society of Manufacturing Engineers Technical Paper FC83-690. Dearborn Mich.: SME, 1984.

Ivaska, Joseph, Jr., *Lubrication Implications for Integrated Pressworking.* Society of Manufacturing Engineers Technical Paper MF85-862. Dearborn, Mich.: SME, 1984.

Nachtman, Elliot S., *A Review of Surface Lubricants Interactions During Metal Forming.* Society of Manufacturing Engineers Technical Paper MS77-338. Dearborn, Mich.: SME 1977.

Smith, David A., *Die Design Handbook.* Third Ed. Dearborn, Mich.: SME 1990.

Smith, David A., *Quick Die Change.* Dearborn, Mich.: SME, 1991.

Wallis, Bernard J., *Transfer Die Technology.* Dearborn, Mich.: Livernois Automation Co., 1991.

Staff, Society of Manufacturing Engineers, *Tool and Manufacturing Engineers Handbook,* Vol. 2, Forming. Dearborn, Mich.: SME 1984.

Staff, Society of Manufacturing Engineers, *Fundamentals of Tool Design,* Third Ed. Dearborn, Mich.: SME, 1991.

Index

A

Air blowoff, 156, 157, 414
Angular
    edge part breakage, 49
    holes, jig grinding, 91
Area
    calculation of, 128, 129

B

Banko patent, 113
Blank
    calculation of diameters, 122-125
    development, 11
        cold, 119
        double, 136
        unsymmetrical, 131
    stages, 19
Bridges and carriers, 154
Buttons
    tapered, 5

C

Cam
    -actuated cam strippers, 62
    curling stages, 59
    forming punches, 58
    guide rails, 55
    piercing, 60, 61
    positive-return, 55
    stages, 55
Carbide progressive dies
    applications, 215
    cemented carbides, 216
        production of, 216
        properties, 217
        steel-bonded, 245-249
    design considerations, 220
    finishing, 223
Centrobaric calculation, 131
Counterbored holes, 8
Countersinking stage, 22

D

Deburring stage, 22
Determining stages, 146
Die
    carts, 435-441
    clearance, 3
    definition, 3
    design, 255
        CAD/CAM, 280
        CAD/CNC, 283
    dimensions, 273
    engineering, 255
    life, 8
    machining, 280
        CNC, 281
    plates, 315-318, 321
Dieing machines, 83
Dies
    blank, 5
    cutoff, 139
    cutoff and form, 147
    cutoff and flange, 148
    compound, 13
    drawing, 10, 151, 154
    form, 147
    lamination, 51
    surface treatment for, 243
    two-per-stroke, 145, 169
        layout, 157
Draw
    beads, 14
    formulas, 121
    radii, 16
    stages, 14
Dressing
    ID radius, 95
    ID wheel, 96
Dwell, 58, 204, 205

E

Electrical discharge machining
    advantage, 313
    cycle, 308
    description, 307
    in carbide, 222
    overcut, 309
    setup, 317
    surface finish, 312
    wear ratios, 311
        corner, 312
        end, 312
        volumetric, 311
    wire-cut, 321
Electronic sensing
    control checklist, 393
    cyclic events, 385
    opposed-mode sensors, 405
    photoelectric, 403-405

Index

productivity, 392
proximity sensors, 395, 396
ready signals, 389
semicyclic events, 389
static events, 383
stop outputs, 391
Electric connection die strip, 174
Embossing
   circular, 20
   clearance, 21
   stages, 20
   V-type, 20

F

Flange lines, 24
Flanging steel, 25
Fracturing, 10

G

Gages, 50
Grinding
   carbide dies, 113
   conical sections, 104
   holes, 90
   jig, 87
      angular holes, 91
      odd-shaped sections, 91
   long radii, 103
   long sections, 102
   operations, 87, 97
   outer diameters, 96
   radius, 99,
     full inside, 99
     full outside, 99
     partial inside, 100
     partial outside, 101
     small outside, 101
   round blank stages, 93
   S curves, 111
   thin stock, 98
   to an internal angle, 104
   warped sections, 98
Grinding chart, 89

H

Hooke's Law, 27

I

Idle station, 206
Inserts, 45

L

Lance stages, 28
Lancing upward, 30

Lifters
   bar-type, 19, 39
   stock guide, 37
Lubrication
   application techniques, 361
   clarification, 366
   cleaning and finishing, 364
   disposal, 366
   material compatibility, 357
   microbiology, 365
   recycling, 366
   selection, 353
   structure, 367
   tooling interaction, 353
   troubleshooting, 371
   water-based, 376

M

Materials
   aluminum bronzes, 251
   antimonial lead, 254
   beryllium coppers, 252
   bismuth alloys, 254
   carbon and low-alloy steels, 226
   cast irons and steels, 229
   cold-rolled steels, 228
   hot-rolled steels, 228
   stainless and maraging steels, 233
   steel-bonded carbides, 245
   tool steel, 234
   zinc-based alloys, 252
Mathematics
   angular bend lines, 335
   angular layout, 332
   compound angles, 340
      calculation of, 345
   compound sine rolls, 331
   hole circles, 332
   locus of radius, 339
   radial bend lines, 337
   sine plates, 329
   sine setups, 330
   trigonometric functions, 327
Metal flow, 11
Microswitches, 47

N

Nitrogen
   die cylinders, 209
   ejection systems, 210
Notch stages, 143

O

Ohio Knife, 315
Overbend, 26 (see also springback)

P

Pilots
  entry, 46
  immovable, 48
  indirect, 293
  retaining, 293
  spring-loaded, 47
  stages, 143
Pins
  ejector, 4
  guide, 97
  kicker, 4
  push-off, 6
Plastic flow, 16
Pocket depth, 9
Press
  selection, 65
  size requirements, 65
  speeds, 67
  systems, 69
  types, 66
Presses
  die, 83
  four-slide, 84
  hydraulic, 80
    advantages, 80
    C-frame, 82
    controls, 82
    limitations, 81
    straightside, 81
  lamination, 84
  mechanical, 69
    adjustable-bed stationary, 71
    controls, 79
    gap-frame (C-frame), 69, 71
      open-back inclinable, 71
      open-back stationary, 71
      permanently inclined, 71
    motor selection, 79
    round-column, 75
    straightside, 69, 75
      advantages, 73
  transfer, 199
Pressworking process planning, 256
Progressive dies
  basic types, 141
  blank, 142
  conventional, 183
  classification, 141
  cut-and-carry, 154, 164
    double-strip, 160
  cutoff and form acute angle, 186
  draw, 151
  extrusion and double flange, 183
  form and cutoff, 185
  stage, 3
  station, 3
  tilting a panel, 191
  unsupported forming, 188
Progressive transfer dies
  butting type, 201
  description, 199
  three-dimensional, 211, 212
  transfer mechanisms, 201
  versus progressive dies, 202
    carrier strip, 207
  walking beam, 210, 211
Pull-out holes, 9
Punch
  as an electrode, 310
  cam forming, 58
  contours, 45
  definition, 3
  heels, 10
  notching, 10
  with shear, 5

Q

Quick die change
  custom-built systems, 427
  dealing with a mix of equiment, 419
  equipping presses for, 426
  grouping presses and dies for, 418
  inventory costs, 256
  planning, 420
  press factors, 421
  types, 429

R

Restrike stages, 26
Right- and left-hand parts, 159

S

Shearing
  in the punch, 23
  punch, with, 5
Slug
  columns, 5
  disposal, cam-piercing dies, 63
Springback, 26, 148, 151
Stamping
  bridges, 178
  design, 177
  double panels, 178
  holes near bend lines, 181
  interlocking parts, 182
  intersecting bend lines, 181
  radii, 180
  slug bridges, 179
  straight line in blank, 180
  symmetrical edges, 182
Steels
  carbon and low-alloy, 226
  cast irons and, 229

cold-rolled, 228
hot-rolled, 228
stainless and maraging, 233
tool, 234
water-hardening, 31
Stock guide
   design, 35
   lifters, 37
Stops
   for double runs, 300
   latch, 297
   overfeed, 306
   pin, 297
   solid, 296
   starting, 298
   trigger, 300
   trim, 302
   underfeed, 306
Strip
   carrier, 207
   computer-aided design, 280
   computer-aided machining, 280
   cut-and-carry, 164
   design, 163
   development, 277
      CNC, 281
   extended members, 172
   for frames, 170
   two-per-stroke, 169, 178
Stripper
   and pilot, 46
   balance, 34

cam-actuated, 61
for cam piercing, 60
keepers, 41
positive, 51
pressure, 44
rubber, 61
spool-type retainer, 42
spring-actuated, 52, 61
springs, 44
support, 32
wear, 31
Surface finish, 312
Surface treatments, 243-245
Switches
   Candy, 390

T

Thinning, 10
T-slots, 427-432
T-tables, 431-434

W

Whirligig
   horizontal, 107
   vertical, 5, 109
Wrinkle
   formation, 17, 34